Quantum Computing Architecture and Hardware for Engineers

Hiu Yung Wong

Quantum Computing Architecture and Hardware for Engineers

Step by Step

Hiu Yung Wong
Department of Electrical Engineering
San Jose State University
San Jose, CA, USA

ISBN 978-3-031-78218-3   ISBN 978-3-031-78219-0 (eBook)
https://doi.org/10.1007/978-3-031-78219-0

© The Editor(s) (if applicable) and The Author(s), under exclusive license to Springer Nature Switzerland AG 2025

This work is subject to copyright. All rights are solely and exclusively licensed by the Publisher, whether the whole or part of the material is concerned, specifically the rights of translation, reprinting, reuse of illustrations, recitation, broadcasting, reproduction on microfilms or in any other physical way, and transmission or information storage and retrieval, electronic adaptation, computer software, or by similar or dissimilar methodology now known or hereafter developed.
The use of general descriptive names, registered names, trademarks, service marks, etc. in this publication does not imply, even in the absence of a specific statement, that such names are exempt from the relevant protective laws and regulations and therefore free for general use.
The publisher, the authors and the editors are safe to assume that the advice and information in this book are believed to be true and accurate at the date of publication. Neither the publisher nor the authors or the editors give a warranty, expressed or implied, with respect to the material contained herein or for any errors or omissions that may have been made. The publisher remains neutral with regard to jurisdictional claims in published maps and institutional affiliations.

This Springer imprint is published by the registered company Springer Nature Switzerland AG
The registered company address is: Gewerbestrasse 11, 6330 Cham, Switzerland

If disposing of this product, please recycle the paper.

To my wife, Michelle, my daughter, Jennifer, and my son, James

Preface

The purpose of this book is to teach quantum computing hardware from an engineer's perspective. Engineers play an important role in quantum computers. However, college and graduate engineering students usually do not have the required physics and mathematics training to understand how quantum computer hardware works. This book provides step-by-step guidance to connect engineers to the quantum world. It is based on the teaching materials I created at San José State University for EE274 Quantum Computing Architectures in Spring 2023.

In this book, quantum computers built on silicon spin qubits and superconducting qubits are discussed in detail. The physics, mathematics, and their connection to microwave electronics are presented based on how they fulfill the five DiVincenzo's criteria. Readers will be able to understand the nuts and bolts of qubit initialization, readout, and manipulation. At the end, I also present a superconducting qubit integrated circuit design example, which is not commonly found elsewhere.

The characteristics of this book are three. Firstly, I try to write the equations and explain the process step-by-step to avoid ambiguity. As an engineer, I believe that this is very useful for serious engineers who want to connect their knowledge to quantum computing hardware. Such a connection is difficult to find in the literature. Secondly, I try to connect and contrast the spin qubits with superconducting qubits so that the readers can understand them in a unified theory. Lastly, simulation programs and design examples along with teaching videos are provided.

I learned quantum computing hardware by myself and I believe I have written the book in a way that the confusions and ambiguities an engineer usually faces are highlighted and answered. I also need to thank Dr. Yaniv Jacob Rosen and Dr. Kristin Beck from the Lawrence Livermore National Laboratory for the collaborations since 2021, through which I learned a lot from both of them and also had the opportunity to practice my knowledge in superconducting qubit hardware.

This book is organized into four parts. Part I gives an overview of quantum computers and the basics of linear algebra, i.e., the Schrödinger equation, the density matrix, and the Bloch sphere. These represent the minimal knowledge required to understand quantum computing hardware. Readers are also encouraged to read my other book, "Introduction to Quantum Computing: From a Layperson to a Programmer in 30 Steps" if one is interested in learning the basics of quantum algorithms.

In Part II, we discuss the physics of spin magnetic moment and magnetic field interaction which is the basics of spin qubits. Besides Larmor precession, we go through the physics and mathematics of Rabi oscillation under a linearly oscillating magnetic field and rotating magnetic field. In the process, we build the mathematical tools and concepts, such as rotating frame and rotating wave approximation which are required to understand qubit operations in many technologies. With this knowledge, we show how a silicon electron qubit can be implemented on a Metal-Oxide-Semiconductor (MOS) technology.

In Part III, we first introduce Lagrangian and Hamiltonian mechanics, which are usually not covered in engineering classes. Then we show how to connect classical mechanics to quantum mechanics through operator promotion. We will discuss the details of a mechanical simple harmonic oscillator and reuse their equations in an LC tank. Then we show that an LC tank is not a suitable qubit due to the lack of anharmonicity and introduce the physics and fabrication of the Josephson junction due to its ability to provide nonlinear inductance for anharmonicity in a superconducting qubit. We go through circuit quantization and discuss the characteristics of the Cooper pair box in its charge qubit and transmon qubit regimes. We highlight the similarities between the superconducting qubit and spin qubit Hamiltonian and show that we can reuse the solutions from the spin qubit after careful treatments. In this process, we share scripts for solving the superconducting qubits.

In Part IV, we discuss the roles of microwave electronics in quantum computers. Then we show the design parameters and methodologies of superconducting integrated quantum chip. Finally, we touch upon errors and decoherence time measurements. After reading this book, readers will be more ready to study more advanced topics such as open quantum systems.

This book is accompanied by many teaching videos and GitHub codes. Readers can find them in Appendix A.

For any comments, please send an email to intro.qc.wong@gmail.com.

San Jose, California, USA Hiu Yung Wong
August 2024

Contents

Part I Quantum Computer Overview and Essential Linear Algebra and Quantum Mechanics

1 Quantum Computer Hardware and Architecture—An Overview 3
   1.1   Introduction 3
   1.2   Quantum Computer as a Classical Electronics-Controlled Wavefunction 4
   1.3   DiVincenzo's Criteria 5
   1.4   Tracing the Signal: A Superconducting Qubit Quantum Computer Example 7
   1.5   Summary 10
   Problems 10
   References 11

2 Linear Algebra—Vectors, States, and Measurement 13
   2.1   Introduction 13
   2.2   Vector Space, Inner Product, and Hilbert Space 14
   2.3   Review of Vector Basics 17
   2.4   Tensor Product 22
   2.5   Summary 23
   Problems 24
   References 24

3 Linear Algebra—Operators, Matrices, and Quantum Gates 25
   3.1   Introduction 25
   3.2   Operators 26
   3.3   Matrices 27
   3.4   Measurement of a Quantum State—Part 2 34
   3.5   Tensor Product of Matrices 35
   3.6   Summary 36
   Problems 37
   Reference 37

4	**Schrödinger Equation and Quantum Gates**		39
	4.1	Introduction	39
	4.2	Schrödinger Equation	40
	4.3	Solving Schrödinger Equation	41
	4.4	Relationship Between Hamiltonian and Quantum Gate	47
	4.5	Basic Quantum Gates	48
	4.6	Entanglement	53
	4.7	Summary	55
	Problems		55
	References		56
5	**Bloch Sphere, Quantum Gates, and Pauli Matrices**		57
	5.1	Introduction	57
	5.2	Bloch Sphere	58
	5.3	Quantum Gate and Bloch Sphere	59
	5.4	Pauli Matrices	60
	5.5	Universal Sets of Gates	64
	5.6	Summary	68
	Problems		68
	References		69
6	**Density Matrix and the Bloch Sphere**		71
	6.1	Introduction	71
	6.2	Real Vector Space of 2×2 Hermitian Matrices	72
	6.3	Density Matrix	76
	6.4	Expectation Value	82
	6.5	Density Matrix, Expectation Value, and Bloch Sphere	83
	6.6	Summary	88
	Problems		88
	Reference		89

Part II Silicon Spin Qubit Architecture and Hardware

7	**Spin Qubit—Preliminary Physics**		93
	7.1	Introduction	93
	7.2	Magnetic Moment, Angular Momentum, and Gyromagnetic Ratio	94
	7.3	Spin, Spin Angular Momentum, and Spin Magnetic Moment	96
	7.4	Interaction Between Magnetic Moment and an External Magnetic Field	98
	7.5	Summary	99
	Problems		100
8	**Spin Qubit—Larmor Precession—Phase Shift Gate**		101
	8.1	Introduction	101
	8.2	Construction of Single-Qubit Gate Hamiltonian Under a Constant Magnetic Field	102

	8.3	Larmor Precession and Phase Shift Gate	104
	8.4	Implementation of Phase Shift Gate	108
	8.5	Summary	110
	Problems		111
	Reference		111
9	**Spin Qubit—Rabi Oscillation**		113
	9.1	Introduction	113
	9.2	Spin Angular Momentum Operator	114
	9.3	Rabi Oscillation	115
	9.4	Spin Resonance and Rotating Wave Approximation (RWA)	120
	9.5	Rabi Oscillation and Rabi Frequency	122
	9.6	Intuitive View of Spin Resonance and Rotating Wave Approximation	125
	9.7	Summary	127
	Problems		127
	References		128
10	**Spin Qubit—Rabi Oscillation Under Rotating Field Using Rotating Frame**		129
	10.1	Introduction	129
	10.2	Experimental Setup	130
	10.3	Setting Up the Hamiltonian	131
	10.4	Transforming to Rotating Frame	133
	10.5	Solving the Schrödinger Equation	134
	10.6	Intuitive Understanding of the Solution	138
	10.7	Another Method to Perform Rotating Frame Transformation	140
	10.8	Summary	142
	Problems		142
11	**Electron Spin Qubit in Semiconductor—Implementation, Initialization, and Readout**		143
	11.1	Introduction	143
	11.2	Why Silicon?	144
	11.3	Spin Qubits in Quantum Dots	145
	11.4	Silicon Spin Qubit	146
	11.5	Qubit Initialization	149
	11.6	Qubit Readout	152
	11.7	Summary	153
	Problems		154
	References		154
12	**Electron Spin Qubit in Semiconductor—1-Qubit and 2-Qubit Gates**		155
	12.1	Introduction	155
	12.2	1-Qubit Gate Implementation	156
	12.3	2-Qubit Gate for Spin Implementation	157

	12.4	Summary	165
	Problems		165
	References		165

Part III Superconducting Qubit Architecture and Hardware

13	**Lagrangian Mechanics and Hamiltonian Mechanics**		169
	13.1	Introduction	169
	13.2	Lagrangian Mechanics	170
	13.3	Hamiltonian Mechanics	174
	13.4	Hamiltonian and Hamilton's Equation	175
	13.5	Summary	179
	Problems		180
	References		180
14	**Quantization of Simple Harmonic Oscillator**		181
	14.1	Introduction	181
	14.2	SHO Hamiltonian and Promotion of Operators	182
	14.3	Annihilation, Creation, and Number Operators	185
	14.4	Energy Eigenstates, Coherent State, and Linkage to Classical SHO	193
	14.5	Summary	198
	Problems		199
	References		199
15	**Quantization of an LC Tank: A Bad Qubit**		201
	15.1	Introduction	201
	15.2	Classical LC Tank	202
	15.3	Generalized Coordinate, Velocity, and Momentum in the LC Tank	206
	15.4	Quantization of an LC Tank	208
	15.5	Summary	213
	Problems		213
	References		214
16	**Superconductor and Josephson Junction**		215
	16.1	Introduction	215
	16.2	Superconductors	216
	16.3	Josephson Junction	220
	16.4	Flux-Tunable Josephson Junction Loop	230
	16.5	Summary	233
	Problems		233
	References		234
17	**Cooper Pair Box Qubit: Hamiltonian**		235
	17.1	Introduction	235
	17.2	Isolated Cooper Pair Box	236

	17.3	Cooper Pair Box Coupled to a Voltage Source	241
	17.4	Summary	247
		Problems	247
		References	247
18	**Cooper Pair Box: Analytical Solution**		**249**
	18.1	Introduction	249
	18.2	General Hamiltonian	250
	18.3	Hamiltonian with Two-Basis-State Approximation	251
	18.4	Eigenenergies in Cooper Pair Box	252
	18.5	Eigenstates of a Cooper Pair Box	254
	18.6	Summary	256
		Problems	256
		References	257
19	**Cooper Pair Box: Numerical Solution**		**259**
	19.1	Introduction	259
	19.2	Anharmonicity	260
	19.3	Python Code for Finding Eigenvalues and Eigenvectors	261
	19.4	Transmon Qubit	265
	19.5	Summary	270
		Problems	270
		References	271
20	**Charge Qubit Dynamics: Precession and 1-Qubit Gate**		**273**
	20.1	Introduction	273
	20.2	Mapping Charge Qubit to Spin Qubit	274
	20.3	One-Qubit Gate	279
	20.4	Summary	286
		Problems	286
		Reference	287
21	**Transmon Qubit: One-Qubit and Two-Qubit Gates**		**289**
	21.1	Introduction	289
	21.2	One-Qubit Gate for Transmon	290
	21.3	Two-Qubit Entanglement Gate for Transmon	297
	21.4	Summary	307
		Problems	308
		References	308
22	**Superconducting Qubit: Readout and Initialization**		**309**
	22.1	Introduction	309
	22.2	Light-Matter Interaction	310
	22.3	Jaynes-Cummings Hamiltonian	312
	22.4	Dispersive Readout	317
	22.5	Microwave Readout Circuit Example	320
	22.6	Qubit Initialization	322

	22.7	Summary	323
	Problems		323
	References		323

Part IV Quantum Computer Design and Implementation

23 Microwave Electronics in Quantum Computers ... 327
- 23.1 Introduction ... 327
- 23.2 Overview ... 328
- 23.3 Qubit Manipulation Path ... 329
- 23.4 Readout Path ... 333
- 23.5 Summary ... 341
- Problems ... 341
- References ... 342

24 Design of an Integrated Superconducting Qubit Chip ... 343
- 24.1 Introduction ... 343
- 24.2 Layout ... 344
- 24.3 Design and Numerical Examples ... 346
- 24.4 Summary ... 353
- Problems ... 354
- References ... 354

25 Errors and Decoherence ... 355
- 25.1 Introduction ... 355
- 25.2 Errors ... 356
- 25.3 Decoherence Times ... 359
- 25.4 Summary ... 362
- Problems ... 362
- References ... 363

A Resources ... 365

Index ... 367

Part I
Quantum Computer Overview and Essential Linear Algebra and Quantum Mechanics

Chapter 1
Quantum Computer Hardware and Architecture—An Overview

1.1 Introduction

In this chapter, we will have an overview of a quantum computer from an engineer's perspective. We will first look at the basic operations of a quantum computer, namely, qubit initialization, qubit manipulation, and qubit readout. We then introduce and discuss DiVincenzo's criteria which set the minimal requirement for a technology to build a quantum computer. Then we will use a superconducting qubit quantum computer as an example to trace the propagation, attenuation, modification, and amplification of the microwave signals to gain a deeper understanding of the operating principles of a quantum computer and the critical role of classical and microwave electronics in a quantum computer. We will also discuss the scaling of a quantum computer.

1.1.1 Learning Outcomes

Understand the basic operations of a quantum computer; be able to describe the role of classical electronics in a quantum computer; appreciate the challenges in achieving a highly scaled quantum computer.

1.1.2 Teaching Videos

- Search for Ch1 in this playlist
    - https://tinyurl.com/3yhze3jn

- Other videos
  - htttps://youtu.be/M93Qtu3J-5M

1.2 Quantum Computer as a Classical Electronics-Controlled Wavefunction

A quantum computer is nothing but a **quantum bit (qubit)** wavefunction controlled by classical electronics (Fig. 1.1). The qubit wavefunction, such as the spin of an electron, the electronic state of an ion, and the excess number of Cooper pairs in a superconducting qubit, is passive. Without interacting with the external environment, it is a "boring" *passive wavefunction*. A quantum computing process is also nothing but the modification of the wavefunction through an appropriate **Hamiltonian**. We will discuss Hamiltonian in future chapters (e.g., Chap. 4). Hamiltonian is the total energy of the system which *governs the evolution (change) of the wavefunction*. An appropriate Hamiltonian is created by applying a suitable laser pulse or microwave pulse with the right *frequency*, right *shape*, right *amplitude*, and right *phase* at the right *time* for a right *duration*. The pulses are generated in classical electronics and it is not surprising that microwave engineering, laser optics, and signal processing play a key role in a quantum computer. Since the pulses need to be short and fast, high-speed digital circuits (such as application-specific integrated circuits (**ASICs**) and field programmable gate arrays (**FPGAs**)) and efficient classical programming and data processing are crucial in a quantum computer.

At the time of writing, *a quantum computer is just a well-organized physics laboratory*.

Figure 1.1 shows that there are three critical interactions between the classical electronics and the qubit wavefunctions ($QB_{n-1}, \cdots, QB_0$), namely, *qubit initialization*, *qubit manipulation*, and *qubit readout*. These are all performed by sending an appropriate pulse to the qubit from the classical electronics.

Qubit initialization is a process to initialize the qubit to a particular state which is usually the ground state, $|0\rangle$. Most quantum computing algorithm starts with all qubits initialized to $|0\rangle$ [1, 2]. Moreover, in algorithms with error corrections, a steady supply of initialized qubits is required. There are two ways to initialize a qubit to $|0\rangle$. One is through **thermalization**. In this process, we let the qubit wait for a long enough time so that it will decay to the ground state by losing energy to its environment. This is typically a long process. The second approach is to measure the qubit and a pulse is applied to reset the qubit to the ground state if it is not at the ground state (i.e., $|1\rangle \rightarrow |0\rangle$). This is called the **active reset**. We will discuss qubit initialization with more details in Chaps. 11 and 22.

Qubit manipulation is equivalent to applying a quantum gate to the qubit. As mentioned, pulses with the appropriate frequency, shape, amplitude, phase, and

1.3 DiVincenzo's Criteria

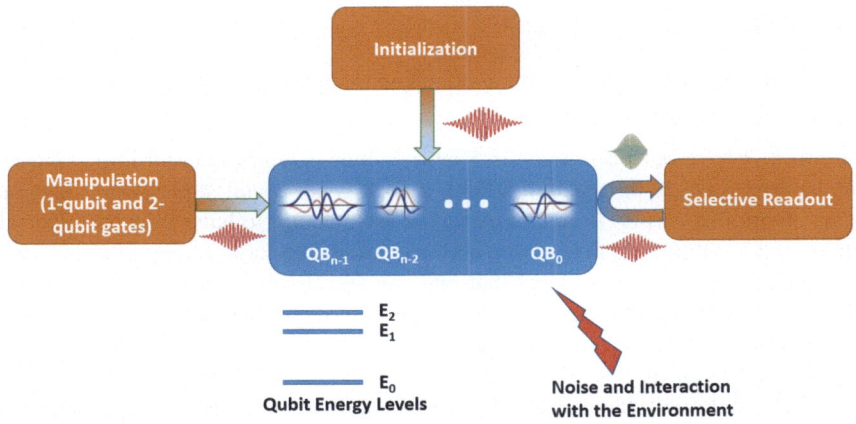

Fig. 1.1 Block diagram of a typical quantum computer

duration are sent to interact with the qubit so that the qubit wavefunction will change. If the pulse only interacts with a single qubit, it is a **1-qubit gate**. If the pulse causes two qubits to interact with each other, it is a **2-qubit gate**. Any quantum gate can be decomposed or approximated by a set of 1-qubit and 2-qubit gates (which is called **a set of universal quantum gates**). This is just like the fact that any classical logic may be implemented by NOT and AND gates. It is possible to have gates for 3 or more qubits. However, they are more difficult to implement and have lower fidelity.

Qubit readout refers to measuring all or some of the qubits. The readout is essential in quantum computing as it is used to measure the final result. In some algorithms, intermediate results need to be measured to determine the next step (e.g. in quantum teleportation, see Chapter 19 in [1]). It is also used to measure ancillary qubits for error correction in the middle of the algorithm. Due to the nature of quantum mechanics, upon measurement, a qubit wavefunction will collapse to the measurement basis states ($|0\rangle$ or $|1\rangle$). This is achieved by applying a pulse to the qubit and observing the change of the reflected pulse which is modulated by the state of the qubit ($|0\rangle$ or $|1\rangle$) in some architectures (such as superconducting qubit). In the silicon electron spin qubit case, the spin state is mapped to a charge state, and the charge state is detected by electronics.

1.3 DiVincenzo's Criteria

In his 2000 paper [3], DiVincenzo suggested that any architecture needed to have the following five criteria to make a useful quantum computer:

1. The ability to initialize the state of the qubits to a simple fiducial state

2. A "universal" set of quantum gates
3. A qubit-specific measurement capability
4. A scalable physical system with well-characterized qubit
5. Long relevant decoherence times

Items 1–3 have been discussed in Sect. 1.2. Item 1 refers to qubit initialization. Initialization needs to have a high fidelity (which is not trivial) and also needs to be fast as mentioned earlier. Item 2 requests the architecture to have a set of universal 1-qubit and 2-qubit gates that can be used to build up (or at least approximate) any quantum circuits. Item 3 requires that one can selectively measure any particular qubits without affecting others. This is required in many circuits such as quantum teleportation and Shor's algorithm [1].

Item 4 refers to the fact that the qubit must be a well-distinguished two-level system. It should only have states $|0\rangle$ and $|1\rangle$ and they need to be well-separated. This is not trivial for some systems. For example, a superconducting qubit can have states $|0\rangle, |1\rangle, |2\rangle, |3\rangle, \cdots$ (Fig. 1.1). If the energy separation between $|0\rangle$ and $|1\rangle$, $E_1 - E_0$, is too close to the energy separation between $|1\rangle$ and $|2\rangle$, $E_2 - E_1$, i.e., $E_2 - E_1 \approx E_1 - E_0$, the qubit state can leave the $|0\rangle / |1\rangle$ Hilbert space easily, resulting in a failure in the computation. We will discuss this in detail in Sect. 19.2.

$|0\rangle$ and $|1\rangle$ need to be well-separated because of two reasons. Firstly, there are thermal and **quantum noises**. When there is noise, they cannot be distinguished well unless their energy separation is well above the noise level. When the noise level is large, the qubit can transit from one state to another due to the noise. The noise will also distort the measurement signal making the state indistinguishable [4]. The second reason is due to the so-called **line-width broadening**. The energy of each level is not sharp. It has a certain spread due to **the uncertainty principle** and its interaction with the environment. Therefore, even if there is no noise, the two levels need to be well separated.

Thermal noise is proportional to kT/q, where k is the Boltzmann constant, T is the temperature, and q is the elementary charge. Figure 1.2 shows the thermal energy as a function of temperature in kelvin (K). The qubit bit energy separation ($E_1 - E_0$) of a typical superconducting qubit is about 0.02 meV which is equivalent to the thermal noise at 0.23 K $=230$ mK. Therefore, to operate a superconducting quantum computer, a **dilution refrigerator** is needed in which the temperature is maintained at about 10 to 20 mK, corresponding to a thermal energy of about 10–20 times lower than the qubit energy separation.

Therefore, besides those mentioned in Sect. 1.2, *cryogenic technologies and electronics* are critical players in quantum computers.

Item 5 is about the **decoherence time** which is a measure of duration a qubit can maintain its state without intentional interaction (such as the initialization, manipulation, and readout in Fig. 1.1). A qubit may lose its information by interacting with the environment. There are two types of decoherence time. The first type is called the T_1-*decoherence time* or the *longitudinal decoherence time*. It is a time constant of how fast an excited state, $|1\rangle$, will decay to a ground state, $|0\rangle$. The second type is called the T_2-*decoherence time* or the *transverse decoherence time*.

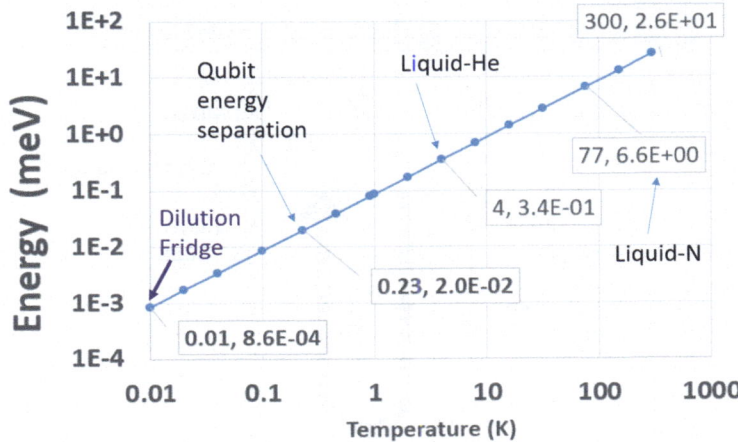

Fig. 1.2 Thermal energy as a function of temperature

It is a time constant of how fast a quantum superposition of $|0\rangle$ and $|1\rangle$ decays to a classical mixture of $|0\rangle$ and $|1\rangle$. In this process, the state loses its phase information too and thus it is also called the **dephasing time**. This will be discussed in detail in Chap. 25.

Item 5 tells us that the qubit needs to have a long decoherence time so that it can perform enough computation (many applications of quantum gates) before it loses its information. Since decoherence is related to noise, we again want to put the qubit at a cryogenic temperature to avoid the thermal noise.

There is a dilemma in DiVincenzo's Criteria. To have a long decoherence time, qubits should be well-isolated from the environment. A qubit that does not have a strong interaction mechanism with the environment usually is well-isolated. However, qubit control requires that the qubit can be coupled to the environment (the control) easily. Usually, these two cannot be achieved at the same time. Therefore, a quantum computer architecture that has a long decoherence time usually has a long gate time (as coupling to the control is weaker). Moreover, we also require dissipative coupling to the environment for initialization if active reset is not preferred.

1.4 Tracing the Signal: A Superconducting Qubit Quantum Computer Example

It would be instructive to look deeper into a quantum computer by tracing how the signals propagate. Here I will choose a superconducting qubit quantum computer which I have worked on. Figure 1.3 shows the schematic of a typical

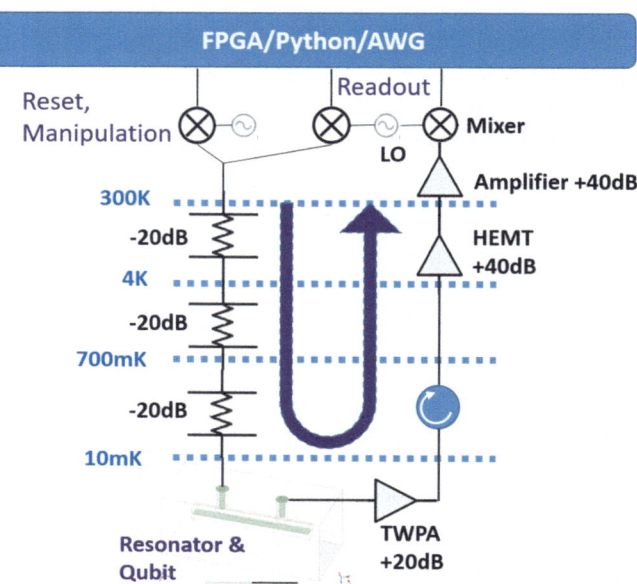

Fig. 1.3 Schematic of a typical superconducting qubit quantum computer

superconducting qubit quantum computer. Every part is classical except the qubit and the **traveling wave parametric amplifier (TWPA)** at 10 mK are quantum.

On the top at room temperature, a high-performance server in conjunction with high-speed electronics such as FPGAs is used to generate signals, analyze data, and make decisions. It controls equipment such as an arbitrary waveform generator (AWG) to generate appropriate pulses (with the right phase, shape, duration, amplitude, and frequency) for qubit initialization, manipulation, and readout. The signal will be up-converted to microwave frequency in the GHz regime through a mixer and a local oscillator (LO) which is required to interact with the qubit as the superconducting qubit operates in the GHz domain (e.g., a qubit energy of 0.02 meV corresponds to about a 5 GHz microwave photon).

The pulses will be generated with a large enough amplitude and go through a chain of attenuators from room temperature to about 10 mK. Note that the manipulation/reset pulses usually have a different frequency (so a different LO is used) than that of the readout pulse. The attenuators are required to attenuate the thermal noise from room temperature so that the thermal noise becomes negligible when it reaches the qubit. At that stage, the thermal noise is smaller than the *quantum noise*. Quantum noise is a result of the *Heisenberg's Uncertainty Principle* and cannot be avoided which poses a fundamental limit.

The pulse does not interact with the qubit directly. It often interacts with the qubit through capacitive or inductive coupling. If it is a manipulation pulse (i.e., a quantum gate) or initialization pulse for active reset, the process is completed.

However, if it is a readout pulse (i.e., a measurement pulse), the reflected or transmitted signal will be detected. In some quantum computers, the readout and manipulation/reset are coupled to the qubit through different paths. Usually, the readout pulse is passed through a resonator capacitively coupled to the qubit. The resonant frequency of the resonator will change depending on the state of the qubit ($|0\rangle$ or $|1\rangle$), and thus the reflected/transmitted pulse is changed accordingly.

When the readout pulse is reflected or transmitted, the signal is very weak and cannot be detected by classical electronics. Therefore, amplification is required. However, every amplifier has a certain noise factor which will be discussed in Chap. 23. When an amplifier amplifies the signal, it also amplifies the noise and thus the **signal-to-noise ratio**, *S/N ratio*, will not be improved. Indeed, it also adds additional noise to the amplified signal due to the noise sources in the amplifier, and thus further degrades the S/N ratio depending on its noise factor. Based on microwave theory [5], to minimize the overall noise factor, it is desirable to have an amplifier with the lowest noise factor at the beginning of the amplification chain. Therefore, a TWPA is added at the beginning. This is a type of **quantum parametric amplifier** [6] and has the minimal possible noise factor. The gain of a TWPA is usually low. Therefore, further amplifications are required. Finally, the signal is downconverted to a lower frequency using a mixer and an LO for signal processing to distinguish $|0\rangle$ and $|1\rangle$ states in classical computers and electronics.

I would also highlight two important aspects related to electrical engineering in this system. Firstly, a *high-electron-mobility transistor* (HEMT) amplifier is commonly used in quantum computers to amplify readout signals. This is because it has a low noise factor. Secondly, superconducting qubits are usually fabricated through *integrated circuit (IC)* technologies. The knowledge in IC can be applied readily to superconducting qubit design except that microwave analysis is required and it is built on materials usually not used in a traditional silicon IC chip. It is also worth noting that a superconducting qubit chip has most of the area occupied by classical microwave components such as capacitors, inductors, transmission lines, and resonators. Readers can see a design example in Chap. 24.

1.4.1 Scaling of Quantum Computers

After understanding what a quantum computer looks like, it is natural to see that the scaling of a quantum computer is not trivial. How can we increase the number of qubits to realize a powerful quantum computer? There are three issues.

Firstly, too many transmission lines need to be wired from room temperature to cryogenic temperature to control each qubit. While multiplexing is possible, this will eventually post a bottleneck to the **throughput** and speed. Throughput refers to the amount of signal that can be processed per unit amount of time. This requires us to miniaturize the classical control electronics and place them in the proximity of the qubits and, thus, need to be cooled down to cryogenic temperature.

Secondly, even if we can miniaturize the control electronics, a typical dilution refrigerator can only have a cooling power of less than 1 mW [7]. A typical classical CPU has a power of 100 W. It is thus almost impossible to put all control electronics in mK-regime. But it is still desirable to bring the electronics as close to the qubit as possible (to increase the throughput) and as cold as possible (to reduce thermal noise). Therefore, cryogenic electronics at 4.2 K (boiling point of liquid He) has been widely studied [8].

Finally, due to decoherence, error-protected qubits are required. In theory, it requires about 100–1000 physical qubits to create an error-protected qubit. This increases the scaling requirement by 100–1000 times to have a useful quantum computer.

1.5 Summary

In this chapter, we give an overview of the operations of a typical quantum computer. A quantum computer has three basic operations, namely, initialization, manipulation, and readout. Any architecture needs to fulfill DiVincenzo's criteria to be a useful quantum computer. Besides being able to perform basic operations, the qubits need to have well-separated levels and long decoherence time. We then use the superconducting qubit as an example to demonstrate the basic operations and constraints of a quantum computer. At the moment of writing, a quantum computer is still just a well-organized sophisticated physics laboratory. To build a quantum computer with many qubits and with error correction is not trivial. This requires a lot of engineering work and engineers are expected to play an important role. In the following chapters, we will review the basics of quantum computing and physics. We will then study the details of how spin qubits and superconducting qubits fulfill DiVincenzo's criteria.

Problems

1.1 Thermal Noise

If we want to operate a superconducting qubit at room temperature, what should be the energy separation between $|0\rangle$ and $|1\rangle$ (i.e., qubit energy) if we want it to be at least 10 times the thermal noise. What is the corresponding frequency? Is this feasible? If you do not know the equations, you can refer to Fig. 1.2 and the discussion in Sect. 1.3. Note that noise energy is proportional to temperature and qubit energy is proportional to the qubit frequency.

References

1. Hiu-Yung Wong. *Introduction to Quantum Computing*. Springer, 2024.
2. M. A. Nielsen and I. L. Chuang. *Quantum Computation and Quantum Information: 10th Anniversary Edition*. Cambridge University Press, 2011.
3. David P. DiVincenzo. The physical implementation of quantum computation. *Fortschritte der Physik*, 48(9–11):771–783, 2000.
4. Hiu Yung Wong, Prabjot Dhillon, Kristin M. Beck, and Yaniv J. Rosen. A simulation methodology for superconducting qubit readout fidelity. *Solid-State Electronics*, 201:108582, 2023.
5. Behzad Razavi. *RF Microelectronics*. Pearson, 2011.
6. C. Macklin, K. O'Brien, D. Hover, M. E. Schwartz, V. Bolkhovsky, X. Zhang, W. D. Oliver, and I. Siddiqi. A near–quantum-limited josephson traveling-wave parametric amplifier. *Science*, 350(6258):307–310, 2015.
7. H. Zu, W. Dai, and A.T.A.M. de Waele. Development of dilution refrigerators–a review. *Cryogenics*, 121:103390, 2022.
8. Tom Jiao, Edwin Antunez, and Hiu Yung Wong. Study of cryogenic mosfet sub-threshold swing using ab initio calculation. *IEEE Electron Device Letters*, 44(10):1604–1607, 2023.

Chapter 2
Linear Algebra—Vectors, States, and Measurement

2.1 Introduction

In this and the next few chapters, we will review linear algebra basics. In this chapter, we will review the definitions of vector space, inner product space, and Hilbert space. Although it is not necessary to understand them thoroughly if you feel comfortable with using the equations and mathematical tools given, it is still very instructive if you want to have a basic idea of the foundation of quantum computing and quantum physics. We will also practice *bra-ket* notation and their representations in row and column vectors. We will study how to represent a vector in different bases. Then, we will discuss the effect of measurement in quantum mechanics and the importance of orthonormal basis and normalized vectors. Finally, we will review how to combine two spaces (of two physical systems) into a larger one using the tensor product.

2.1.1 Learning Outcomes

Know how to represent vectors in different bases; be more familiar with *bra-ket* operations; able to perform tensor products.

2.1.2 Teaching Videos

- Search for Ch2 in this playlist
  - https://tinyurl.com/3yhze3jn

- Other videos
- https://youtu.be/Xireluiir9Y
- https://youtu.be/cEpjouPIuQY
- https://youtu.be/Wrmigi645J4

2.2 Vector Space, Inner Product, and Hilbert Space

A **vector space** (or **linear space**), in general, has a set of **vectors**, $\{\vec{u}, \vec{v}, \vec{w}, \cdots\}$, and a set of **scalars**, $\{a, b, c, \cdots\}$. There is an **addition operation**, $+$, for the vectors, resulting in another vector in the space (e.g., $\vec{w} = \vec{u} + \vec{v}$). The vector can be scaled by the scalars (e.g., $a\vec{v}$). They need to obey the following **vector axioms**:

$$\vec{u} + (\vec{v} + \vec{w}) = (\vec{u} + \vec{v}) + \vec{w},$$
$$\vec{u} + \vec{v} = \vec{v} + \vec{u},$$
$$\vec{u} + \vec{0} = \vec{u},$$
$$\vec{v} + \vec{-v} = \vec{0},$$
$$a(b\vec{u}) = (ab)\vec{u},$$
$$1\vec{u} = \vec{u},$$
$$a(\vec{u} + \vec{v}) = a\vec{u} + a\vec{v},$$
$$(a+b)\vec{u} = a\vec{u} + b\vec{u}. \tag{2.1}$$

We will not discuss the details and interested readers can refer to standard textbooks or Wikipedia [2]. However, we will emphasize a few things. $\vec{-v}$ refers to the requirement that an *inverse* vector must exist for $\vec{v}$. Therefore $\vec{-v}$ is just $-\vec{v}$. Moreover, there exists a **zero** vector $\vec{0}$ and there should be a **multiplicative identity**, 1, in the scalar set.

If the scalars are only real numbers, then the space is called a **real vector space**. If they include complex numbers, the corresponding space is called a **complex vector space**.

We can further require that a certain vector space *has an operation called the inner product defined for its vectors*. Then it is called an **inner product space**. The inner product between two vectors $\vec{u}$ and $\vec{v}$ is written as $\langle \vec{u} | \vec{v} \rangle$ *which is a scalar*. Before we continue to study the properties of inner products, let us discuss how our 3D space behaves as an inner product space.

Example 2.1 Euclidean vector space is an inner product space. The 3D space we live in is a Euclidean vector space. Discuss how it fulfills the definition of an inner product space.

2.2 Vector Space, Inner Product, and Hilbert Space

For any position vectors in the space, $\vec{v}$ and $\vec{w}$, we already know from our daily experience that they obey Eq. (2.1) with real number scalars. So it is a vector space.

We can set $\hat{x}$, $\hat{y}$, and $\hat{z}$ as its basis vectors (see Chapter 3 in [1]) and they are **orthonormal** (to be discussed in Sect. 2.3.4). Then for any position vectors in the space, $\vec{v}$ and $\vec{w}$, they can be written as a linear combination of the basis vectors,

$$\vec{v} = v_1\hat{x} + v_2\hat{y} + v_3\hat{z},$$
$$\vec{w} = w_1\hat{x} + w_2\hat{y} + w_3\hat{z}, \qquad (2.2)$$

where v_1, v_2, v_3, w_1, w_2, and w_3 are real numbers.

We define the inner product as

$$\langle v|w \rangle = \begin{pmatrix} v_1 & v_2 & v_3 \end{pmatrix} \begin{pmatrix} w_1 \\ w_2 \\ w_3 \end{pmatrix},$$
$$= v_1 w_1 + v_2 w_2 + v_3 w_3. \qquad (2.3)$$

It is also helpful to state that the inner product in a real 3D space is equivalent to

$$\langle v|w \rangle = \vec{v} \cdot \vec{w} = |\vec{v}||\vec{w}|\cos\theta, \qquad (2.4)$$

where θ is the angle between the two vectors. And it should be noted that $\langle v\ w \rangle$ is just the $bra - ket$ notation for the **dot product**, $\vec{v} \cdot \vec{w}$, that we usually use.

Based on our daily experience, it is easy to understand that the inner product in a real 3D space satisfies the following equations:

$$\vec{v} \cdot \vec{w} = \vec{w} \cdot \vec{v},$$
$$\vec{v} \cdot \vec{v} \geq 0,$$
$$\vec{u} \cdot (a\vec{v} + b\vec{w}) = a\vec{u} \cdot \vec{v} + b\vec{u} \cdot \vec{w}, \qquad (2.5)$$

where the second equation just tells us that any vector has a zero or positive length. ∎

Now, we formally define a general **Euclidean vector space** as a *finite inner product space* with *real* scalars. A *finite* inner product space means that the number of basis vectors (e.g., 3 in our 3D real space) is finite instead of infinite.

Based on our experience with the real 3D Euclidean vector space, we can generalize the criteria of the inner product in Eq. (2.5) to any inner product space with complex scalars and with an arbitrary number of dimensions. For simplicity, we will remove the arrow in the vector notation now. Since $\langle \vec{u}|\vec{v} \rangle = \vec{u} \cdot \vec{v}$ can be a complex number, the definition of an inner product is an operation that obeys the following rules:

$$\langle v|w\rangle = \langle w|v\rangle^*,$$
$$\langle v|v\rangle \geq 0,$$
$$\langle u|(a|v\rangle + b|w\rangle) = a\langle u|v\rangle + b\langle u|w\rangle. \quad (2.6)$$

We can use the inner product to define another quantity called **norm**, which is

$$\|v\| = \sqrt{\langle v|v\rangle}, \quad (2.7)$$

and is just the length of the vector and sometimes we write it as $|\vec{v}|$ in the real 3D space (e.g., Eq. (2.4)).

We then define the **distance** between two vectors, $\vec{u}$ and $\vec{v}$, as

$$d(u,v) = \|u - v\| = \sqrt{\langle u-v|u-v\rangle}, \quad (2.8)$$

which is just the *norm* or the length of their difference.

I introduce *norm* and *distance* because I would like to give a more formal definition of **Hilbert Space**. A Hilbert space is a *real* or *complex* inner product space that is *complete* with respect to its distance. The definition of "completeness" here is *different* from the completeness of basis vector (e.g., Section 10.4 in [1]). It is pretty involving mathematically. We can approximately explain it as the following. There is something called the **Cauchy sequence**. This sequence has its elements eventually become arbitrarily close to the next one (converged). For an inner product space, we can create many Cauchy sequences of its *vectors*. Each Cauchy sequence will converge to a vector (in other words, its distance to other elements will converge to 0). If the converged vector *is also a vector in the inner product space*, i.e., the inner product space contains that converged vector, then it is said to be complete. Since we measure the convergence using distance, it is said to be complete with respect to its distance.

Note that any finite-dimensional inner product space is complete. Therefore, a Euclidean vector space is a Hilbert space because it is complete.

Figure 2.1 summarizes the properties of the spaces discussed.

	Has vectors and scalars which obey (2.1)	Has inner product defined that obeys (2.6)	Is complete?	Has infinite dimensions?	Has complex scalars?
Vector Space	✓	Maybe	Maybe	Maybe	Maybe
Inner Product Space	✓	✓	Maybe	Maybe	Maybe
Hilbert Space	✓	✓	✓	Maybe	Maybe
Euclidean Vector Space	✓	✓	✓	✗	✗

Fig. 2.1 Summary of the properties of various spaces discussed in this section

2.3 Review of Vector Basics

Here we will succinctly list the basic concepts in linear algebra that are relevant to quantum computing. For a more detailed discussion, please see [1].

2.3.1 Basis Vectors/States, Bra-ket Notation, and Representation of Vectors/States

A state in a quantum system is represented by a vector. In other words, every vector in a Hilbert space is also a state. We may choose some vectors to be the **basis vectors** and every vector in the space can be represented as a linear combination of the basis vectors. For example, in real 3D space, a vector $\vec{v}$ can be expressed as the combination of the unit vectors (as basis vectors), $\hat{x}$, $\hat{y}$, and $\hat{z}$, as

$$\vec{v} = v_1\hat{x} + v_2\hat{y} + v_3\hat{z} = \begin{pmatrix} v_1 \\ v_2 \\ v_3 \end{pmatrix}, \tag{2.9}$$

where v_1, v_2, and v_3 are scalars.

To write a vector, besides putting an arrow on top of the variable (e.g., $\vec{v}$),

$$\vec{v} = \alpha \hat{v}_x + \beta \hat{v}_y, \tag{2.10}$$

we may also use the *bra-ket* notation in which the variable is written in a *ket*,

$$|v\rangle = \alpha |v_x\rangle + \beta |v_y\rangle. \tag{2.11}$$

It should be noted that anything inside the *ket* is just a name. As long as it is not ambiguous, we may write anything that is convenient. For example, if $|v\rangle$ is the vector for a photon polarization state, and $|v_x\rangle$ and $|v_y\rangle$ are the horizontal and vertical polarization states of a photon, respectively, we may write Eq. (2.11) as,

$$|\text{Final Polarization}\rangle = \alpha |\text{Horizontal Polarization}\rangle +$$
$$\beta |\text{Vertical Polarization}\rangle. \tag{2.12}$$

If a different basis vector set is chosen, the representation will be different. For example, in Fig. 2.2, $|V_1\rangle$ can be expressed as a linear combination of the basis vectors in one of the two bases, $|0\rangle / |1\rangle$ or $|+\rangle / |-\rangle$.

$$|V_1\rangle = 1|0\rangle + 0|1\rangle = |0\rangle = \begin{pmatrix} 1 \\ 0 \end{pmatrix}, \begin{matrix} \to |0\rangle \\ \to |1\rangle \end{matrix} \tag{2.13}$$

Fig. 2.2 Vector $|V_1\rangle$ can be represented in two different bases, $|0\rangle/|1\rangle$ or $|+\rangle/|-\rangle$

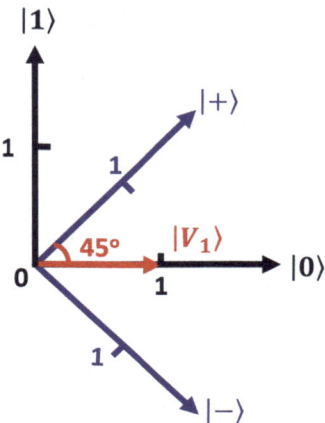

$$= \frac{1}{\sqrt{2}}(|+\rangle + |-\rangle) = \frac{1}{\sqrt{2}}\begin{pmatrix}1\\1\end{pmatrix}\cdot\begin{matrix}\rightarrow |+\rangle\\ \rightarrow |-\rangle\end{matrix} \quad (2.14)$$

It should be noted that *it has different column vectors in different bases*. This is because each row represents the amount of basis vector, indicated after the right arrows, the vector contains. Also, we have used the *bra − ket* notation. It should be emphasized that while we label the basis vectors as $|0\rangle$, $|1\rangle$, $|+\rangle$, and $|-\rangle$, they are just some given notations. They are NOT the $|0\rangle$, $|1\rangle$, $|+\rangle$, and $|-\rangle$ in a qubit space, even though they are related through the same equations.

Example 2.2 Change of Basis: Derive Eq. (2.14) from Eq. (2.13).

Here we assume that we already know that $|0\rangle = \frac{1}{\sqrt{2}}(|+\rangle + |-\rangle)$ and $|1\rangle = \frac{1}{\sqrt{2}}(|+\rangle - |-\rangle)$. We could have derived them using trigonometry in Fig. 2.2 but we assume the readers know how to do it. Then,

$$|V_1\rangle = 1|0\rangle + 0|1\rangle,$$
$$= 1\left(\frac{1}{\sqrt{2}}(|+\rangle + |-\rangle)\right) + 0\left(\frac{1}{\sqrt{2}}(|+\rangle - |-\rangle)\right),$$
$$= \frac{1}{\sqrt{2}}(|+\rangle + |-\rangle). \quad (2.15)$$

∎

In Sect. 3.3.5, we will give a transformation matrix to facilitate basis transformation and discuss its properties.

2.3.2 High-Dimensional Vectors

For an n-dimensional space, it has a basis of n basis vectors. Every vector, $\vec{a} = |a\rangle$, in that space, can be expressed as a linear combination of the basis vectors as

$$\vec{a} = a_0 \hat{x}_0 + a_1 \hat{x}_1 + \cdots + a_{n-1} \hat{x}_{n-1}, \tag{2.16}$$

or in *bra-ket* notation, it is,

$$\begin{aligned}|a\rangle &= a_0 |x_0\rangle + a_1 |x_1\rangle + \cdots + a_{n-1} |x_{n-1}\rangle, \\ &= a_0 |0\rangle + a_1 |1\rangle + \cdots + a_{n-1} |n-1\rangle, \\ &= \begin{pmatrix} a_0 \\ a_1 \\ \vdots \\ a_{n-1} \end{pmatrix},\end{aligned} \tag{2.17}$$

where in line 2, we changed the names of the basis states to emphasize that whatever is put inside the *ket* is just a name. As long as it is not confusing, it does not matter how we name it. When we write the vector in a column form, we have assumed that the basis vectors are orthonormal, which will be discussed soon in Sect. 2.3.4.

Each vector has a corresponding vector in the *bra*-space (**dual correspondence**). This is similar to the fact that every object has an image in the mirror. The *bra* of $|b\rangle$ is written as $\langle b|$. And to construct the *bra* version of $|b\rangle$ in matrix form, we need to perform **conjugate transpose**. That is to swap the rows and columns and then apply complex conjugate to each element. Therefore, the column vector has a row vector in its *bra* version. For example, vector $|b\rangle$, which is expressed as

$$|b\rangle = b_0 |x_0\rangle + b_1 |x_1\rangle + \cdots + b_{n-1} |x_{n-1}\rangle = \begin{pmatrix} b_0 \\ b_1 \\ \vdots \\ b_{n-1} \end{pmatrix},$$

(2.18)

has its *bra* version expressed as

$$\begin{aligned}\langle b| &= b_0^* \langle x_0| + b_1^* \langle x_1| + \cdots + b_{n-1}^* \langle x_{n-1}|, \\ &= \begin{pmatrix} b_0^* & b_1^* & \cdots & b_{n-1}^* \end{pmatrix}.\end{aligned} \tag{2.19}$$

The *bra-ket* notation is very useful in linear algebra. For example, the inner product of two vectors, $|b\rangle$ and $|a\rangle$, is just the multiplication between the *bra* of

$|b\rangle$ and the *ket* of $|a\rangle$,

$$\langle b|a\rangle = a_0 b_0^* + a_1 b_1^* + \cdots + a_{n-1} b_{n-1}^*, \qquad (2.20)$$

which is the same as how we wrote it in Eq. (2.3).

Example 2.3 For $|a\rangle = \begin{pmatrix} 3i+2 \\ 0 \\ 5 \\ 4-2i \end{pmatrix}$ and $|b\rangle = \begin{pmatrix} 2 \\ i \\ 0 \\ 2i \end{pmatrix}$, find $\langle a|b\rangle$.

$$\langle a|b\rangle = \begin{pmatrix} -3i+2 & 0 & 5 & 4+2i \end{pmatrix} \begin{pmatrix} 2 \\ i \\ 0 \\ 2i \end{pmatrix},$$

$$= (-6i+4) + 0 + 0 + (8i-4) = 2i.$$

∎

2.3.3 *Measurement of a Quantum State*

Measurement is not a part of linear algebra. However, I would like to interject this topic so that we can understand the following sections better. The measurement of a quantum state results in the **collapse of the state** to one of its basis states. That means that the measurement outcome is one of the basis states and the original quantum state no longer exists. The process is completely random except that the probability it will collapse to a certain basis vector is the square of the magnitude of the corresponding coefficient (Fig. 2.3). For example, for $|\Psi\rangle = \alpha|0\rangle + \beta|1\rangle$, the probability it will collapse to $|0\rangle$ is

$$\text{Prob}(|0\rangle) = \alpha\alpha^* = |\alpha|^2, \qquad (2.21)$$

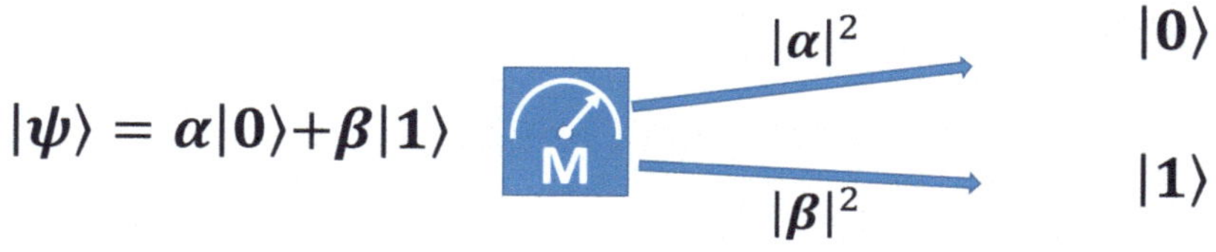

Fig. 2.3 Upon measurement, a state will collapse to one of the basis states with a probability equal to the square of the magnitude of the corresponding coefficient

2.3 Review of Vector Basics

and the probability it will collapse to $|1\rangle$ is

$$\text{Prob}(|1\rangle) = \beta\beta^* = |\beta|^2. \qquad (2.22)$$

2.3.4 Orthonormal Basis and Vector Normalization

An **orthonormal basis** is a basis with orthonormal basis vectors. Let it be an n-dimensional space and thus it has n basis vectors, $|x_0\rangle, |x_1\rangle, \cdots, |x_{n-1}\rangle$. If each basis vector is normalized (with a length of 1) and orthogonal to the others (with 0 overlap or inner product with other basis vectors), it is called an orthonormal basis. This can be written as

$$\langle x_i | x_j \rangle = \begin{cases} 0 & if\ i \neq j, \text{(orthogonal)} \\ 1 & if\ i = j, \text{(normalized)} \end{cases}$$
$$= \delta_{ij}, \qquad (2.23)$$

where we use the **Kronecker delta** in the last line. Note that when $i = j$, $\langle x_i\ x_j \rangle = \langle x_i | x_i \rangle$ and this is just the square of the norm of the basis vector, $|x_i\rangle$ (Eq. (2.7)). If it is one, then it means that the length is also one. Working on an orthonormal basis provides a lot of convenience in calculations due to the fact that $\langle x_i | x_j \rangle$ results in either 0 or 1. We can also thus write the vector in a column or row form as in Eqs. (2.18) and (2.19).

For example, if a vector $|V\rangle$ represented in an orthonormal basis, $|x_i\rangle$, is given by

$$|V\rangle = a_0 |x_0\rangle + \cdots + a_{n-1}|x_{n-1}\rangle, \qquad (2.24)$$

to find its norm squared, $\|v\|^2$ (Eq. (2.7)), we have

$$\langle v|v \rangle = (a_0^* a_1^* \cdots a_{n-1}^*) \begin{pmatrix} a_0 \\ a_1 \\ \vdots \\ a_{n-1} \end{pmatrix},$$
$$= a_0^* a_0 + a_1^* a_1 + \cdots + a_{n-1}^* a_{n-1},$$
$$= |a_0|^2 + |a_1|^2 + \cdots + |a_{n-1}|^2. \qquad (2.25)$$

If $\langle v|v \rangle = 1$, then vector $|v\rangle$ is a **normalized vector**. As shown in Eq. (2.25), this means that a normalized vector has the sum of the coefficient modulus squared equals one. Recalling that upon the measurement of a quantum state, the probability is the corresponding coefficient modulus squared (Eqs. (2.21) and (2.22)), then a quantum state must be normalized so that the probability of collapsing to *any* of the basis states is one. In other words, this ensures the sum of the probabilities of measuring one of the basis states to be one. Therefore, any quantum state must be a normalized vector.

2.4 Tensor Product

We can combine two or more vector spaces through **tensor product**. For a more detailed discussion, please refer to Chapters 11 and 12 in [1]. What is the meaning of *combing two vector spaces* and why do we want to do that? For example, we can describe the spin of an electron using a 2D Hilbert space. It has two basis vectors, $|0\rangle_1$ and $|1\rangle_1$. Here I used subscript 1 to indicate that this is the vector space belonging to the first electron. The state of any possible spin of the electron is a vector, $|\psi_1\rangle$, in this Hilbert space and is a linear combination of the basis states,

$$|\psi_1\rangle = \alpha_1 |0\rangle_1 + \beta_1 |1\rangle_1. \tag{2.26}$$

Similarly, if there is a second electron, its basis vectors are $|0\rangle_2$ and $|1\rangle_2$. Its state is given by,

$$|\psi_2\rangle = \alpha_2 |0\rangle_2 + \beta_1 |1\rangle_2. \tag{2.27}$$

If we want to describe the two electrons together or treat the two electrons as a *single* physical system, then the tensor product is the mathematical tool for us to do so.

As aligned with our common sense, the new system must have a larger vector space. Here, *let me emphasize* that a larger space is *NOT* obtained through a simple extension of a lower-dimension one to a higher-dimension one (e.g., adding a time dimension to the 3D space to become a 4D space-time). It is a result of the tensor product of two lower-dimension spaces. The basis states will be expanded as a tensor product, $\otimes$, of the lower space basis states. The number of the new basis states is the number of the permutations of the lower space basis states. For example, the two-electron system as a whole has four basis states, $|0\rangle_1 \otimes |0\rangle_2$, $|0\rangle_1 \otimes |1\rangle_2$, $|1\rangle_1 \otimes |0\rangle_2$, and $|1\rangle_1 \otimes |1\rangle_2$. We may also omit $\otimes$ by writing it as $|0\rangle_1 |0\rangle_2$, $|0\rangle_1 |1\rangle_2$, $|1\rangle_1 |0\rangle_2$, and $|1\rangle_1 |1\rangle_2$. And if we agree with each other that the first (second) number refers to the first (second) electron, we can also succinctly write it as $|00\rangle$, $|01\rangle$, $|10\rangle$, and $|11\rangle$.

With the new space, we also expect that a state in the combined system must be a linear combination of the new basis vectors. This can be seen clearly by considering the tensor product of the two-electron system. The following demonstrates how to perform tensor products without explaining the background. Readers can treat it as a result of the definitions and should appreciate its similarity to a regular algebraic product.

2.5 Summary

For example, if the first electron is in state $|\psi_1\rangle$ and the second electron is in state $|\psi_2\rangle$, then the state of the whole system, $|\psi\rangle$, is obtained through the tensor product of $|\psi_1\rangle$ and $|\psi_2\rangle$ (Eqs. (2.26) and (2.27)):

$$\begin{aligned}|\psi\rangle &= |\psi_1\rangle \otimes |\psi_2\rangle, \\ &= (\alpha_1|0\rangle_1 + \beta_1|1\rangle_1) \otimes (\alpha_2|0\rangle_2 + \beta_2|1\rangle_2), \\ &= \alpha_1\alpha_2|0\rangle_1|0\rangle_2 + \alpha_1\beta_2|0\rangle_1|1\rangle_2 + \beta_1\alpha_2|1\rangle_1|0\rangle_2 + \beta_1\beta_2|1\rangle_1|1\rangle_2, \\ &= \alpha_1\alpha_2|00\rangle + \alpha_1\beta_2|01\rangle + \beta_1\alpha_2|10\rangle + \beta_1\beta_2|11\rangle.\end{aligned} \quad (2.28)$$

We can also do this in matrix form,

$$\begin{aligned}|\psi\rangle &= |\psi_1\rangle \otimes |\psi_2\rangle, \\ &= \begin{pmatrix}\alpha_1 \\ \beta_1\end{pmatrix} \otimes \begin{pmatrix}\alpha_2 \\ \beta_2\end{pmatrix}, \\ &= \begin{pmatrix}\alpha_1\begin{pmatrix}\alpha_2 \\ \beta_2\end{pmatrix} \\ \beta_1\begin{pmatrix}\alpha_2 \\ \beta_2\end{pmatrix}\end{pmatrix}, \\ &= \begin{pmatrix}\alpha_1\alpha_2 \\ \alpha_1\beta_2 \\ \beta_1\alpha_2 \\ \beta_1\beta_2\end{pmatrix} \cdot \begin{matrix}|00\rangle \\ |01\rangle \\ |10\rangle \\ |11\rangle\end{matrix}\end{aligned} \quad (2.29)$$

where in the last line, the corresponding basis states are indicated. The same methodology is used for higher-dimensional spaces.

If more than 2 spaces need to be combined, we can do this one after another.

2.5 Summary

We review the basic properties of vectors in various vector spaces. In quantum computing, we will work in the Hilbert space. Therefore, the inner product which is an important component of the Hilbert space plays an important role in all calculations. We also discuss the measurement of a quantum state. Although measurement is not a part of linear algebra, it requires that all quantum state needs to be normalized. We also practice how to combine two subsystems into a larger one using tensor product. In the next chapter, we will discuss more advanced linear algebra. We will discuss matrices and operators and their applications in quantum computing.

Problems

2.1 Vector Space

a is a scalar and $|W\rangle$ is a vector. Given that,

$$|aW\rangle = a\,|W\rangle, \tag{2.30}$$

using also Eq. (2.1), prove the following equations:

$$\langle V|aW\rangle = a\,\langle V|W\rangle. \tag{2.31}$$

$$\langle aV|W\rangle = a^*\,\langle V|W\rangle. \tag{2.32}$$

2.2 Orthonormal Basis

Prove Eq. (2.25) using *bra-ket* notation (e.g., Eq. (2.24)) instead of using matrix form.

2.3 Tensor Product

Find the tensor product of $|a\rangle$ and $|b\rangle$ in Example 2.3.

References

1. Hiu-Yung Wong. *Introduction to Quantum Computing*. Springer, 2024.
2. Vector space. https://en.wikipedia.org/wiki/Vector_space. Accessed:2024-01-08.

Chapter 3
Linear Algebra—Operators, Matrices, and Quantum Gates

3.1 Introduction

Vectors are the fundamentals in a Hilbert space. They represent the state of a physical system corresponding to that Hilbert space. But a state or a vector is *passive*. What is interesting is the application of an operator to rotate a vector in the space or, in other words, to transform the state. An operator corresponds to a matrix. In this chapter, we will review the concepts of eigenvalue, eigenvector, Hermitian matrix, and unitary matrix. We will also review how to construct a projection operator and a unitary transformation matrix. Then we will revisit the meaning of a measurement in a quantum system using the new knowledge we have learned. Finally, we will discuss how to perform a tensor product for matrices.

3.1.1 Learning Outcomes

Understand the definitions of Hermitian matrix, unitary matrix, and project matrix; able to transform vectors and matrices from one basis to another; appreciate that applying an operator to a vector is to rotate the vector in its Hilbert space.

3.1.2 Teaching Videos

- Search for Ch3 in this playlist
  - https://tinyurl.com/3yhze3jn

- Other videos
    - https://youtu.be/Wrmigi645J4
    - https://youtu.be/Z_fXDssH2JA

3.2 Operators

An **operator** in a Hilbert space maps the vectors in one space to another space. However, sometimes, the second space is the same as the first one. In this case, an operator can be considered to be rotating a vector in its space. Here we will consider the operators that map the vectors to the same space. For example, an operator M applied to vector $|\alpha\rangle$ may give another vector $|\alpha'\rangle$,

$$|\alpha'\rangle = M |\alpha\rangle. \tag{3.1}$$

If the vector is represented as a column vector, it is natural that the operator must be a matrix.

The operators are linear and observe the distribution law,

$$M(C_\alpha |\alpha\rangle + C_\beta |\beta\rangle) = C_\alpha M |\alpha\rangle + C_\beta M |\beta\rangle, \tag{3.2}$$

where C_α and C_β are complex scalars and $|\beta\rangle$ is another vector.

3.2.1 Dual Correspondence

We introduced the concept of **dual correspondence** in Chap. 2 that every vector $|\alpha\rangle$ in the *ket* space has a corresponding vector $\langle\alpha|$ in the *bra* space. Therefore, there is a corresponding *bra* version for $|\alpha'\rangle$, i.e., $\langle\alpha'|$, in Eq. (3.1),

$$\langle\alpha'| \quad \Leftrightarrow \quad |\alpha'\rangle. \tag{3.3}$$

Is there a dual correspondence for the *ket*-space operator M? Yes, it is the **adjoint** of M, i.e.. $M^\dagger$. The adjoint of a matrix is just the **conjugate transpose** of that matrix. This is the same as how we find the *bra* version of a vector (e.g., Eq. (2.19)). Therefore, Eq. (3.3) can be rewritten as,

$$\langle\alpha| M^\dagger \quad \Leftrightarrow \quad M |\alpha\rangle. \tag{3.4}$$

Note that the operator is applied from the **right** of the *bra* vector. This can be easily appreciated by recognizing that a *bra* vector is a row vector instead of a column vector.

The dual equation of Eq. (3.2) is thus,

$$(C_\alpha^* \langle\alpha| + C_\beta^* \langle\beta|) M^\dagger = C_\alpha^* \langle\alpha| M^\dagger + C_\beta^* \langle\beta| M^\dagger. \tag{3.5}$$

Note that complex conjugate is taken for scalars C_α and C_β in the *bra* version.

Based on the first line in Eq. (2.6), we can further derive that

$$\langle\beta|\alpha'\rangle = \langle\alpha'|\beta\rangle^*,$$
$$\langle\beta| M |\alpha\rangle = \langle\alpha| M^\dagger |\beta\rangle^*. \tag{3.6}$$

These are some useful equations we will use in quantum mechanics.

3.3 Matrices

We mentioned that operators can be represented as matrices when the vectors are represented as row or column vectors. Therefore, it is important to understand some of the important properties of matrices.

3.3.1 Eigenvalues and Eigenvectors

A matrix maps (transforms) a vector to another vector. For a given matrix, there is a set of vectors to which it only scales by a scalar when it is applied. These vectors are called the **eigenvectors** of the matrix. The corresponding amounts it scales are the **eigenvalues** of the matrix. For example, if $|i\rangle$ is an eigenvector of M, then

$$M |i\rangle = \lambda_i |i\rangle, \tag{3.7}$$

where λ_i is a scalar and the eigenvalue of M, corresponding to the eigenvector, $|i\rangle$.

For an $n \times n$ matrix, it has n eigenvalues (counting multiplicities) over the complex field (the eigenvalues can be complex or real). For the same operator, it can be represented in a different matrix form if a different basis is chosen. For some matrices (**diagonalizable** matrices), if the eigenvectors are chosen to be the basis states (**eigenbasis**), then the matrix is a diagonal matrix with the eigenvalues along the diagonal,

$$M = \begin{pmatrix} \lambda_0 & 0 & \cdots & 0 \\ 0 & \lambda_1 & \cdots & 0 \\ \vdots & \vdots & \ddots & \vdots \\ 0 & 0 & \cdots & \lambda_{n-1} \end{pmatrix}_{\text{In } M\text{'s eigenbasis}}. \tag{3.8}$$

The process of finding the eigenbasis so that the matrix is in a diagonal form is called the **diagonalization**. Diagonalization is a very important tool in solving the Schrödinger equation. Note again that *not all matrices are diagonalizable*. Readers are encouraged to refer to Section 9.2 in [1] to review how to find the eigenvalues and eigenvectors and, thus, the diagonalization of a matrix.

If the eigenvectors and eigenvalues are given, we can also construct the matrix from the eigenvectors and eigenvalues using this equation,

$$M = \sum_{i=0}^{n-1} \lambda_i |i\rangle \langle i|. \tag{3.9}$$

This is trivial if the matrix is in the eigenbasis which has the form of Eq. (3.8). This is still true in general and can be proved by using the basis transformation to be discussed in Sect. 3.3.5.

3.3.2 Hermitian Matrix

We discussed earlier that the adjoint of an operator M is written as $M^\dagger$. When it is written as a matrix, the adjoint of M is its conjugate transpose. If the adjoint of a matrix equals itself, it is also called a **self-adjoint** or **Hermitian** matrix.

Example 3.1 Show that $\sigma_y = \begin{pmatrix} 0 & -i \\ i & 0 \end{pmatrix}$ is Hermitian.

$$\sigma_y^\dagger = \begin{pmatrix} 0 & -i \\ i & 0 \end{pmatrix}^{T*},$$

$$= \begin{pmatrix} 0 & i \\ -i & 0 \end{pmatrix}^*,$$

$$= \begin{pmatrix} 0 & -i \\ i & 0 \end{pmatrix} = \sigma_y. \tag{3.10}$$

Therefore, it is Hermitian. Here σ_y^{T*} refers to applying a transpose operation followed by complex conjugation to σ_y. ∎

3.3.3 Projection Operator

A projection operator, P, is an operator that satisfies the following equation:

$$P = PP, \tag{3.11}$$

3.3 Matrices

which means that applying it twice is the same as applying it once (**idempotent**). For our purpose, we want to be more specific on what it does. Therefore, we will label it as $P_{|v\rangle}$ to indicate that it can be an operator to extract the $|v\rangle$ component from any vectors. $|v\rangle$ needs to be a *normalized* vector. Therefore, $\langle v|v\rangle = 1$ (see Eq. (2.25) and after). For example, $P_{|v\rangle}|\alpha\rangle$ should give us the $|v\rangle$ component in $|\alpha\rangle$. To construct $P_{|v\rangle}$, we can use this equation,

$$P_{|v\rangle} = |v\rangle\langle v|. \tag{3.12}$$

Let us try two examples to understand better.

Example 3.2 Show $P_{|v\rangle} = P_{|v\rangle}P_{|v\rangle}$.

$$\begin{aligned}P_{|v\rangle}P_{|v\rangle} &= (|v\rangle\langle v|)(|v\rangle\langle v|), \\ &= |v\rangle(\langle v|v\rangle)\langle v|, \\ &= |v\rangle\langle v|, \\ &= P_{|v\rangle}.\end{aligned} \tag{3.13}$$

Therefore, as long as $|v\rangle$ is a *normalized* vector, $P_{|v\rangle}$ satisfies the definition of a projection operator. ∎

Example 3.3 In a 1-qubit system, a general state can be written as $|\psi\rangle = \alpha|0\rangle + \beta|1\rangle$. Find the $|0\rangle$ component of $|\psi\rangle$ using the corresponding projection operator.

It is trivial that the $|0\rangle$ component of $|\psi\rangle$ is $\alpha|0\rangle$ from the given expression. Let us use the projection operator to find it, too. Firstly, we recall that the column and row representations of $|0\rangle$ and $\langle 0|$ are,

$$|0\rangle = \begin{pmatrix} 1 \\ 0 \end{pmatrix},$$

$$\langle 0| = \begin{pmatrix} 1 & 0 \end{pmatrix}. \tag{3.14}$$

Therefore, the projection operator for $|0\rangle$ is

$$\begin{aligned}P_{|0\rangle} &= |0\rangle\langle 0|, \\ &= \begin{pmatrix} 1 \\ 0 \end{pmatrix}\begin{pmatrix} 1 & 0 \end{pmatrix}, \\ &= \begin{pmatrix} 1 & 0 \\ 0 & 0 \end{pmatrix}.\end{aligned} \tag{3.15}$$

We may use two methods to find the answer. Firstly, by using *bra-ket* notation, we have,

$$P_{|0\rangle} |\psi\rangle = |0\rangle \langle 0| (\alpha |0\rangle + \beta |1\rangle),$$
$$= |0\rangle (\alpha \langle 0|0\rangle + \beta \langle 0|1\rangle),$$
$$= |0\rangle \alpha = \alpha |0\rangle. \tag{3.16}$$

We may also use the matrix method and we have,

$$P_{|0\rangle} |\psi\rangle = \begin{pmatrix} 1 & 0 \\ 0 & 0 \end{pmatrix} \begin{pmatrix} \alpha \\ \beta \end{pmatrix},$$
$$= \begin{pmatrix} \alpha \\ 0 \end{pmatrix},$$
$$= \alpha \begin{pmatrix} 1 \\ 0 \end{pmatrix} = \alpha |0\rangle. \tag{3.17}$$

Both methods give the same result as expected. ■

3.3.4 Unitary Matrix

A **unitary matrix**, U, is a matrix that satisfies the following equations:

$$UU^\dagger = U^\dagger U = I,$$
$$U^\dagger = U^{-1}. \tag{3.18}$$

Unlike a Hermitian matrix which is equal to its adjoint, a unitary matrix has its **inverse** equal to its adjoint. The most important property of a unitary matrix is that it preserves the inner product of two vectors when both vectors are transformed by the same unitary matrix. For example, after the transformation, vectors $|g\rangle$ and $|f\rangle$ become $|g'\rangle = U |g\rangle$ and $|f'\rangle = U |f\rangle$, respectively. The inner product of the new vectors is

$$\langle g'|f'\rangle = (\langle g| U^\dagger)(U |f\rangle),$$
$$= \langle g| (U^\dagger U) |f\rangle,$$
$$= \langle g| I |f\rangle,$$
$$= \langle g|f\rangle, \tag{3.19}$$

3.3 Matrices

where Eqs. (3.4) and (3.18) are used in line 1 and line 3, respectively. Since a unitary matrix preserves the inner product of two vectors, it also *preserves the norm* of any vector as the norm of a vector is just the square root of the inner product of the vector to itself (Eq. (2.7)). Therefore, later we will see that a quantum gate must be unitary so that the state vector norm is not changed after each operation and keeps normalized.

It should also be noted that when a unitary matrix is written in matrix form, each of its columns is a normalized vector and is orthogonal to other columns. This is the same for the rows. This means that if the matrix is,

$$U = \begin{pmatrix} b_{0,0} & b_{0,1} & \cdots & b_{0,n-1} \\ b_{1,0} & b_{1,1} & \cdots & b_{1,n-1} \\ \vdots & \vdots & \ddots & \vdots \\ b_{n-1,0} & b_{n-1,1} & \cdots & b_{n-1,n-1} \end{pmatrix},$$

$$= \begin{pmatrix} |v_0\rangle & |v_1\rangle & \cdots & |v_{n-1}\rangle \end{pmatrix}, \tag{3.20}$$

where we have set

$$|v_i\rangle = \begin{pmatrix} b_{0,i} \\ b_{1,i} \\ \vdots \\ b_{n-1,i} \end{pmatrix}, \tag{3.21}$$

then we have

$$\langle v_i | v_j \rangle = \delta_{i,j}. \tag{3.22}$$

3.3.5 *Transformation of Basis*

Sometimes we want to work on a different basis for convenience. Then we need to perform an appropriate transformation of the vectors and matrices. For example, in Fig. 3.1, vector $|V\rangle$ might be originally represented in the old basis $|0\rangle / |1\rangle$ as $|V\rangle = \alpha_0 |0\rangle + \alpha_1 |1\rangle$. We want to find its representation in a new basis $|0'\rangle / |1'\rangle$ and it might be $|V\rangle = \alpha'_0 |0'\rangle + \alpha'_1 |1'\rangle$. We have done something similar in Fig. 2.2. Here, we want to show an equation to help us perform the transformation.

Suppose an n-dimensional vector is represented in a vector form with the basis vectors in the old basis being $|0\rangle, |1\rangle, \cdots, |n-1\rangle$. Now we want to represent it in a new basis with basis vectors $|0'\rangle, |1'\rangle, \cdots, |n-1'\rangle$. The transformation matrix

Fig. 3.1 Representation of vector $|V\rangle$ in the new basis $|0'\rangle/|1'\rangle$ is the same as the representation of vector $|V\rangle_{new}$ in the old basis $|0\rangle/|1\rangle$

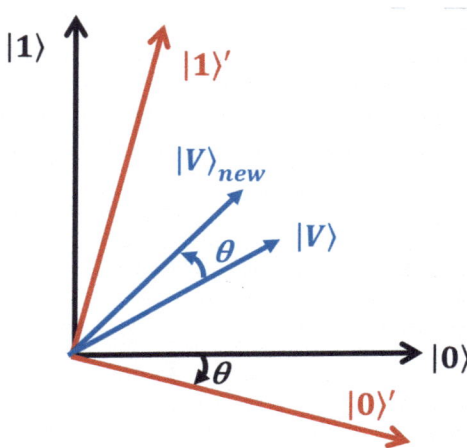

to represent a vector in the new basis is given by

$$U = \begin{pmatrix} \langle 0'|0\rangle & \langle 0'|1\rangle & \cdots & \langle 0'|n-1\rangle \\ \langle 1'|0\rangle & \langle 1'|1\rangle & \cdots & \langle 1'|n-1\rangle \\ \vdots & \vdots & \ddots & \vdots \\ \langle n-1'|0\rangle & \langle n-1'|1\rangle & \cdots & \langle n-1'|n-1\rangle \end{pmatrix}. \tag{3.23}$$

By using this matrix, we can find the coefficients of the vector in the new basis through matrix multiplication.

$$\begin{pmatrix} \alpha'_0 \\ \alpha'_1 \\ \vdots \\ \alpha'_{n-1} \end{pmatrix} = \begin{pmatrix} \langle 0'|0\rangle & \langle 0'|1\rangle & \cdots & \langle 0'|n-1\rangle \\ \langle 1'|0\rangle & \langle 1'|1\rangle & \cdots & \langle 1'|n-1\rangle \\ \vdots & \vdots & \ddots & \vdots \\ \langle n-1'|0\rangle & \langle n-1'|1\rangle & \cdots & \langle n-1'|n-1\rangle \end{pmatrix} \begin{pmatrix} \alpha_0 \\ \alpha_1 \\ \vdots \\ \alpha_{n-1} \end{pmatrix}. \tag{3.24}$$

We may better appreciate the meaning of this equation if we realize that the i-th row of the left-hand side (α'_i), which represents the amount of $|i'\rangle$ component in the new basis, is the sum of the amount of each component in the old basis (e.g., α_j) weighted by their overlaps (commons) with $|i'\rangle$, i.e., $\langle i'|j\rangle$.

Equation (3.24) also reveals another important thing. As discussed, a matrix applying to a vector is also a transformation of the vector. Therefore, the left-hand side can also be regarded as a new vector $|V\rangle_{new}$ after a certain operation *in the old basis*. What is this operation? As shown in Fig. 3.1, if the new basis can be obtained by rotating the old basis clockwise by an angle, θ, the operation is equivalent to a counterclockwise rotation of the vector by an angle, θ, in the old basis. In the figure, it can be seen that the representation of vector $|V\rangle$ in the new basis $|0'\rangle/|1'\rangle$ is the same as the representation of vector $|V\rangle_{new}$ in the old basis $|0\rangle/|1\rangle$. In general,

3.3 Matrices

when we represent a vector in a new basis formed by a transformation U^{-1} of the old basis, it is the same as transforming the vector in the old basis by its inverse, i.e., U.

Example 3.4 For the problem in Fig. 2.2, construct the transformation matrix to convert the representation of $|V_1\rangle$ in the old $|0\rangle / |1\rangle$ basis to the new $|+\rangle / |-\rangle$ basis.

Firstly, we recognize that the old basis has basis vectors $|0\rangle$ and $|1\rangle$. The new basis has basis vectors $|0'\rangle = |+\rangle$ and $|1'\rangle = |-\rangle$. Therefore, the transformation matrix is

$$U = \begin{pmatrix} \langle 0'|0\rangle & \langle 0'|1\rangle \\ \langle 1'|0\rangle & \langle 1'|1\rangle \end{pmatrix} = \begin{pmatrix} \langle +|0\rangle & \langle +|1\rangle \\ \langle -|0\rangle & \langle -|1\rangle \end{pmatrix},$$

$$= \begin{pmatrix} \frac{1}{\sqrt{2}} & \frac{1}{\sqrt{2}} \\ \frac{1}{\sqrt{2}} & -\frac{1}{\sqrt{2}} \end{pmatrix} = \frac{1}{\sqrt{2}} \begin{pmatrix} 1 & 1 \\ 1 & -1 \end{pmatrix}. \tag{3.25}$$

Now, if we apply U to $|V_1\rangle = |0\rangle$ as given in the first line of Eq. (2.15), we get

$$U|V_1\rangle = \frac{1}{\sqrt{2}} \begin{pmatrix} 1 & 1 \\ 1 & -1 \end{pmatrix} \begin{pmatrix} 1 \\ 0 \end{pmatrix} = \frac{1}{\sqrt{2}} \begin{pmatrix} 1 \\ 1 \end{pmatrix}, \tag{3.26}$$

which has the same coefficients as those in the last line of Eq. (2.15). As discussed, $U|V_1\rangle$ is also the vector formed after rotating $|V_1\rangle$ counterclockwise by $45°$ in the old basis. ∎

If two vectors $|g\rangle$ and $|f\rangle$ are represented in a new basis, we expect that their inner product will not change. Since representing the vectors in a new basis is equivalent to transforming the vectors in the old basis, i.e., $|g'\rangle = U|g\rangle$ and $|f'\rangle = U|f\rangle$, then it means that *the transformation matrix must be unitary* in order to preserve their inner product (See Eq. (3.19)). Therefore, it also obeys Eq. (3.18), i.e., $U^\dagger U = I$.

Similar to vectors, when the basis is changed, matrices also need to be transformed accordingly. A matrix M is transformed to M' through

$$M' = UMU^\dagger. \tag{3.27}$$

This is also called the **similarity transformation**. It is not difficult to see that this makes sense. Assume $|w\rangle = M|v\rangle$. This means that vector $|v\rangle$ is transformed by an operator M to another vector $|w\rangle$ and they are all in the same old basis. Now if we want to work on a new basis by applying the basis transformation matrix U, we have,

$$U|w\rangle = U(M|v\rangle),$$
$$= UMI|v\rangle,$$

$$= UM(U^\dagger U)|v\rangle,$$
$$= (UMU^\dagger)(U|v\rangle), \qquad (3.28)$$

which clearly shows that $|w\rangle$ in the new basis ($U|w\rangle$) equals the operator in the new basis (($UMU^\dagger$), Eq. (3.27)) multiplying $|v\rangle$ in the new basis ($U|v\rangle$). This preserves the relationship between the vectors in the old basis, i.e., $|w\rangle = M|v\rangle$.

3.4 Measurement of a Quantum State—Part 2

We have discussed some of the basics of measurement in Sect. 2.3.3. When a quantum state is measured, the state will collapse to one of the basis states. For example, if we perform a spin measurement on a spin qubit, the quantum state will collapse to either spin-up, $|\uparrow\rangle$, or spin-down, $|\downarrow\rangle$. Experimentally, we will also obtain a real number in the measurement (e.g., $\frac{1}{2}$ or $-\frac{1}{2}$).

In general, depending on what we are measuring, the basis states it will collapse to and the real values measured are the eigenvectors and eigenvalues, respectively, of a Hermitian matrix, A. For example, if we are performing a spin measurement of an electron, this measurement corresponds to the Hermitian matrix, $\frac{1}{2}\sigma_z = \begin{pmatrix} \frac{1}{2} & 0 \\ 0 & -\frac{1}{2} \end{pmatrix}$, which has eigenvalues of $\frac{1}{2}$ and $-\frac{1}{2}$. It should be noted that the corresponding Hermitian matrix is **NOT an operator to perform the measurement**. It is only that its eigenvectors are the states it will collapse to and its eigenvalues are the numbers being measured experimentally.

It should also be clear to the readers why the corresponding operators must be Hermitian. This is because the Hermitian matrix has real eigenvalues which are what will be measured experimentally.

We mentioned that the probability of a state $|\Psi\rangle = \alpha|0\rangle + \beta|1\rangle$ collapsing to one of the basis states (which are the eigenstates of the corresponding Hermitian matrix) is the square of the modulus of the corresponding coefficient. Here, we will give a more versatile definition of the probability, $Prob(|i\rangle)$, it will collapse to basis state $|i\rangle$. That is,

$$Prob(|i\rangle) = \langle\Psi|P_{|i\rangle}|\Psi\rangle, \qquad (3.29)$$

where the projection operator to $|i\rangle$ is used.

Example 3.5 Derive Eq. (2.21) using Eq. (3.29).

$$Prob(|0\rangle) = \langle\Psi|P_{|0\rangle}|\Psi\rangle,$$
$$= (\alpha^*\langle 0| + \beta^*\langle 1|)(|0\rangle\langle 0|)(\alpha|0\rangle + \beta|1\rangle),$$
$$= \alpha^*\langle 0|0\rangle\langle 0|\alpha|0\rangle,$$

$$= \alpha^*\alpha = |\alpha|^2, \tag{3.30}$$

where from line 2 to line 3, we have used the fact that $\langle 0|1\rangle = 0$. ∎

3.4.1 Expectation Value in a Measurement

If A is the Hermitian matrix corresponding to a measurement and has eigenvectors $|0\rangle$ and $|1\rangle$, then the **expectation value** or the average value obtained by performing the measurement on many identically prepared state $|\Psi\rangle$ is the sum of the eigenvalues (λ_0, λ_1) of each eigenvector weighted by the probability of the eigenvector to which $|\Psi\rangle$ will collapse. Therefore, the expectation value of A (or the average measured value) for the given state $|\Psi\rangle$ is

$$\begin{aligned}\langle A\rangle &= Prob(|0\rangle)\lambda_0 + Prob(|1\rangle)\lambda_1,\\ &= \langle\Psi|\,P_{|0\rangle}\,|\Psi\rangle\,\lambda_0 + \langle\Psi|\,P_{|1\rangle}\,|\Psi\rangle\,\lambda_1,\\ &= \langle\Psi|0\rangle\langle 0|\Psi\rangle\,\lambda_0 + \langle\Psi|1\rangle\langle 1|\Psi\rangle\,\lambda_1,\\ &= \langle\Psi|\,(|0\rangle\langle 0|\,\lambda_0 + |1\rangle\langle 1|\,\lambda_1)\,|\Psi\rangle,\\ &= \langle\Psi|\,A\,|\Psi\rangle.\end{aligned} \tag{3.31}$$

In the last line, we used the fact that working in A's eigenbasis, A is a diagonal matrix with the eigenvalues along the diagonal which is $A = |0\rangle\langle 0|\,\lambda_0 + |1\rangle\langle 1|\,\lambda_1$ (see Eqs. (3.8) and (3.9)).

3.5 Tensor Product of Matrices

In Sect. 2.4, we discussed how to construct a larger space by combining smaller spaces using the **tensor product**. The state/vector of the combined system can be described by the tensor product of the states/vectors of the smaller systems (Eq. (2.28)). Note that it can also be a linear combination of the tensor products if they are **entangled** which will be discussed later. We also need to create an operator for the combined system so that it is equivalent to the individual operators in the subsystems. For example, if M_1 is applied to $|\psi_1\rangle$ and M_2 is applied to $|\psi_2\rangle$, what is the equivalent operator M applied to state of the combined system, i.e., $|\psi_1\rangle\otimes|\psi_2\rangle$?

We construct M using a tensor product of M_1 and M_2,

$$M = M_1 \otimes M_2. \tag{3.32}$$

As a result, we have

$$\begin{aligned} M |\psi\rangle &= M(|\psi_1\rangle \otimes |\psi_2\rangle), \\ &= (M_1 \otimes M_2)(|\psi_1\rangle \otimes |\psi_2\rangle), \\ &= (M_1 |\psi_1\rangle) \otimes (M_2 |\psi_2\rangle). \end{aligned} \qquad (3.33)$$

Note that the operator in each subsystem only applies to the state in that system. For example, a magnetic pulse to rotate the spin state of electron 1 (M_1) is only physically applied to electron 1 and should not have an effect on electron 2. If it has, this is already an operator in the combined system.

When the operators are expressed in their matrix form, we follow the approach in Eq. (2.29) to perform the tensor product.

Example 3.6 If $M_1 = \begin{pmatrix} a & b \\ c & d \end{pmatrix}$ and $M_2 = \begin{pmatrix} e & f \\ g & h \end{pmatrix}$, find $M = M_1 \otimes M_2$.

$$\begin{aligned} M &= \begin{pmatrix} a & b \\ c & d \end{pmatrix} \otimes \begin{pmatrix} e & f \\ g & h \end{pmatrix}, \\ &= \begin{pmatrix} a \begin{pmatrix} e & f \\ g & h \end{pmatrix} & b \begin{pmatrix} e & f \\ g & h \end{pmatrix} \\ c \begin{pmatrix} e & f \\ g & h \end{pmatrix} & d \begin{pmatrix} e & f \\ g & h \end{pmatrix} \end{pmatrix}, \\ &= \begin{pmatrix} ae & af & be & bf \\ ag & ah & bg & bh \\ ce & cf & de & df \\ cg & ch & dg & dh \end{pmatrix}. \end{aligned} \qquad (3.34)$$

∎

3.6 Summary

In this chapter, we have reviewed some fundamental concepts of matrix. A Hermitian matrix has real eigenvalues. Therefore, all measurements must be corresponding to a Hermitian matrix. However, it is emphasized that the Hermitian matrix is not an operator that results in a measurement. It is only that its eigenvectors are the collapsed states after a measurement and its eigenvalues are the experimentally measured values. A unitary matrix preserves the inner products of vectors and preserves the vector norms. Therefore, the transformation matrix for basis changing must be unitary. We also learn how to create the operators of a combined system using a tensor product of the operators in the subsystems. Now, we have reviewed most of the essential basic linear algebra and we can start studying the physics of quantum computers, namely, the Schrödinger equation in the next chapter.

Problems

3.1 Dual Correspondence
Prove Eq. (3.6) by using Eq. (2.6).

3.2 Adjoint Matrix
Find the adjoint matrix of $\begin{pmatrix} i & 1 \\ 0 & i \end{pmatrix}$. Is it Hermitian? Is it unitary?

3.3 Transformation
How is a general vector $|\Psi\rangle = \begin{pmatrix} \alpha \\ \beta \end{pmatrix}$ transformed in the example in Fig. 2.2?

3.4 Tensor Product
Transform $|0\rangle_1 \otimes |0\rangle_2$ using the matrices in Eq. (3.34). Firstly, transform each qubit individually in its own space and then find the combined vector using a tensor product. Secondly, transform $|0\rangle_1 \otimes |0\rangle_2$ in the combined space using the corresponding matrix in the combined space. Show that both methods give the same result.

3.5 Diagonal Matrix
Show that this is a diagonal matrix by performing appropriate substitutions: $A = |0\rangle\langle 0| \lambda_0 + |1\rangle\langle 1| \lambda_1$. See also Eq. (3.31).

Reference

1. Hiu-Yung Wong. *Introduction to Quantum Computing*. Springer, 2024.

Chapter 4
Schrödinger Equation and Quantum Gates

4.1 Introduction

To realize an operator or a quantum gate, we need to set up the hardware so that it has the appropriate energy landscape, which is called the Hamiltonian. A quantum state will then evolve by following the Schrödinger equation of the given Hamiltonian. In this chapter, we will first study how to solve the Schrödinger equation using matrix mechanics with both diagonal and non-diagonal Hamiltonians. Then we will discuss how a quantum gate can be generated for a given Hamiltonian. We will then review a few important 1-qubit quantum gates. We will also discuss the CNOT gate, which is a 2-qubit entanglement gate, and demonstrate how to use it to create an entanglement state by combining it with other 1-qubit gates.

4.1.1 Learning Outcomes

Understand the meaning of the Schrödinger equation; be able to solve the Schrödinger equation in matrix form for different types of Hamiltonians; be familiar with the basic gates and entanglement creation.

4.1.2 Teaching Videos

- Search for Ch4 in this playlist
    - https://tinyurl.com/3yhze3jn

- Other videos
    - https://youtu.be/wyenXYGu5lo
    - https://youtu.be/DvJPM3ACkNw
    - https://youtu.be/Wrmigi645J4
    - https://youtu.be/tKx-JZg0qYk

4.2 Schrödinger Equation

The **Schrödinger equation** is the *governing equation* in quantum mechanics. It is difficult to solve. However, it is relatively easy if it is applied to a 1-qubit system, which is the case in most parts of this book. The Schrödinger equation is given as

$$i\hbar \frac{\partial}{\partial t} |\psi\rangle = \boldsymbol{H} |\psi\rangle, \qquad (4.1)$$

where i is the imaginary number, $\sqrt{-1}$, t is time, and $\hbar$ is the **reduced Planck constant**. $\hbar = \frac{h}{2\pi}$ with the **Planck constant**, $h = 6.626 \times 10^{-34} J \cdot s$. $\boldsymbol{H}$ is the **Hamiltonian** *of the system that we are investigating*. The Hamiltonian is the total energy of the system, which is the sum of the potential and kinetic energies. We will discuss it more in depth in Chap. 13. Here, we assume that $\boldsymbol{H}$ is given. Also, we write $\boldsymbol{H}$ in boldface because we treat it as a matrix here. In the following chapters, we will start writing it as an operator after we have learned the necessary knowledge. $|\psi\rangle$ is the state of the system.

Let us first descriptively understand what the Schrödinger equation tries to tell us. It says that the rate of change of the state ($\frac{\partial |\psi\rangle}{\partial t}$) is proportional to (scaled by $i\hbar$) the Hamiltonian multiplied by the state ($\boldsymbol{H} |\psi\rangle$).

Recalling that we represent a state as a vector, the Hamiltonian must be a matrix. Writing and solving the Schrödinger equation in this way is called the **matrix mechanics** as proposed by Heisenberg in contrast to Schrödinger's wave formulation. For finite (discrete) Hilbert spaces such as those of qubit systems, matrix mechanics is often more convenient.

Let us now consider a 1-qubit system. A single qubit is a 2D Hilbert space with complex scalars with two basis states, $|0\rangle$ and $|1\rangle$. Again, $|0\rangle$ and $|1\rangle$ are just the labels of the basis states and it does not matter what the underlying physics is. A general state in the system, $|\psi\rangle$, can be represented as a linear combination of the basis states

$$|\psi\rangle = \alpha |0\rangle + \beta |1\rangle = \begin{pmatrix} \alpha \\ \beta \end{pmatrix}, \qquad (4.2)$$

where α and β are complex scalars. If α and β are determined, then $|\psi\rangle$ is determined. Therefore, solving the Schrödinger equation for $|\psi\rangle$ in the 1-qubit system is equivalent to finding α and β.

4.3 Solving Schrödinger Equation

Since it is a 2D space, the matrix must be 2×2 in size. We assume the Hamiltonian to be

$$H = \begin{pmatrix} H_{00} & H_{01} \\ H_{10} & H_{11} \end{pmatrix}, \quad (4.3)$$

where H_{00}, H_{01}, H_{10}, and H_{11} are complex numbers. Then Eq. (4.1) becomes

$$i\hbar \frac{\partial}{\partial t} \begin{pmatrix} \alpha \\ \beta \end{pmatrix} = \begin{pmatrix} H_{00} & H_{01} \\ H_{10} & H_{11} \end{pmatrix} \begin{pmatrix} \alpha \\ \beta \end{pmatrix}. \quad (4.4)$$

To find α and β, we perform scalar multiplication on the left-hand side and matrix multiplication on the right-hand side,

$$\begin{pmatrix} i\hbar \frac{\partial \alpha}{\partial t} \\ i\hbar \frac{\partial \beta}{\partial t} \end{pmatrix} = \begin{pmatrix} H_{00} & H_{01} \\ H_{10} & H_{11} \end{pmatrix} \begin{pmatrix} \alpha \\ \beta \end{pmatrix},$$

$$= \begin{pmatrix} H_{00}\alpha + H_{01}\beta \\ H_{10}\alpha + H_{11}\beta \end{pmatrix}. \quad (4.5)$$

The vectors on the left and right are the same and so do their coefficients. Therefore, we obtain two simultaneous *differential equations*,

$$i\hbar \frac{\partial \alpha}{\partial t} = H_{00}\alpha + H_{01}\beta, \quad (4.6)$$

$$i\hbar \frac{\partial \beta}{\partial t} = H_{10}\alpha + H_{11}\beta. \quad (4.7)$$

To solve Eqs. (4.6) and (4.7), we need to solve a second-order differential equation by substituting one into another. We will study two cases to understand how to solve them in general.

4.3 Solving Schrödinger Equation

In general, a **matrix differential equation** with the following form,

$$i\hbar \frac{\partial}{\partial t} |\psi(t)\rangle = H |\psi\rangle; \quad |\psi(t=0)\rangle = |\psi_0\rangle, \quad (4.8)$$

has a general solution of

$$|\psi(t)\rangle = e^{\frac{H}{i\hbar}t} |\psi_0\rangle = e^{-i\frac{H}{\hbar}t} |\psi_0\rangle, \quad (4.9)$$

when $\boldsymbol{H}$ is a *constant matrix* (independent of time). If it is time-dependent, more sophisticated equations are needed and readers can refer to Chapter 2 in [2]. Therefore, if we know how to perform **matrix exponential**, we can obtain the solution, too.

4.3.1 Diagonal Hamiltonian

If the given Hamiltonian is diagonalized, then H_{01} and H_{10} are zero. A Hamiltonian (or, in general, an operator) is diagonal if the basis being used is the eigenbasis of the Hamiltonian (see Sect. 3.3.1). The equations to be solved become

$$i\hbar \frac{\partial \alpha}{\partial t} = H_{00}\alpha, \tag{4.10}$$

$$i\hbar \frac{\partial \beta}{\partial t} = H_{11}\beta. \tag{4.11}$$

We can see that now α and β are **decoupled**, and each equation contains only one variable and can be solved independently. Note that α and β refer to the amount of each basis state ($|0\rangle$ and $|1\rangle$) that $|\psi\rangle$ has. Therefore, **non-zero off-diagonal elements** *enable the coupling between different basis states*. For example, even if $\beta = 0$ at $t = 0$, eventually, β will become non-zero if there are non-zero off-diagonal elements which enable the coupling.

The solutions to the equations are

$$\alpha(t) = \alpha_0 e^{-i\frac{H_{00}}{\hbar}t}, \tag{4.12}$$

$$\beta(t) = \beta_0 e^{-i\frac{H_{11}}{\hbar}t}, \tag{4.13}$$

where α_0 and β_0 are constants and they are the initial values of $\alpha(t)$ and $\beta(t)$ at $t = 0$. One may substitute Eqs. (4.12) and (4.13) into Eqs. (4.10) and (4.11), respectively, to show that they are indeed the solutions.

Therefore, the state (vector) of the 1-qubit system changes as a function of time when it has a diagonal Hamiltonian $\boldsymbol{H} = \begin{pmatrix} H_{00} & 0 \\ 0 & H_{11} \end{pmatrix}$ as

$$|\psi\rangle = \begin{pmatrix} \alpha(t) \\ \beta(t) \end{pmatrix} = \begin{pmatrix} \alpha_0 e^{-i\frac{H_{00}}{\hbar}t} \\ \beta_0 e^{-i\frac{H_{11}}{\hbar}t} \end{pmatrix}. \tag{4.14}$$

We may also check this by using Eq. (4.9). When the Hamiltonian is *diagonal*, we can exponentiate it easily by only exponentiating the diagonal elements (the proof

4.3 Solving Schrödinger Equation

will be given in Example 4.1). That is,

$$e^{-i\frac{H}{\hbar}t} = e^{-i\frac{\begin{pmatrix} H_{00} & 0 \\ 0 & H_{11} \end{pmatrix}}{\hbar}t},$$

$$= \begin{pmatrix} e^{-i\frac{H_{00}}{\hbar}t} & 0 \\ 0 & e^{-i\frac{H_{11}}{\hbar}t} \end{pmatrix}. \quad (4.15)$$

Therefore, Eq. (4.9) becomes

$$|\psi(t)\rangle = e^{-i\frac{H}{\hbar}t}|\psi_0\rangle,$$

$$\begin{pmatrix} \alpha(t) \\ \beta(t) \end{pmatrix} = \begin{pmatrix} e^{-i\frac{H_{00}}{\hbar}t} & 0 \\ 0 & e^{-i\frac{H_{11}}{\hbar}t} \end{pmatrix} \begin{pmatrix} \alpha_0 \\ \beta_0 \end{pmatrix},$$

$$\begin{pmatrix} \alpha(t) \\ \beta(t) \end{pmatrix} = \begin{pmatrix} \alpha_0 e^{-i\frac{H_{00}}{\hbar}t} \\ \beta_0 e^{-i\frac{H_{11}}{\hbar}t} \end{pmatrix}, \quad (4.16)$$

which is the same as the solution in Eq. (4.14).

Example 4.1 Prove Eq. (4.15).

We prove this by using the Taylor expansion of $e^{-i\frac{H}{\hbar}t}$ and the definition of the zero exponent of a matrix, H,

$$H^0 = I. \quad (4.17)$$

The Taylor series of $e^{-i\frac{H}{\hbar}t}$ is

$$e^{-i\frac{H}{\hbar}t} = \sum_{k=0}^{\infty} \frac{(-i\frac{H}{\hbar}t)^k}{k!},$$

$$= \sum_{k=0}^{\infty} \frac{1}{k!} \begin{pmatrix} -i\frac{H_{00}}{\hbar}t & 0 \\ 0 & -i\frac{H_{11}}{\hbar}t \end{pmatrix}^k,$$

$$= \sum_{k=0}^{\infty} \frac{1}{k!} \begin{pmatrix} (-i\frac{H_{00}}{\hbar}t)^k & 0 \\ 0 & (-i\frac{H_{11}}{\hbar}t)^k \end{pmatrix},$$

$$= \begin{pmatrix} \sum_{k=0}^{\infty} \frac{1}{k!}(-i\frac{H_{00}}{\hbar}t)^k & 0 \\ 0 & \sum_{k=0}^{\infty} \frac{1}{k!}(-i\frac{H_{11}}{\hbar}t)^k \end{pmatrix},$$

$$= \begin{pmatrix} e^{-i\frac{H_{00}}{\hbar}t} & 0 \\ 0 & e^{-i\frac{H_{11}}{\hbar}t} \end{pmatrix}. \quad (4.18)$$

where in line 2, Eq. (4.3) is used to substitute H. In line 3, we use the fact that when a diagonal matrix multiplies itself, it is the same as each diagonal element multiplies itself. In line 4, we just use the definition of matrix summation, and in line 5, the Taylor series of number exponential is used. ∎

4.3.2 Non-diagonal Hamiltonian

If the Hamiltonian is not diagonal (i.e., at least one of the off-diagonal elements is non-zero), then we cannot use Eq. (4.15). We need to solve the system of linear equations in Eqs. (4.6) and (4.7). This is tedious. However, if we already know the eigenvalues and eigenvectors of H, we can work on its eigenbasis to find the solutions and then transform it back to the basis we are interested in. By the way, since H is the total energy of the system, its eigenvalues are also called the **eigenenergies**.

We had discussed how to construct a general transformation matrix in Sect. 3.3.5. Assume we are in an old basis with basis states $|0\rangle$ and $|1\rangle$. In this basis, H is *not* diagonal. We can work in the eigenbasis of H (the new basis with basis vectors $|0'\rangle$ and $|1'\rangle$) by creating a transformation matrix, U, based on Eq. (3.23)

$$U = \begin{pmatrix} \langle 0'|0\rangle & \langle 0'|1\rangle \\ \langle 1'|0\rangle & \langle 1'|1\rangle \end{pmatrix}. \tag{4.19}$$

Then Eq. (4.1) becomes

$$U i\hbar \frac{\partial}{\partial t} |\psi\rangle = U H I |\psi\rangle,$$

$$i\hbar \frac{\partial}{\partial t} U |\psi\rangle = U H U^\dagger U |\psi\rangle. \tag{4.20}$$

This is like how we derived Eq. (3.28) by applying U, which is time independent, from the left and using the identity, $U^\dagger U = I$. More specifically, we can set $|\psi'\rangle = U|\psi\rangle$ and $H' = UHU^\dagger$ which is *diagonal* and we can solve

$$i\hbar \frac{\partial}{\partial t} |\psi'\rangle = H' |\psi'\rangle, \tag{4.21}$$

as how we did in the diagonal Hamiltonian case in Eq. (4.16). After obtaining, $|\psi'\rangle$, we can get $|\psi\rangle$ by using

$$|\psi\rangle = U^\dagger |\psi'\rangle. \tag{4.22}$$

Of course, the difficulty is to find the eigenvectors of H which is computationally intensive when the matrix is large.

4.3.3 Using Taylor Expansion

In principle, we can also calculate matrix exponential using Taylor expansion. Sometimes, an analytical closed form can be found. In the following example, while the matrix can be diagonalized (see Problem 4.2), we use Taylor expansion to calculate the matrix exponential.

Example 4.2 This example will be used later when we try to construct an **iSWAP gate** for superconducting transmon qubits in Chap. 21. Given the following Hamiltonian, find $e^{-i\frac{H}{\hbar}t}$.

$$H = \begin{pmatrix} 0 & 0 & 0 & 0 \\ 0 & 0 & g & 0 \\ 0 & g & 0 & 0 \\ 0 & 0 & 0 & 0 \end{pmatrix}. \quad (4.23)$$

We will use the Taylor series of $e^{-i\frac{H}{\hbar}t}$.

$$e^{-i\frac{H}{\hbar}t} = \sum_{k=0}^{\infty} \frac{(-i\frac{H}{\hbar}t)^k}{k!},$$

$$= \sum_{k=0}^{\infty} \frac{1}{k!} \begin{pmatrix} 0 & 0 & 0 & 0 \\ 0 & 0 & \frac{-igt}{\hbar} & 0 \\ 0 & \frac{-igt}{\hbar} & 0 & 0 \\ 0 & 0 & 0 & 0 \end{pmatrix}^k. \quad (4.24)$$

Let us first study $\begin{pmatrix} 0 & 0 & 0 & 0 \\ 0 & 0 & \frac{-igt}{\hbar} & 0 \\ 0 & \frac{-igt}{\hbar} & 0 & 0 \\ 0 & 0 & 0 & 0 \end{pmatrix}^k$. When $k = 0$, it is just I as given in Eq. (4.17).

Also, we note that

$$\begin{pmatrix} 0 & 0 & 0 & 0 \\ 0 & 0 & \frac{-igt}{\hbar} & 0 \\ 0 & \frac{-igt}{\hbar} & 0 & 0 \\ 0 & 0 & 0 & 0 \end{pmatrix}^2 = \begin{pmatrix} 0 & 0 & 0 & 0 \\ 0 & 0 & \frac{-igt}{\hbar} & 0 \\ 0 & \frac{-igt}{\hbar} & 0 & 0 \\ 0 & 0 & 0 & 0 \end{pmatrix} \begin{pmatrix} 0 & 0 & 0 & 0 \\ 0 & 0 & \frac{-igt}{\hbar} & 0 \\ 0 & \frac{-igt}{\hbar} & 0 & 0 \\ 0 & 0 & 0 & 0 \end{pmatrix},$$

$$= \begin{pmatrix} 0 & 0 & 0 & 0 \\ 0 & -\left(\frac{gt}{\hbar}\right)^2 & 0 & 0 \\ 0 & 0 & -\left(\frac{gt}{\hbar}\right)^2 & 0 \\ 0 & 0 & 0 & 0 \end{pmatrix}. \tag{4.25}$$

Similarly,

$$\begin{pmatrix} 0 & 0 & 0 & 0 \\ 0 & 0 & \frac{-igt}{\hbar} & 0 \\ 0 & \frac{-igt}{\hbar} & 0 & 0 \\ 0 & 0 & 0 & 0 \end{pmatrix}^3 = \begin{pmatrix} 0 & 0 & 0 & 0 \\ 0 & 0 & i\left(\frac{gt}{\hbar}\right)^3 & 0 \\ 0 & i\left(\frac{gt}{\hbar}\right)^3 & 0 & 0 \\ 0 & 0 & 0 & 0 \end{pmatrix}, \tag{4.26}$$

$$\begin{pmatrix} 0 & 0 & 0 & 0 \\ 0 & 0 & \frac{-igt}{\hbar} & 0 \\ 0 & \frac{-igt}{\hbar} & 0 & 0 \\ 0 & 0 & 0 & 0 \end{pmatrix}^4 = \begin{pmatrix} 0 & 0 & 0 & 0 \\ 0 & \left(\frac{gt}{\hbar}\right)^4 & 0 & 0 \\ 0 & 0 & \left(\frac{gt}{\hbar}\right)^4 & 0 \\ 0 & 0 & 0 & 0 \end{pmatrix}, \tag{4.27}$$

and

$$\begin{pmatrix} 0 & 0 & 0 & 0 \\ 0 & 0 & \frac{-igt}{\hbar} & 0 \\ 0 & \frac{-igt}{\hbar} & 0 & 0 \\ 0 & 0 & 0 & 0 \end{pmatrix}^5 = \begin{pmatrix} 0 & 0 & 0 & 0 \\ 0 & 0 & -i\left(\frac{gt}{\hbar}\right)^5 & 0 \\ 0 & -i\left(\frac{gt}{\hbar}\right)^5 & 0 & 0 \\ 0 & 0 & 0 & 0 \end{pmatrix}. \tag{4.28}$$

We see that it has off-diagonal terms $(\frac{-igt}{\hbar})^k$ when k is odd, and it has diagonal terms $(\frac{-igt}{\hbar})^k$ when k is even. Therefore, the Taylor expansion can be written as

$$e^{-i\frac{H}{\hbar}t} = \mathbf{I} + \sum_{k=1}^{\infty} \frac{1}{k!} \begin{pmatrix} 0 & 0 & 0 & 0 \\ 0 & 0 & \frac{-igt}{\hbar} & 0 \\ 0 & \frac{-igt}{\hbar} & 0 & 0 \\ 0 & 0 & 0 & 0 \end{pmatrix}^k,$$

4.4 Relationship Between Hamiltonian and Quantum Gate

$$= \frac{1}{0!}\begin{pmatrix}1&0&0&0\\0&1&0&0\\0&0&1&0\\0&0&0&1\end{pmatrix} + \frac{1}{1!}\begin{pmatrix}0&0&0&0\\0&0&\frac{-igt}{\hbar}&0\\0&\frac{-igt}{\hbar}&0&0\\0&0&0&0\end{pmatrix} + \frac{1}{2!}\begin{pmatrix}0&0&0&0\\0&-\left(\frac{gt}{\hbar}\right)^2&0&0\\0&0&-\left(\frac{gt}{\hbar}\right)^2&0\\0&0&0&0\end{pmatrix}$$

$$+ \frac{1}{3!}\begin{pmatrix}0&0&0&0\\0&0&i\left(\frac{gt}{\hbar}\right)^3&0\\0&i\left(\frac{gt}{\hbar}\right)^3&0&0\\0&0&0&0\end{pmatrix} + \frac{1}{4!}\begin{pmatrix}0&0&0&0\\0&\left(\frac{gt}{\hbar}\right)^4&0&0\\0&0&\left(\frac{gt}{\hbar}\right)^4&0\\0&0&0&0\end{pmatrix}$$

$$+ \begin{pmatrix}0&0&0&0\\0&0&-i\left(\frac{gt}{\hbar}\right)^5&0\\0&-i\left(\frac{gt}{\hbar}\right)^5&0&0\\0&0&0&0\end{pmatrix} + \cdots ,$$

$$= \begin{pmatrix}1&0&0&0\\0&1-\frac{1}{2!}\left(\frac{gt}{\hbar}\right)^2+\frac{1}{4!}\left(\frac{gt}{\hbar}\right)^4-\cdots&-i\frac{1}{1!}\frac{gt}{\hbar}+i\frac{1}{3!}\left(\frac{gt}{\hbar}\right)^3-i\frac{1}{5!}\left(\frac{gt}{\hbar}\right)^5+\cdots&0\\0&-i\frac{1}{1!}\frac{gt}{\hbar}+i\frac{1}{3!}\left(\frac{gt}{\hbar}\right)^3-i\frac{1}{5!}\left(\frac{gt}{\hbar}\right)^5+\cdots&1-\frac{1}{2!}\left(\frac{gt}{\hbar}\right)^2+\frac{1}{4!}\left(\frac{gt}{\hbar}\right)^4-\cdots&0\\0&0&0&1\end{pmatrix},$$

$$= \begin{pmatrix}1&0&0&0\\0&\cos\frac{gt}{\hbar}&-i\sin\frac{gt}{\hbar}&0\\0&-i\sin\frac{gt}{\hbar}&\cos\frac{gt}{\hbar}&0\\0&0&0&1\end{pmatrix}, \tag{4.29}$$

where in the first line, we singled out $k = 0$ and summed from $k = 1$ to $k = \infty$. In the last line, we used the Taylor expansions of $\cos\frac{gt}{\hbar}$ and $\sin\frac{gt}{\hbar}$. ∎

4.4 Relationship Between Hamiltonian and Quantum Gate

A **quantum gate** is used to transform a quantum state (e.g., $|\psi_{in}\rangle$) to another (e.g., $|\psi_{out}\rangle$). Therefore, a quantum gate is a matrix, U, when the states are represented as column vectors. It is a 2×2 matrix for a 1-qubit system and it is a $2^n \times 2^n$ matrix for an n-qubit system. Therefore,

$$|\psi_{out}\rangle = U|\psi_{in}\rangle . \tag{4.30}$$

To implement a quantum gate (i.e., to implement U), we need to apply an appropriate Hamiltonian so that the initial state of the system will change to the desired state. Equation (4.9) describes how the initial state $|\psi_0\rangle$ changes to the final state at t, $|\psi(t)\rangle$. Note again this is only true if the Hamiltonian is time-independent. If we let $|\psi_0\rangle = |\psi_{in}\rangle$ and $|\psi(t)\rangle = |\psi_{out}\rangle$, then Eq. (4.9) is equivalent to Eq. (4.30) if

$$U = e^{-i\frac{H}{\hbar}t}. \tag{4.31}$$

Note that H can be diagonal or non-diagonal. Also, since H is the energy operator and its eigenvalues (or eigenenergies) are real, then it is also Hermitian with $H = H^\dagger$. That means

$$\begin{aligned} UU^\dagger, \\ &= e^{\frac{-iHt}{\hbar}} (e^{\frac{-iHt}{\hbar}})^\dagger, \\ &= e^{\frac{-iHt}{\hbar}} e^{\frac{iH^\dagger t}{\hbar}}, \\ &= e^{\frac{-iHt}{\hbar}} e^{\frac{iHt}{\hbar}}, \\ &= I. \end{aligned} \tag{4.32}$$

Therefore, any quantum gate, U, is *unitary* and *reversible*. We need to be careful not to be confused with the roles of the Hamiltonian and quantum gate in quantum state manipulation.

What I have shown is the quantum gate of a 1-qubit system. The idea is the same for multiple qubits.

4.5 Basic Quantum Gates

In this section, we will first review the definitions and properties of some basic quantum gates and their matrix form. Readers may refer to Chapters 15 to 18 in [1] for more detailed discussions.

It should be noted that (1) all quantum gates are defined based on how they transform the basis states, (2) every quantum gate has a corresponding matrix once the basis is chosen, and (3) all quantum gates must be **unitary** as discussed in Sect. 4.4.

4.5 Basic Quantum Gates

4.5.1 Identity Gate

The **identity gate**, I, as its name implies, keeps a vector unchanged. In physics, applying an identity gate to a state means leaving the state as it is without applying any interaction (H is a zero matrix in Eq. (4.1)). While it seems that it is trivial and redundant, it plays an important role when one needs to construct an operator for a larger space formed by a tensor product of smaller ones, in which one qubit goes through a non-trivial operator and one qubit is left unchanged.

An identity gate is defined as,

$$I \left|0\right\rangle = \left|0\right\rangle,$$
$$I \left|1\right\rangle = \left|1\right\rangle. \qquad (4.33)$$

Therefore, when it is applied to a general state $\left|\Psi\right\rangle = \alpha \left|0\right\rangle + \beta \left|1\right\rangle$, we have,

$$\begin{aligned} I \left|\Psi\right\rangle &= I(\alpha \left|0\right\rangle + \beta \left|1\right\rangle), \\ &= I\alpha \left|0\right\rangle + I\beta \left|1\right\rangle, \\ &= \alpha I \left|0\right\rangle + \beta I \left|1\right\rangle, \\ &= \alpha \left|0\right\rangle + \beta \left|1\right\rangle, \\ &= \left|\Psi\right\rangle. \end{aligned} \qquad (4.34)$$

The matrix of I is

$$I = \begin{pmatrix} 1 & 0 \\ 0 & 1 \end{pmatrix}. \qquad (4.35)$$

4.5.2 NOT Gate

The **NOT** gate, U_{NOT}, is defined as,

$$U_{NOT} \left|0\right\rangle = \left|1\right\rangle,$$
$$U_{NOT} \left|1\right\rangle = \left|0\right\rangle. \qquad (4.36)$$

Therefore, when it is applied to a general state $\left|\Psi\right\rangle = \alpha \left|0\right\rangle + \beta \left|1\right\rangle$, we have,

$$\begin{aligned} U_{NOT} \left|\Psi\right\rangle &= U_{NOT}(\alpha \left|0\right\rangle + \beta \left|1\right\rangle), \\ &= \alpha U_{NOT} \left|0\right\rangle + \beta U_{NOT} \left|1\right\rangle, \\ &= \alpha \left|1\right\rangle + \beta \left|0\right\rangle. \end{aligned} \qquad (4.37)$$

The matrix of U_{NOT} is

$$U_{NOT} = \begin{pmatrix} 0 & 1 \\ 1 & 0 \end{pmatrix}. \tag{4.38}$$

4.5.3 Phase Shift Gate

A **phase shift gate**, $U_{PS,\Phi}$, has a **gate parameter**, Φ. This parameter determines how much *relative* phase shift will be applied to a quantum state. A phase shift gate is defined as,

$$U_{PS,\Phi} |0\rangle = |0\rangle,$$
$$U_{PS,\Phi} |1\rangle = e^{i\Phi} |1\rangle. \tag{4.39}$$

Therefore, when it is applied to a general state $|\Psi\rangle = \alpha |0\rangle + \beta |1\rangle$, we have,

$$\begin{aligned} U_{PS,\Phi} |\Psi\rangle &= U_{PS,\Phi} (\alpha |0\rangle + \beta |1\rangle), \\ &= \alpha U_{PS,\Phi} |0\rangle + \beta U_{PS,\Phi} |1\rangle, \\ &= \alpha |0\rangle + \beta e^{i\Phi} |1\rangle. \end{aligned} \tag{4.40}$$

The matrix of $U_{PS,\Phi}$ is

$$U_{PS,\Phi} = \begin{pmatrix} 1 & 0 \\ 0 & e^{i\Phi} \end{pmatrix}. \tag{4.41}$$

If we set Φ to π, $\pi/2$, and $\pi/4$, we will obtain **Z**, **S**, and **T** gates, respectively. That is,

$$Z = \begin{pmatrix} 1 & 0 \\ 0 & e^{i\pi} \end{pmatrix} = \begin{pmatrix} 1 & 0 \\ 0 & -1 \end{pmatrix}. \tag{4.42}$$

$$S = \begin{pmatrix} 1 & 0 \\ 0 & e^{i\frac{\pi}{2}} \end{pmatrix} = \begin{pmatrix} 1 & 0 \\ 0 & i \end{pmatrix}. \tag{4.43}$$

$$T = \begin{pmatrix} 1 & 0 \\ 0 & e^{i\frac{\pi}{4}} \end{pmatrix}. \tag{4.44}$$

4.5 Basic Quantum Gates

It is important to recognize that $U_{PS,\Phi}$ only changes the **relative phase** between the coefficients of the two basis states, $|0\rangle$ and $|1\rangle$. It does *not* change the square of the modulus of the coefficients because $|e^{i\phi}| = 1$. Therefore, it does not change the probability of the state collapsing to each basis state (see Eqs. (2.21) and (2.22)).

4.5.4 Hadamard Gate

The **Hadamard gate**, H, is a quantum gate to create **superposition**. Although I am using the same symbol as the Hamiltonian, we should not be confused. A Hadamard gate is defined as,

$$H|0\rangle = \frac{1}{\sqrt{2}}(|0\rangle + |1\rangle) = |+\rangle,$$

$$H|1\rangle = \frac{1}{\sqrt{2}}(|0\rangle - |1\rangle) = |-\rangle, \quad (4.45)$$

where we have defined $|+\rangle = \frac{1}{\sqrt{2}}(|0\rangle + |1\rangle)$ and $|-\rangle = \frac{1}{\sqrt{2}}(|0\rangle - |1\rangle)$.

Therefore, when it is applied to a general state $|\Psi\rangle = \alpha|0\rangle + \beta|1\rangle$, we have,

$$H|\Psi\rangle = H(\alpha|0\rangle + \beta|1\rangle),$$
$$= \alpha H|0\rangle + \beta H|1\rangle,$$
$$= \alpha|+\rangle + \beta|-\rangle. \quad (4.46)$$

The matrix of H is

$$I = \frac{1}{\sqrt{2}}\begin{pmatrix} 1 & 1 \\ 1 & -1 \end{pmatrix}. \quad (4.47)$$

It should be noted that H equals its inverse H^{-1}; therefore,

$$HH = HH^{-1} = I. \quad (4.48)$$

4.5.5 CNOT Gate

The **CNOT** gate, U_{XOR}, is also called the **XOR** gate. CNOT stands for **controlled NOT**. It is a 2-qubit gate. This means that it is a 4×4 matrix. The Hamiltonian used to generate the gate (see Eq. (4.31)) is also a 4×4 matrix in the 4D space with basis states $|00\rangle$, $|01\rangle$, $|10\rangle$, and $|11\rangle$ (see Sect. 2.4)). A CNOT gate has one **control qubit** and one **target qubit**. Their meanings can be understood better by first looking at the definition of U_{XOR},

$$U_{XOR} |00\rangle = |0, 0 \oplus 0\rangle = |0, 0\rangle = |00\rangle ,$$
$$U_{XOR} |01\rangle = |0, 0 \oplus 1\rangle = |0, 1\rangle = |01\rangle ,$$
$$U_{XOR} |10\rangle = |1, 1 \oplus 0\rangle = |1, 1\rangle = |11\rangle ,$$
$$U_{XOR} |11\rangle = |1, 1 \oplus 1\rangle = |1, 0\rangle = |10\rangle , \quad (4.49)$$

which can be summarized as

$$U_{XOR} |ab\rangle = |a, a \oplus b\rangle , \quad (4.50)$$

with a and b taking the value of 0 or 1. $\oplus$ is the *classical* XOR operation, which explains why this gate is also called an XOR gate. We can also understand its definition from a "control-target" point of view. When it is applied to a basis *ket*, if the **most significant bit (MSB)** is 0 in the *ket*, the **least significant bit (LSB)** is unchanged. If the MSB is 1 in the *ket*, the LSB is flipped which is equivalent to having received a *classical* NOT operation. Therefore, the MSB is a control qubit, the LSB is a target qubit, and the operation is a controlled-NOT operation. When it is applied to a general state $|\Psi\rangle = \alpha |00\rangle + \beta |01\rangle + \gamma |10\rangle + \delta |11\rangle$, we have,

$$\begin{aligned} U_{XOR} |\Psi\rangle &= U_{XOR}(\alpha |00\rangle + \beta |01\rangle + \gamma |10\rangle + \delta |11\rangle), \\ &= \alpha U_{XOR} |00\rangle + \beta U_{XOR} |01\rangle + \gamma U_{XOR} |10\rangle + \delta U_{XOR} |11\rangle , \\ &= \alpha |00\rangle + \beta |01\rangle + \gamma |11\rangle + \delta |10\rangle , \\ &= \alpha |00\rangle + \beta |01\rangle + \delta |10\rangle + \gamma |11\rangle , \end{aligned} \quad (4.51)$$

which can be visualized in the top circuit in Fig. 4.1.

The matrix of U_{XOR} is

$$U_{XOR} = \begin{pmatrix} 1 & 0 & 0 & 0 \\ 0 & 1 & 0 & 0 \\ 0 & 0 & 0 & 1 \\ 0 & 0 & 1 & 0 \end{pmatrix} . \quad (4.52)$$

We may also swap the role of the MSB and LSB such that the LSB is the control qubit and the MSB is the target qubit (the bottom circuit in Fig. 4.1). In this case, we have

$$U_{XOR} |00\rangle = |0 \oplus 0, 0\rangle = |0, 0\rangle = |00\rangle ,$$
$$U_{XOR} |01\rangle = |0 \oplus 1, 1\rangle = |1, 1\rangle = |11\rangle ,$$
$$U_{XOR} |10\rangle = |1 \oplus 0, 0\rangle = |1, 0\rangle = |10\rangle ,$$

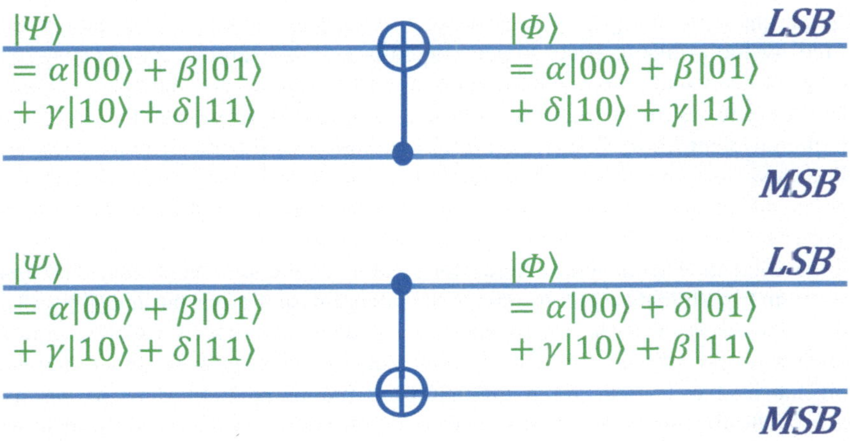

4.6 Entanglement

Fig. 4.1 Operation of a CNOT gate. Top: the MSB is the control qubit. Bottom: the LSB is the control qubit

$$U_{XOR}|11\rangle = |1 \oplus 1, 1\rangle = |0, 1\rangle = |01\rangle. \quad (4.53)$$

Therefore,

$$\begin{aligned}
U_{XOR}|\Psi\rangle &= U_{XOR}(\alpha|00\rangle + \beta|01\rangle + \gamma|10\rangle + \delta|11\rangle), \\
&= \alpha U_{XOR}|00\rangle + \beta U_{XOR}|01\rangle + \gamma U_{XOR}|10\rangle + \delta U_{XOR}|11\rangle, \\
&= \alpha|00\rangle + \beta|11\rangle + \gamma|10\rangle + \delta|01\rangle, \\
&= \alpha|00\rangle + \delta|01\rangle + \gamma|10\rangle + \beta|11\rangle,
\end{aligned} \quad (4.54)$$

and,

$$U_{XOR} = \begin{pmatrix} 1 & 0 & 0 & 0 \\ 0 & 0 & 0 & 1 \\ 0 & 0 & 1 & 0 \\ 0 & 1 & 0 & 0 \end{pmatrix}. \quad (4.55)$$

4.6 Entanglement

Entanglement, in addition to superposition, is what makes quantum computing powerful. A CNOT gate is an entanglement gate that can be used to create entangled states. In any quantum hardware, we need to be able to implement at least one entanglement gate. Other commonly used entanglement gates are controlled-phase shift gates and iSWAP gates. For different quantum computing architectures, different entanglement gates are used because the gates that can be generated are limited by the available Hamiltonian (Eq. (4.31)). For example, an electron spin qubit in silicon may use a controlled-phase shift gate as its entanglement gate (Chap. 12) and a superconducting qubit may use an iSWAP gate (Chap. 21). They are all equivalent to the CNOT gate after combining with some 1-qubit gates. We will discuss them individually in the following chapters. Here, we will review how to create an entanglement using the CNOT gate and other 1-qubit gates. The readers can refer to Chapters 13 and 14 in [1] for more details.

Figure 4.2 shows a quantum circuit for creating an entanglement state. The time flows from the left to the right. It starts with both qubits at the ground state $|0\rangle_A |0\rangle_B$. It then goes through a Hadamard gate for its MSB after which a CNOT gate is applied with the MSB as the control qubit. The figure has already shown how the qubit state evolves from the left to the right using bra-ket notation. Here, we will use matrix multiplication to obtain the same result. Of course, in matrix multiplication, the process goes from the right to the left in the *equation*. The output is

$$U_{XOR}(H \otimes I)|00\rangle,$$

$$= \begin{pmatrix} 1 & 0 & 0 & 0 \\ 0 & 1 & 0 & 0 \\ 0 & 0 & 0 & 1 \\ 0 & 0 & 1 & 0 \end{pmatrix} \frac{1}{\sqrt{2}} \begin{pmatrix} 1 & 1 \\ 1 & -1 \end{pmatrix} \otimes \begin{pmatrix} 1 & 0 \\ 0 & 1 \end{pmatrix} \begin{pmatrix} 1 \\ 0 \\ 0 \\ 0 \end{pmatrix},$$

$$= \begin{pmatrix} 1 & 0 & 0 & 0 \\ 0 & 1 & 0 & 0 \\ 0 & 0 & 0 & 1 \\ 0 & 0 & 1 & 0 \end{pmatrix} \frac{1}{\sqrt{2}} \begin{pmatrix} 1 & 0 & 1 & 0 \\ 0 & 1 & 0 & 1 \\ 1 & 0 & -1 & 0 \\ 0 & 1 & 0 & -1 \end{pmatrix} \begin{pmatrix} 1 \\ 0 \\ 0 \\ 0 \end{pmatrix},$$

$$= \begin{pmatrix} 1 & 0 & 0 & 0 \\ 0 & 1 & 0 & 0 \\ 0 & 0 & 0 & 1 \\ 0 & 0 & 1 & 0 \end{pmatrix} \frac{1}{\sqrt{2}} \begin{pmatrix} 1 \\ 0 \\ 1 \\ 0 \end{pmatrix},$$

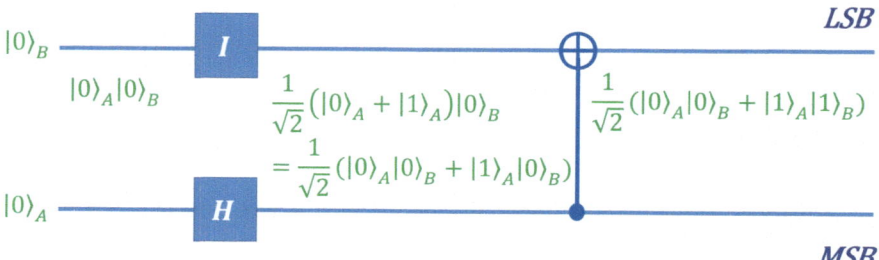

Fig. 4.2 An entanglement circuit formed by a 1-qubit Hadamard gate and a CNOT gate

$$= \frac{1}{\sqrt{2}} \begin{pmatrix} 1 \\ 0 \\ 0 \\ 1 \end{pmatrix},$$

$$= \frac{1}{\sqrt{2}}(|00\rangle + |11\rangle), \tag{4.56}$$

resulting in an entangled state.

4.7 Summary

We learn how to solve the Schrödinger equation using matrix formulation. We need to be careful when performing matrix exponential when the matrix is non-diagonal. We show that the Hamiltonian and the time the Hamiltonian is applied would determine the effective quantum gate. Therefore, the creation of quantum gates is nothing but **Hamiltonian engineering**. We review some of the fundamental 1-qubit gates. We also review the CNOT gate which can be used to create entanglement. It is important to note that the control qubit can be either the MSB or LSB. Finally, the CNOT gate might not be the native entanglement gate in some technologies. However, it is equivalent to them after combining with some 1-qubit quantum gates and we will discuss them in the following chapters.

Problems

4.1 The Schrödinger Equation
Show that Eqs. (4.12) and (4.13) are the solutions of Eqs. (4.10) and (4.11), respectively.

4.2 Matrix Exponential through Diagonalization
Find $e^{-i\frac{H}{\hbar}t}$ for the H in Eqs. (4.23) by first diagonalizing H. Then calculate the matrix exponential in the diagonal form. Using the transformation matrix to convert it back to the original basis. Note that you need to find the eigenvectors in this process. You will find that two of the eigenvalues are 0 and you want to pick the two corresponding eigenvectors to be orthonormal to others.

4.3 Phase Shift Gates
Using matrix multiplication, show that $T^4 = S^2 = Z$.

4.4 Entanglement Circuit
Create an entanglement circuit and use matrix multiplication to prove the function of the circuit when the control qubit is the LSB.

References

1. Hiu-Yung Wong. *Introduction to Quantum Computing*. Springer, 2024.
2. J.J. Sakurai. *Modern Quantum Mechanics*. Addison-Wesley, 1993.

Chapter 5
Bloch Sphere, Quantum Gates, and Pauli Matrices

5.1 Introduction

In this chapter, we will introduce the Bloch sphere to which a 2D complex space of a single qubit is mapped. The Bloch sphere is embedded in the real 3D space. As a result, the manipulation of a qubit by a quantum gate is equivalent to a rotation on the Bloch sphere. We will show the rotation matrix of an arbitrary quantum gate and its decomposition. Then we will discuss Pauli matrices and their properties. Pauli matrices are the generators of rotations. Understanding the properties of the Pauli matrices helps us derive many important equations. Finally, we will discuss the universal sets of quantum gates which can be used to implement all quantum gates.

5.1.1 Learning Outcomes

Understand the nature of the Bloch sphere and its relationship to the real 3D space; be able to construct the rotation matrix of any given rotation on the Bloch sphere; understand the properties of Pauli matrices.

5.1.2 Teaching Videos

- Search for Ch5 in this playlist
  - https://tinyurl.com/3yhze3jn

- Other videos
  - https://youtu.be/IRoYYJM8Gq8
  - https://youtu.be/MLkDyY91_GU
  - https://youtu.be/JR2jRCeTHDc
  - https://youtu.be/MTmQKP_9iJ0

5.2 Bloch Sphere

A single-qubit state, $|\Psi\rangle$, can be expressed as a linear combination of the two basis states, $|0\rangle$ and $|1\rangle$. Therefore, $|\Psi\rangle = \alpha|0\rangle + \beta|1\rangle$. α and β are complex numbers. Each complex number has a magnitude and a phase ($\alpha = |\alpha|e^{i\delta_\alpha}$ and $\beta = |\beta|e^{i\delta_\beta}$). Therefore, we need to fix four real numbers in order to fix α and β and, thus, $|\Psi\rangle$. As a result, $|\Psi\rangle$ has four **degrees of freedom (DOFs)**. So, it is impossible to visualize the 2D complex Hilbert space of a 1-qubit system in our real 3D space.

However, since a physical state vector must be normalized (Sect. 2.3.4), the DOF is reduced to 3 due to the constraint that $|\alpha|^2 + |\beta|^2 = 1$. We can set $|\alpha| = \cos\frac{\theta}{2}$ and $|\beta| = \sin\frac{\theta}{2}$ as $(\cos\frac{\theta}{2})^2 + (\sin\frac{\theta}{2})^2 = 1$ so that we use the parameter θ instead of $|\alpha|$ and $|\beta|$. We can then write the qubit state as

$$\begin{aligned}|\Psi\rangle &= \alpha|0\rangle + \beta|1\rangle, \\ &= |\alpha|e^{i\delta_\alpha}|0\rangle + |\beta|e^{i\delta_\beta}|1\rangle, \\ &= \cos\frac{\theta}{2}e^{i\delta_\alpha}|0\rangle + \sin\frac{\theta}{2}e^{i\delta_\beta}|1\rangle, \\ &= e^{i(\delta_\alpha+\delta_\beta)/2}(\cos\frac{\theta}{2}e^{i(\delta_\alpha-\delta_\beta)/2}|0\rangle + \sin\frac{\theta}{2}e^{i(-\delta_\alpha+\delta_\beta)/2}|1\rangle), \\ &= e^{i(\delta_\alpha+\delta_\beta)/2}(\cos\frac{\theta}{2}e^{-i\phi/2}|0\rangle + \sin\frac{\theta}{2}e^{i\phi/2}|1\rangle), \\ &= e^{i\gamma}(\cos\frac{\theta}{2}e^{-i\phi/2}|0\rangle + \sin\frac{\theta}{2}e^{i\phi/2}|1\rangle). \end{aligned} \quad (5.1)$$

We have performed some variable changes and substitutions. For example, we introduced $\gamma = (\delta_\alpha + \delta_\beta)/2$ and $\phi = \delta_\beta - \delta_\alpha$. Now, $|\Psi\rangle$ is described by three *real* parameters, γ, θ, and ϕ. $e^{i\gamma}$ gives the global phase of the state. It is not important when the qubit is isolated because the inner product or the expectation value of the state will not change. For example, if a phase of $e^{i\gamma'}$ is added to state $|\Psi\rangle$ so it becomes $|\Psi'\rangle = e^{i\gamma'}|\Psi\rangle$ and we perform a measurement corresponding to an observable M, the expectation value (Eq. (3.31)) is

$$\langle\Psi'|M|\Psi'\rangle = \langle\Psi|e^{-i\gamma'}Me^{i\gamma'}|\Psi\rangle = \langle\Psi|M|\Psi\rangle. \quad (5.2)$$

5.3 Quantum Gate and Bloch Sphere

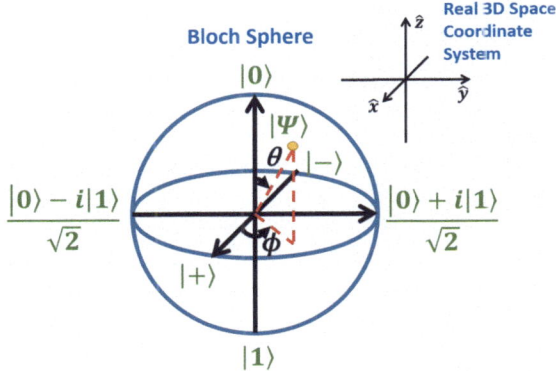

Fig. 5.1 Bloch sphere embedded in real 3D space

This is because complex conjugate is applied to the scalar phase associated with the *bra* (Eq. (3.5)) which cancels the extra phase from the *ket*.

So, we will ignore γ, and now the qubit is described by two parameters, θ, and ϕ, with 2 DOFs,

$$|\Psi\rangle = \cos\frac{\theta}{2}e^{-i\phi/2}|0\rangle + \sin\frac{\theta}{2}e^{i\phi/2}|1\rangle. \tag{5.3}$$

This equation happens to map to the surface of a unit sphere, **embedded** in our real 3D space, with θ corresponding to the **polar angle** and ϕ corresponding to the **azimuthal angle** on the equatorial plane. This allows us to visualize the hyperspace that the qubit resides in. But we need to remind ourselves that the qubit does **not** reside in the 3D space! This unit sphere is called the **Bloch sphere** (Fig. 5.1). Every point on the surface of the Bloch sphere corresponds to an infinite number of equivalent 1-qubit states which are different by only a global phase, $e^{i\gamma}$.

5.3 Quantum Gate and Bloch Sphere

Now with the introduction of the Bloch sphere, we can use it to help us understand better how a quantum gate transforms a state. Since every 1-qubit state resides on the surface of the Bloch sphere, the application of a quantum gate is just a rotation of the state on the Bloch sphere from one point to another. An arbitrary quantum gate, U, corresponds to a 2×2 unitary matrix (Eq. (4.31)). Each element in the matrix is a complex number. Therefore, there are 8 DOFs as there are eight real parameters in the matrix (cf. Sect. 5.2). Since U is unitary, its column vectors ($|v_0\rangle$ and $|v_1\rangle$) must be orthonormal (Eq. (3.22)) which imposes more constraints due to the three equations, namely, $\langle v_0|v_0\rangle = 1$, $\langle v_1|v_1\rangle = 1$, and $\langle v_0|v_1\rangle = 0$. The first two make sure the two column vectors are normalized. Since they are about the norm of the vector, they only have real numbers in the equations. So, each of them

Fig. 5.2 Decomposition of any 1-qubit gate into three Euler rotations with the global phase ignored (This is also called the "Z-Y" decomposition of a single-qubit gate)

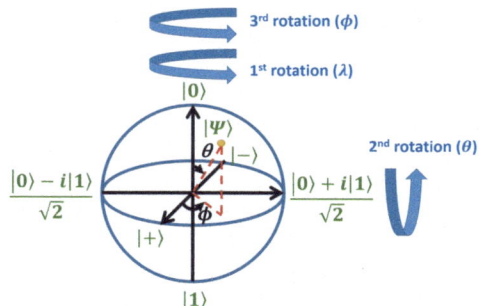

reduces the DOFs by one. For the third equation, it has both real and imaginary parts. Therefore, it is equivalent to two real equations (as one needs to equate the real and imaginary parts separately). Therefore, it reduces the DOFs by 2. As a result of being unitary, U only has 4 DOFs and can be described by four parameters. One possible representation of an arbitrary 1-qubit quantum gate is, thus,

$$U = U_{\theta,\phi,\lambda,\alpha} = e^{i\alpha} \begin{pmatrix} \cos\frac{\theta}{2} & -e^{i\lambda}\sin\frac{\theta}{2} \\ e^{i\phi}\sin\frac{\theta}{2} & e^{i(\lambda+\phi)}\cos\frac{\theta}{2} \end{pmatrix}, \quad (5.4)$$

where it is completely described by four real parameters, θ, ϕ, λ, and α.

This matrix corresponds to a series of three **Euler rotations** on the Bloch sphere and a global phase shift. Firstly, it rotates the state about the z-axis by λ. Then it rotates the state about the y-axis by θ followed by a rotation about the z-axis again by ϕ. Then it has an additional phase shift which is $e^{i\alpha}e^{i\frac{\lambda+\phi}{2}}$ (not $e^{i\alpha}$). We will prove this in Sect. 5.5. However, since a state loses its global phase information on the Bloch sphere and a global phase shift does not affect the observables, we do not need to worry about the actual value of the global phase shift. Figure 5.2 illustrates the rotations corresponding to an arbitrary 1-qubit gate without considering the global phase. This is also called the "Z-Y" decomposition of a single-qubit gate. One may also perform "X-Y" decomposition (see page 176 in [1]).

5.4 Pauli Matrices

Pauli matrices are very important in quantum computing. This is because Pauli matrices are proportional to the **spin angular momentum** of a spin qubit. Angular momentum is the **generator** of rotations (see Chapter 3 in [2]). The rotations turn out to be the rotations on the Bloch sphere embedded in the 3D space about some given axes. We will show this later. And even if it is not a spin qubit (such as a superconducting qubit), the problem can be transformed to use the same framework.

5.4 Pauli Matrices

There are three Pauli matrices. Since they are the operators for a 1-qubit 2D Hilbert space, they are 2×2 matrices,

$$\sigma_x = \sigma_1 = \begin{pmatrix} 0 & 1 \\ 1 & 0 \end{pmatrix}. \tag{5.5}$$

$$\sigma_y = \sigma_2 = \begin{pmatrix} 0 & -i \\ i & 0 \end{pmatrix}. \tag{5.6}$$

$$\sigma_z = \sigma_3 = \begin{pmatrix} 1 & 0 \\ 0 & -1 \end{pmatrix}. \tag{5.7}$$

We also label σ_x, σ_y, and σ_z as σ_1, σ_2, and σ_3, respectively, because this is convenient for indexing.

They have the following basic properties. Firstly, they are Hermitian. This is also expected when I tell you that Pauli matrices are proportional to the spin angular momentum of a spin qubit, which is an observable (see Sect. 3.4). Therefore,

$$\sigma_i = \sigma_i^\dagger \quad (i = 1, 2, 3). \tag{5.8}$$

Secondly,

$$\sigma_i{}^2 = \sigma_i \sigma_i = I \quad (i = 1, 2, 3). \tag{5.9}$$

5.4.1 Commutation Relation

Pauli matrices do *not* commute with each other. Together with Eq. (5.9), it has the following **commutation** property:

$$[\sigma_l, \sigma_m] = \sigma_l \sigma_m - \sigma_m \sigma_l = \sum_n 2i \epsilon_{lmn} \sigma_n. \tag{5.10}$$

ϵ_{lmn} is the **Levi-Civita** symbol. Sometimes we can also write Eq. (5.10) without the summation but require n to be *different from l and m*. The indices l, m, and n can be any of the values of 1, 2, or 3, such as ϵ_{223}, ϵ_{132}, ϵ_{123}, etc. If two of the indices are the same (e.g., $l = m = 2$ in ϵ_{223}), then it is zero (i.e., $\epsilon_{223} = 0$). Otherwise, it is -1 if lmn can be obtained by an odd number of pair exchanges in "123." It is 1 if lmn can be obtained by an even number of pair exchanges in "123." For example, ϵ_{132} is obtained by 1 (odd number) pair exchange in 123 by swapping "2" and "3." Therefore, $\epsilon_{132} = -1$. But ϵ_{123} is obtained by 0 (even number) pair exchanges in "123." Therefore, $\epsilon_{123} = 1$.

Example 5.1 Find ϵ_{xyz}, ϵ_{xzy}, ϵ_{zxy}, and ϵ_{yxz}.

Firstly, x, y, and z correspond to 1, 2, and 3, respectively (Eq. (5.5) to Eq. (5.7)). Therefore,

$$\epsilon_{xyz} = \epsilon_{123} = 1,$$
$$\epsilon_{xzy} = \epsilon_{132} = -1,$$
$$\epsilon_{zxy} = \epsilon_{312} = 1,$$
$$\epsilon_{yxz} = \epsilon_{213} = -1, \qquad (5.11)$$

where the question is structured in a way such that each line has an extra pair of exchange from the previous line. ∎

Example 5.2 Find $[\sigma_x, \sigma_y]$.

Based on Eq. (5.10) and by using x, y, and z directly,

$$[\sigma_x, \sigma_y] = \sigma_x\sigma_y - \sigma_y\sigma_x,$$
$$= 2i\epsilon_{xyz}\sigma_z,$$
$$= 2i\sigma_z. \qquad (5.12)$$

We may further show that this is true by using matrix multiplications.

$$\sigma_x\sigma_y - \sigma_y\sigma_x = \begin{pmatrix} 0 & 1 \\ 1 & 0 \end{pmatrix}\begin{pmatrix} 0 & -i \\ i & 0 \end{pmatrix} - \begin{pmatrix} 0 & -i \\ i & 0 \end{pmatrix}\begin{pmatrix} 0 & 1 \\ 1 & 0 \end{pmatrix},$$
$$= \begin{pmatrix} i & 0 \\ 0 & -i \end{pmatrix} - \begin{pmatrix} -i & 0 \\ 0 & i \end{pmatrix},$$
$$= \begin{pmatrix} 2i & 0 \\ 0 & -2i \end{pmatrix},$$
$$= 2i\sigma_z. \qquad (5.13)$$

∎

When $l = m$, then $\epsilon_{lmn} = 0$ and $[\sigma_l, \sigma_m] = \mathbf{0}$. That means it commutes with itself as $\sigma_l\sigma_m = \sigma_l\sigma_l = \sigma_m\sigma_l$.

5.4.2 Anti-commutation Relation

Pauli matrices anti-commute with the other. That means $\sigma_l\sigma_m + \sigma_m\sigma_l = \mathbf{0}$ if $l \neq m$. If $l = m$, then it is just two times of $\mathbf{I}$ due to Eq. (5.9). Their anti-commutation

5.4 Pauli Matrices

relation can be summarized as,

$$\{\sigma_l, \sigma_m\} = \sigma_l\sigma_m + \sigma_m\sigma_l = 2\delta_{lm}\mathbf{I}, \tag{5.14}$$

where we have used the **Kronecker delta** symbol (Eq. (2.23)).

Example 5.3 Find $\{\sigma_x, \sigma_y\}$ using matrix multiplication.

$$\begin{aligned}\{\sigma_x, \sigma_y\} &= \sigma_x\sigma_y + \sigma_y\sigma_x, \\ &= \begin{pmatrix} 0 & 1 \\ 1 & 0 \end{pmatrix}\begin{pmatrix} 0 & -i \\ i & 0 \end{pmatrix} + \begin{pmatrix} 0 & -i \\ i & 0 \end{pmatrix}\begin{pmatrix} 0 & 1 \\ 1 & 0 \end{pmatrix}, \\ &= \begin{pmatrix} i & 0 \\ 0 & -i \end{pmatrix} + \begin{pmatrix} -i & 0 \\ 0 & i \end{pmatrix}, \\ &= \begin{pmatrix} 0 & 0 \\ 0 & 0 \end{pmatrix}, \\ &= \mathbf{0}. \end{aligned} \tag{5.15}$$

■

5.4.3 Trace Properties

Pauli matrices are **traceless**. The **trace** of a matrix is the sum of the diagonal elements. A traceless matrix is a matrix with zero trace,

$$\text{tr}(\sigma_i) = 0 \quad (i = 1, 2, 3). \tag{5.16}$$

Example 5.4 Show that σ_z is traceless.

$$\text{tr}(\sigma_z) = \text{tr}\begin{pmatrix} 1 & 0 \\ 0 & -1 \end{pmatrix} = 1 + (-1) = 0. \tag{5.17}$$

Therefore, σ_z is traceless. ■

How about the trace of the product of Pauli matrices, $\text{tr}(\sigma_l\sigma_m)$? Again, l and m can be any of x, y, and z (or 1, 2, and 3). It is given by the following equation:

$$\text{tr}(\sigma_l\sigma_m) = 2\delta_{lm}. \tag{5.18}$$

This means that any product of two different Pauli matrices is *traceless*. Let us prove Eq. (5.18).

Example 5.5 Prove Eq. (5.18).

To be more instructive, we will prove by considering $l = m$ and $l \neq m$ separately. Firstly, we express it in terms of the commutator and anti-commutator,

$$\sigma_l \sigma_m = \frac{[\sigma_l, \sigma_m] + \{\sigma_l, \sigma_m\}}{2}. \tag{5.19}$$

When $l = m$, $\delta_{lm} = 1$ and $\epsilon_{lmn} = 0$. Based on Eqs. (5.10) and (5.14),

$$\sigma_l \sigma_m = \frac{\mathbf{0} + 2\mathbf{I}}{2} = \mathbf{I} = \begin{pmatrix} 1 & 0 \\ 0 & 1 \end{pmatrix}. \tag{5.20}$$

Therefore, $\mathrm{tr}(\sigma_l \sigma_m) = \mathrm{tr}(\mathbf{I}) = 1 + 1 = 2$ and Eq. (5.18) is correct.

Now consider when $l \neq m$. Then $\delta_{lm} = 0$.

$$\sigma_l \sigma_m = \frac{2i\epsilon_{lmn}\sigma_n + \mathbf{0}}{2} = \frac{2i\epsilon_{lmn}\sigma_n}{2}. \tag{5.21}$$

This does not give us the answer but we only care about the trace. Therefore,

$$\mathrm{tr}(\sigma_l \sigma_m) = \mathrm{tr}(\frac{2i\epsilon_{lmn}\sigma_n}{2}),$$
$$= \frac{2i\epsilon_{lmn}}{2}\mathrm{tr}(\sigma_n),$$
$$= 0. \tag{5.22}$$

where in line 3, we have used the fact that Pauli matrices are traceless (Eq. (5.16)). And again, Eq. (5.18) is correct. ∎

5.5 Universal Sets of Gates

Now we will discuss the idea of universal sets of quantum gates based on [1] with variations. This is similar to classical logic. In classical logic, any logical operation can be broken down into some universal sets of gates. For example, the set of NOT gate and AND gate (so only 2) can construct any logic. Constructing the circuit using a set of universal gates makes logic synthesis easier and also allows optimization on a limited set of gates. This is the same for quantum computing. If we can decompose all gates into a finite number of gates, efforts can be spent on optimizing those gates (e.g., to achieve the best microwave pulse shapes corresponding to the gates in the set). Moreover, we also want to use the gates that work with error correction.

We already know that every 1-qubit gate may be described by four parameters through Eq. (5.4). We claim that it can be decomposed into three rotations in Fig. 5.2 followed by a global phase shift. It turns out that spin angular momentum is the

5.5 Universal Sets of Gates

generator of rotation on the Bloch sphere. We will not discuss the meaning of "generator." Readers can refer to Chapter 3 in [2]. We will take it for granted. Since the Pauli matrices are proportional to spin angular momenta about the $\hat{x}$, $\hat{y}$, and $\hat{z}$ axes, respectively, they can be used to create (generate) the corresponding rotation matrices. The rotation matrices about $\hat{l}$, $R_l(\theta)$, for $l = x, y, z$, are

$$R_x(\theta) = e^{-i\theta\sigma_x/2} = \cos\frac{\theta}{2}I - i\sin\frac{\theta}{2}\sigma_x = \begin{pmatrix} \cos\frac{\theta}{2} & -i\sin\frac{\theta}{2} \\ -i\sin\frac{\theta}{2} & \cos\frac{\theta}{2} \end{pmatrix}, \quad (5.23)$$

$$R_y(\theta) = e^{-i\theta\sigma_y/2} = \cos\frac{\theta}{2}I - i\sin\frac{\theta}{2}\sigma_y = \begin{pmatrix} \cos\frac{\theta}{2} & -\sin\frac{\theta}{2} \\ \sin\frac{\theta}{2} & \cos\frac{\theta}{2} \end{pmatrix}, \quad (5.24)$$

$$R_z(\theta) = e^{-i\theta\sigma_z/2} = \cos\frac{\theta}{2}I - i\sin\frac{\theta}{2}\sigma_z = \begin{pmatrix} e^{-i\frac{\theta}{2}} & 0 \\ 0 & e^{i\frac{\theta}{2}} \end{pmatrix}. \quad (5.25)$$

The exponential terms, $e^{-i\theta\sigma_l/2}$, in Eq. (5.23) to Eq. (5.25) have the form of $e^{-i\frac{H}{\hbar}t}$ as in Eq. (4.31). Therefore, we need to find a Hamiltonian that is proportional to the Pauli matrices to perform rotations and this will be discussed when we study the actual implementation.

It can be shown that the matrix in Eq. (5.4) can be decomposed into

$$\begin{aligned} U_{\theta,\phi,\lambda,\alpha} &= e^{i\alpha}\begin{pmatrix} \cos\frac{\theta}{2} & -e^{i\lambda}\sin\frac{\theta}{2} \\ e^{i\phi}\sin\frac{\theta}{2} & e^{i(\lambda+\phi)}\cos\frac{\theta}{2} \end{pmatrix}, \\ &= e^{i\alpha}e^{i\frac{\lambda+\phi}{2}}R_z(\phi)R_y(\theta)R_z(\lambda), \\ &= e^{i\alpha'}R_z(\phi)R_y(\theta)R_z(\lambda). \end{aligned} \quad (5.26)$$

Therefore, to implement any 1-qubit gate, we only need to be able to perform a rotation about the $\hat{z}$ and $\hat{y}$ for an arbitrary angle and implement a global phase shifter. The global phase shifter can be denoted as an operator as $e^{i\alpha'}I$, where α' is the total global phase in Eq. (5.26).

However, we also need to perform entanglement operations for 2 or more qubits. We need a 2-qubit CNOT gate, U_{XOR} (Eq. (4.49)). Moreover, in order to implement controlled operations, R_x is also required (see Chapter 4 in [1]). Therefore, one of the **universal sets of quantum gates** is $\{e^{i\alpha'}I, R_x(\theta_x), R_y(\theta_y), R_z(\theta_z), U_{XOR}\}$.

Although there are only five types of gates in this set of quantum gates, four of them are *continuous* due to the continuous parameters, $\alpha', \theta_x, \theta_y$, and θ_z. So, strictly speaking, we still need to be able to implement infinite numbers of gates, although the number of types is limited to 5.

It is also possible to derive a universal set of quantum gates without continuous parameters. This set contains $\{H, S, T, U_{XOR}\}$ (see also Eqs. (4.43) and (4.44)). However, they are *not* exact. But they can infinitely approximate any quantum gates.

5.5.1 Some Useful Mathematics

It will be instructive to show Eqs. (5.26) and (5.23) to (5.25) are correct, through which we will practice some important skills.

Example 5.6 Prove Eq. (5.26).

This is just a simple matrix multiplication and we will work backward.

$$e^{i\alpha} e^{i\frac{\lambda+\phi}{2}} R_z(\phi) R_y(\theta) R_z(\lambda),$$

$$= e^{i\alpha} e^{i\frac{\lambda+\phi}{2}} \begin{pmatrix} e^{-i\frac{\phi}{2}} & 0 \\ 0 & e^{i\frac{\phi}{2}} \end{pmatrix} \begin{pmatrix} \cos\frac{\theta}{2} & -\sin\frac{\theta}{2} \\ \sin\frac{\theta}{2} & \cos\frac{\theta}{2} \end{pmatrix} \begin{pmatrix} e^{-i\frac{\lambda}{2}} & 0 \\ 0 & e^{i\frac{\lambda}{2}} \end{pmatrix},$$

$$= e^{i\alpha} e^{i\frac{\lambda+\phi}{2}} \begin{pmatrix} e^{-i\frac{\phi}{2}} & 0 \\ 0 & e^{i\frac{\phi}{2}} \end{pmatrix} \begin{pmatrix} e^{-i\frac{\lambda}{2}}\cos\frac{\theta}{2} & -e^{i\frac{\lambda}{2}}\sin\frac{\theta}{2} \\ e^{-i\frac{\lambda}{2}}\sin\frac{\theta}{2} & e^{i\frac{\lambda}{2}}\cos\frac{\theta}{2} \end{pmatrix},$$

$$= e^{i\alpha} e^{i\frac{\lambda+\phi}{2}} \begin{pmatrix} e^{-i\frac{\lambda+\phi}{2}}\cos\frac{\theta}{2} & -e^{i\frac{\lambda-\phi}{2}}\sin\frac{\theta}{2} \\ e^{-i\frac{\lambda-\phi}{2}}\sin\frac{\theta}{2} & e^{i\frac{\lambda+\phi}{2}}\cos\frac{\theta}{2} \end{pmatrix},$$

$$= e^{i\alpha} \begin{pmatrix} \cos\frac{\theta}{2} & -e^{i\lambda}\sin\frac{\theta}{2} \\ e^{i\phi}\sin\frac{\theta}{2} & e^{i(\lambda+\phi)}\cos\frac{\theta}{2} \end{pmatrix}. \qquad (5.27)$$

∎

Example 5.7 Prove Eq. (5.24).

We will first show $e^{-i\theta\sigma_y/2} = \cos\frac{\theta}{2} I - i\sin\frac{\theta}{2}\sigma_y$. We first express the matrix exponential in the form similar to the sin and cos functions ($\sin x = \frac{e^{ix} - e^{-ix}}{2i}$ and $\cos x = \frac{e^{ix} + e^{-ix}}{2}$),

$$e^{-i\theta\sigma_y/2} = \frac{e^{-i\theta\sigma_y/2} + e^{i\theta\sigma_y/2}}{2} + \frac{e^{-i\theta\sigma_y/2} - e^{i\theta\sigma_y/2}}{2}, \qquad (5.28)$$

which is composed of a sum term and a difference term. Now, we will use the Taylor expansion of matrix exponential as in Eq. (4.18),

$$e^{-i\theta\sigma_y/2} = \sum_{k=0}^{\infty} \frac{(-i\theta\sigma_y/2)^k}{k!}, \qquad (5.29)$$

$$e^{i\theta\sigma_y/2} = \sum_{k=0}^{\infty} \frac{(i\theta\sigma_y/2)^k}{k!}. \qquad (5.30)$$

These two equations are the same except that the terms with odd k have different signs in Eqs. (5.29) and (5.30). Therefore, for the sum term in Eq. (5.28), only even

5.5 Universal Sets of Gates

k terms are left. So,

$$\frac{e^{-i\theta\sigma_y/2} + e^{i\theta\sigma_y/2}}{2}$$

$$= \frac{1}{2}\left[\frac{2(-i\theta\sigma_y/2)^0}{0!} + \frac{2(-i\theta\sigma_y/2)^2}{2!} + \frac{2(-i\theta\sigma_y/2)^4}{4!} + \cdots\right],$$

$$= \left[\frac{(-i\theta/2)^0}{0!} + \frac{(-i\theta/2)^2}{2!} + \frac{(-i\theta/2)^4}{4!} + \cdots\right]I,$$

$$= \left[\frac{(\theta/2)^0}{0!} - \frac{(\theta/2)^2}{2!} + \frac{(\theta/2)^4}{4!} + \cdots\right]I,$$

$$= \cos\frac{\theta}{2}I, \tag{5.31}$$

where we have used the fact that $\sigma_y\sigma_y = I$ from line 2 to line 3 (Eq. (5.9)). As a result, any even power of σ_y is I.

And for the difference term in Eq. (5.28), only odd k terms are left. So,

$$\frac{e^{-i\theta\sigma_y/2} - e^{i\theta\sigma_y/2}}{2},$$

$$= \frac{1}{2}\left[\frac{2(-i\theta\sigma_y/2)^1}{1!} + \frac{2(-i\theta\sigma_y/2)^3}{3!} + \frac{2(-i\theta\sigma_y/2)^5}{5!} + \cdots\right],$$

$$= \left[\frac{(-i\theta/2)^1}{1!} + \frac{(-i\theta/2)^3}{3!} + \frac{(-i\theta/2)^5}{5!} + \cdots\right]\sigma_y,$$

$$= -i\left[\frac{(\theta/2)^1}{1!} - \frac{(\theta/2)^3}{3!} + \frac{(\theta/2)^5}{5!} + \cdots\right]\sigma_y,$$

$$= -i\sin\frac{\theta}{2}\sigma_y, \tag{5.32}$$

where we used $\sigma_y\sigma_y = I$ again. But since each term has an old power of σ_y, they are all evaluated to be σ_y, instead of I. Therefore, we have proved $e^{-i\theta\sigma_y/2} = \cos\frac{\theta}{2}I - i\sin\frac{\theta}{2}\sigma_y$.

Now, we will prove $\cos\frac{\theta}{2}I - i\sin\frac{\theta}{2}\sigma_y = \begin{pmatrix} \cos\frac{\theta}{2} & -\sin\frac{\theta}{2} \\ \sin\frac{\theta}{2} & \cos\frac{\theta}{2} \end{pmatrix}$. This is straightforward by just performing matrix addition,

$$\cos\frac{\theta}{2}I - i\sin\frac{\theta}{2}\sigma_y,$$

$$= \cos\frac{\theta}{2}\begin{pmatrix} 1 & 0 \\ 0 & 1 \end{pmatrix} - i\sin\frac{\theta}{2}\begin{pmatrix} 0 & -i \\ i & 0 \end{pmatrix},$$

$$= \begin{pmatrix} \cos\frac{\theta}{2} & 0 \\ 0 & \cos\frac{\theta}{2} \end{pmatrix} + \begin{pmatrix} 0 & -\sin\frac{\theta}{2} \\ \sin\frac{\theta}{2} & 0 \end{pmatrix},$$

$$= \begin{pmatrix} \cos\frac{\theta}{2} & -\sin\frac{\theta}{2} \\ \sin\frac{\theta}{2} & \cos\frac{\theta}{2} \end{pmatrix}. \tag{5.33}$$

∎

5.6 Summary

If we ignore the global phase, due to the normalization requirement, a 1-qubit state can be described by two real parameters. As a result, we can map the 2D complex space to the Bloch sphere surface which can be embedded in the real 3D space. The Bloch sphere provides a lot of convenience for our understanding of qubit manipulation. For example, any 1-qubit gate can be described by four real parameters. If the global phase is ignored again, an arbitrary 1-qubit quantum gate can be decomposed into a rotation about the $\hat{z}$ followed by a rotation about $\hat{y}$ and then followed by a third rotation about $\hat{z}$. These rotations can be generated using Pauli matrices. We have learned some important properties of the Pauli matrices. We have also discussed that we may use $\{e^{i\alpha'}I, R_x(\theta_x), R_y(\theta_y), R_z(\theta_y), U_{XOR}\}$ as a universal set of quantum gates. However, some of the gates have continuous parameters. If we allow approximations, $\{H, S, T, U_{XOR}\}$ can be used as a universal set of quantum gates instead.

Problems

5.1 Qubit Gate Matrix
Prove U in Eq. (5.4) is unitary.

5.2 Pauli Matrices
Prove Eqs. (5.8) and (5.9).

5.3 Rotation Matrices
Prove Eqs. (5.23) and (5.25). See Sect. 5.5.1.

5.4 Single Qubit State Representation
Compare Eq. (5.3) to Eq. (1.4) in [1] and argue that they are equivalent.

5.5 Rotation Matrix Representation
Compare Eq. (5.4) to Eqs. (1.17) and (4.12) in [1] and argue that they are equivalent.

References

1. M. A. Nielsen and I. L. Chuang. *Quantum Computation and Quantum Information: 10th Anniversary Edition*. Cambridge University Press, 2011.
2. J.J. Sakurai. *Modern Quantum Mechanics*. Addison-Wesley, 1993.

Chapter 6
Density Matrix and the Bloch Sphere

6.1 Introduction

In this chapter, we will first study the real vector space of 2×2 Hermitian matrices. This is a vector space although its elements are matrices. We will show that any 2×2 Hermitian matrix can be represented as a linear combination of the Pauli matrices and the identity matrix. We will then introduce the inner product of matrices, which is just a natural extension of the concept we learned in a regular vector space. Then we will discuss the concept of density matrix, which is very useful for describing mixed states due to the lack of information or the system being entangled with the external environment. Finally, we will show how to find the expectation values of an operator using a density matrix. Particularly, when the operator is a linear combination of the Pauli matrices, its expectation value is just the projection of the state (including mixed state) on the corresponding normalized Bloch vector on the Bloch sphere.

6.1.1 Learning Outcomes

Know how to perform the decomposition of a 2×2 Hermitian matrix into Pauli matrices and identity matrix; know how to construct density matrices for pure and mixed states; appreciate the meaning of the density matrix of a mixed state and its representation on the Bloch sphere.

6.1.2 Teaching Videos

- Search for Ch6 in this playlist
  - https://tinyurl.com/3yhze3jn
- Other videos
  - https://youtu.be/JR2jRCeTHDc
  - https://youtu.be/KEkOh9IDRI0
  - https://youtu.be/4ns5_BaF4xY
  - https://youtu.be/bUl5-zLR5r4

6.2 Real Vector Space of 2 × 2 Hermitian Matrices

In this section, we will study one special vector space, the **real vector space of** 2×2 **Hermitian matrices**, because of the following reasons. Firstly, we want to understand the relationship between the identity matrix and the Pauli matrices, which is important in the equations of **density matrix**. Secondly, it reinforces our understanding of vector space. Finally, we will take this opportunity to introduce the concept of the **inner product of matrices**.

As the name implies, the vector space we are interested in has Hermitian matrices as it *vector elements*. A vector defining a vector space does not need to be a usual column vector. As defined in Chap. 2, a vector space is defined as long as its vectors satisfy Eq. (2.1). The vector space should contain an addition operation and be also defined over a set of scalar. In the real vector space of Hermitian matrices, we define the addition operation as how we perform a typical matrix addition, and the scalars are limited to real numbers. Readers are encouraged to check that Eq. (2.1) is satisfied.

Since any of its vectors, M, is a 2×2 Hermitian matrix, then $M = M^\dagger$. It has only four degrees of freedom (DOFs). This is because its two diagonal elements are real and its off-diagonal elements are complex conjugate to each other. It is determined by four real numbers, a, b, c, and d, as

$$M = \begin{pmatrix} a & b+ic \\ b-ic & d \end{pmatrix} = M^\dagger. \tag{6.1}$$

Therefore, this is a 4D real space (as the scalars are defined to be real). And it turns out that its basis can be formed by the Pauli matrices and the identity matrix which are,

$$I = \sigma_0 = \begin{pmatrix} 1 & 0 \\ 0 & 1 \end{pmatrix},$$

6.2 Real Vector Space of 2 × 2 Hermitian Matrices

$$\sigma_x = \sigma_1 = \begin{pmatrix} 0 & 1 \\ 1 & 0 \end{pmatrix},$$

$$\sigma_y = \sigma_2 = \begin{pmatrix} 0 & -i \\ i & 0 \end{pmatrix},$$

$$\sigma_z = \sigma_3 = \begin{pmatrix} 1 & 0 \\ 0 & -1 \end{pmatrix}. \tag{6.2}$$

Note that we label I as σ_0. σ_0 does not have all the properties Pauli matrices have. For example, it is not traceless. It does not follow the commutation and anti-commutation relations with the Pauli matrices. But together with the Pauli matrices, they form the basis to express any M,

$$\begin{aligned} M &= a'_0 I + a'_1 \sigma_x + a'_2 \sigma_y + a'_3 \sigma_z, \\ &= a'_0 \sigma_0 + a'_1 \sigma_1 + a'_2 \sigma_2 + a'_3 \sigma_3, \\ &= \sum_{i=0}^{3} a'_i \sigma_i. \end{aligned} \tag{6.3}$$

Again, a'_0 to a'_3 are real based on the definition of the space. We will also define two vectors to facilitate future derivations. Firstly, we define a **Pauli vector** which is composed of the Pauli matrices,

$$\vec{\sigma} = \begin{pmatrix} \sigma_1 \\ \sigma_2 \\ \sigma_3 \end{pmatrix} = \begin{pmatrix} \begin{pmatrix} 0 & 1 \\ 1 & 0 \end{pmatrix} \\ \begin{pmatrix} 0 & -i \\ i & 0 \end{pmatrix} \\ \begin{pmatrix} 1 & 0 \\ 0 & -1 \end{pmatrix} \end{pmatrix}. \tag{6.4}$$

Note that the Pauli vector is a column vector but it has matrices as its elements! As long as it works as a vector, why not? But note that it is not in the Hermitian matrix space that we are discussing. It is just defined for convenience. We will also define a regular vector for the coefficients,

$$\vec{a}' = \begin{pmatrix} a'_1 \\ a'_2 \\ a'_3 \end{pmatrix}. \tag{6.5}$$

Therefore, Eq. (6.3) may be expressed as,

$$\begin{aligned} M &= \sum_{i=0}^{3} a'_i \sigma_i, \\ &= a'_0 \sigma_0 + \left\langle \vec{a'} \middle| \vec{\sigma} \right\rangle, \\ &= a'_0 I + \vec{a'} \cdot \vec{\sigma}, \end{aligned} \quad (6.6)$$

where if you perform an inner product in the second line, you will recover the first line. In the last line, I try to rewrite it in a common (non-*bra-ket*) notation.

6.2.1 Trace of 2 × 2 Hermitian Matrices

Let us now study the **trace** properties of 2×2 Hermitian matrices. Firstly,

$$\begin{aligned} \text{tr}(M) &= \text{tr}(a'_0 I) + \text{tr}(\vec{a'} \cdot \vec{\sigma}), \\ &= a'_0 \text{tr}(I) + 0 = 2a'_0, \end{aligned} \quad (6.7)$$

where we used the fact that the trace operation is linear and the Pauli matrices are traceless (Eq. (5.16)). By rearranging the terms, we also find that

$$a'_0 = \frac{\text{tr}(M)}{2} = \frac{\text{tr}(\sigma_0 M)}{2}, \quad (6.8)$$

where we have used the fact that $\sigma_0 = I$. Another property we want to know is the trace when it is multiplied by a Pauli matrix, σ_k, with $k = 1, 2, 3$ (not including 0). That is,

$$\text{tr}(\sigma_k M) = \text{tr}(a'_0 \sigma_k \sigma_0 + a'_1 \sigma_k \sigma_1 + a'_2 \sigma_k \sigma_2 + a'_3 \sigma_k \sigma_3). \quad (6.9)$$

This looks difficult but it is very straightforward. Since the trace operation is linear, we can do it term by term. The first term has a zero trace because σ_0 is just the identity matrix. So the first term is just a Pauli matrix (which is traceless) scaled by a'_0. For the other terms, they are the products of two Pauli matrices. Only one of them has two identical Pauli matrices and the other two have different Pauli matrices. For example, if $k = 2$, they become $a'_1 \sigma_2 \sigma_1 + a'_2 \sigma_2 \sigma_2 + a'_3 \sigma_2 \sigma_3$. Based on Eq. (5.18), the trace of the product of two different Pauli matrices is 0 and that of the same Pauli matrices is 2. So only $2a'_2$ is left. Therefore, we have

$$\text{tr}(\sigma_k M) = 2a'_k, \quad (6.10)$$

6.2 Real Vector Space of 2 × 2 Hermitian Matrices

and thus,

$$a'_k = \frac{\text{tr}(\sigma_k M)}{2}. \tag{6.11}$$

Combing Eqs. (6.8) and (6.11), we can write

$$\text{tr}(\sigma_i M) = 2a'_i,$$
$$a'_i = \frac{\text{tr}(\sigma_i M)}{2}, \tag{6.12}$$

for $i = 0, 1, 2, 3$. This is actually related to the *inner product* of the matrices which we will discuss next.

6.2.2 Inner Product of Matrices

The real vector space of 2 × 2 Hermitian matrices is a vector space. We can further define an **inner product** to make it an **inner product space** (see Chap. 2). As a reminder, the 2 × 2 Hermitian matrices are the *vectors* in this space. Therefore, the inner product must be defined over the Hermitian matrices (which are the "vectors" in this vector space). The inner product of matrices A and B are defined as,

$$\langle A|B\rangle = \text{tr}(A^\dagger B). \tag{6.13}$$

Therefore, the trace operator has the effect of finding the "overlap" between two matrices. This is also called the Hilbert-Schmidt inner product or trace inner product. It also satisfies the properties in Eq. (2.5).

Let us recall how to find the amount of a component in a regular vector. Assume the vector is $|\psi\rangle = \alpha|0\rangle + \beta|1\rangle$; to find how much $|0\rangle$ it has, we will perform an inner product in this way,

$$\langle 0|\psi\rangle = \alpha\langle 0|0\rangle + \beta\langle 0|1\rangle = \alpha, \tag{5.14}$$

where we used the fact that the basis vectors are orthonormal. We also note that the *bra* version of $|0\rangle$ is used. For a general Hermitian matrix, we have $M = a'_0\sigma_0 + a'_1\sigma_1 + a'_2\sigma_2 + a'_3\sigma_3$. Therefore, we expect that we can find the amount of each component by using the inner product. From Eq. (6.12), we have

$$a'_i = \frac{\text{tr}(\sigma_i M)}{2},$$
$$= \frac{\text{tr}(\sigma_i^\dagger M)}{2},$$

$$= \frac{\langle \sigma_i | M \rangle}{2}, \qquad (6.15)$$

where we have used $\sigma_i = \sigma_i^\dagger$ as they are Hermitian. It can be seen that the amount of each component can be obtained by performing an inner product and scaling it by half.

Example 6.1 Show that σ_x is orthogonal to σ_y.

We will show this by finding their inner product. That is,

$$\begin{aligned}\langle \sigma_x | \sigma_y \rangle &= \mathrm{tr}(\sigma_x^\dagger \sigma_y), \\ &= \mathrm{tr}(\sigma_x \sigma_y), \\ &= 0, \end{aligned} \qquad (6.16)$$

where we have used Eq. (5.8) in line 2 and Eq. (5.18) in line 3. Since their inner product is zero, they are orthogonal to each other. ∎

In general, we have (see Problem 6.5),

$$\begin{aligned}\langle \sigma_i | \sigma_j \rangle &= \mathrm{tr}(\sigma_i^\dagger \sigma_j), \\ &= 2\delta_{ij}.\end{aligned} \qquad (6.17)$$

This shows that the basis vectors, σ_i, for $i = 0, 1, 2, 3$, are **orthogonal** to each other. They are also normalized to two (instead of one). We will not discuss its implications. But I hope this can reinforce our understanding of basis vectors and linear space.

6.3 Density Matrix

6.3.1 Pure and Mixed States

Before introducing **density matrix**, let us clarify a few important concepts regarding quantum states in a mixture and quantum states of which we do not have complete knowledge.

Every quantum system must be in a certain state, $|\psi\rangle$. This is called a **pure state**. We need to emphasize that it can be one of the basis states, e.g., $|\downarrow\rangle$ or $|\uparrow\rangle$ in a 1-qubit system. It can also be a superposition state, $\alpha |\downarrow\rangle + \beta |\uparrow\rangle$. They are on the **Bloch sphere surface**. These are all pure states (left of Fig. 6.1).

Example 6.2 What is the probability of measuring $|\downarrow\rangle$ and $|\uparrow\rangle$ for state $|\psi\rangle = \frac{1}{\sqrt{4}} |\downarrow\rangle + \sqrt{\frac{3}{4}} |\uparrow\rangle$?

6.3 Density Matrix

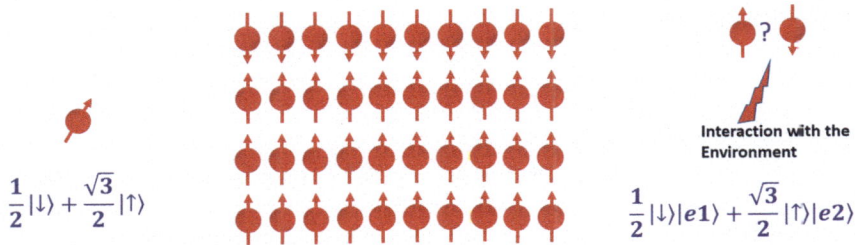

Fig. 6.1 Left: An electron in a superposition state $\frac{1}{\sqrt{4}}|\downarrow\rangle + \sqrt{\frac{3}{4}}|\uparrow\rangle$. Middle: 40 electrons with 10 in state $|\downarrow\rangle$ and 30 in state $|\uparrow\rangle$. Right: An electron entangled with the environment to form a pure state $\frac{1}{\sqrt{4}}|\downarrow\rangle|e1\rangle + \sqrt{\frac{3}{4}}|\uparrow\rangle|e2\rangle$. We cannot say if the electron is $|\uparrow\rangle$ or $|\downarrow\rangle$

This is a *pure state*. Based on Eqs. (2.21) and (2.22), $Prob(|\downarrow\rangle) = \frac{1}{4} = 0.25$ and $Prob(|\uparrow\rangle) = \frac{3}{4} = 0.75$. ∎

Even though every quantum system must be in a definite state (pure state), when the system is complex, we might not have complete information. For example, there might be 40 or more electrons (middle of Fig. 6.1). We do not know each electron's exact state. But we might know statistically, 25% of them are $|\downarrow\rangle$ and 75% of them are $|\uparrow\rangle$. This is just an example. It is possible to have some of them in a superposition state. This system is an **ensemble of pure states**. If we randomly pick an electron and measure its state, we expect $Prob(|\downarrow\rangle) = 25\% = 0.25$ and $Prob(|\uparrow\rangle) = 75\% = 0.75$. While the measurement outcome statistics are the same as the pure state, these two systems are very different.

When the quantum system (e.g., one electron) interacts with the environment, it becomes a larger system. If we describe the larger system as a whole, it is still a well-defined pure state. However, if we only want to describe the quantum system (i.e., the electron), then the quantum system is no longer in a pure state if it has entangled with the environment (right of Fig. 6.1). For example, the pure state of the larger system is $\frac{1}{\sqrt{4}}|\downarrow\rangle|e1\rangle + \sqrt{\frac{3}{4}}|\uparrow\rangle|e2\rangle$, where $|e1\rangle$ and $|e2\rangle$ are the basis states belonging to the environment. But if we try to measure the electron spin, we still have $Prob(|\downarrow\rangle) = \frac{1}{4} = 0.25$ and $Prob(|\uparrow\rangle) = \frac{3}{4} = 0.75$.

The last two cases are **mixed-state** cases.

6.3.2 Density Matrix Definition

Since we do not have complete information to describe the mixed states, the density matrix is thus used and it can capture the essential information of the mixed states succinctly. Assume there is a quantum system with *k pure* states, $|\psi_i\rangle$ for $i = 0, 1, \cdots, k-1$, and each has a measurement probability of P_i. The density matrix of the system, ρ, is defined as,

$$\rho = \sum_{i=0}^{k-1} P_i |\psi_i\rangle \langle \psi_i|,$$
$$= P_0 |\psi_0\rangle \langle \psi_0| + P_1 |\psi_1\rangle \langle \psi_1| \cdots . \quad (6.18)$$

This is not difficult for a 1-qubit system which is in a 2D complex space.

Example 6.3 Find the density matrix for the pure state and mixed state in Fig. 6.1.

For the pure state case, $|\psi\rangle = \frac{1}{\sqrt{4}}|\downarrow\rangle + \sqrt{\frac{3}{4}}|\uparrow\rangle$. Therefore, $k = 1$, $|\psi_0\rangle = |\psi\rangle$, and $P_0 = 1$. We sort the basis vectors so that $|\downarrow\rangle$ goes first. So, $|\downarrow\rangle = \begin{pmatrix} 1 \\ 0 \end{pmatrix}$ and $|\uparrow\rangle = \begin{pmatrix} 0 \\ 1 \end{pmatrix}$. Thus,

$$\rho = P_0 |\psi_0\rangle \langle \psi_0|,$$
$$= 1 \times |\psi\rangle \langle \psi|,$$
$$= \left(\frac{1}{\sqrt{4}}|\downarrow\rangle + \sqrt{\frac{3}{4}}|\uparrow\rangle \right) \times (\frac{1}{\sqrt{4}} \langle \downarrow| + \sqrt{\frac{3}{4}} \langle \uparrow|),$$
$$= \frac{1}{\sqrt{4}}\frac{1}{\sqrt{4}}|\downarrow\rangle\langle\downarrow| + \frac{1}{\sqrt{4}}\sqrt{\frac{3}{4}}|\downarrow\rangle\langle\uparrow| + \sqrt{\frac{3}{4}}\frac{1}{\sqrt{4}}|\uparrow\rangle\langle\downarrow| + \sqrt{\frac{3}{4}}\sqrt{\frac{3}{4}}|\uparrow\rangle\langle\uparrow|,$$
$$= \frac{1}{4}\begin{pmatrix}1\\0\end{pmatrix}(1\ 0) + \sqrt{\frac{3}{16}}\begin{pmatrix}1\\0\end{pmatrix}(0\ 1) + \sqrt{\frac{3}{16}}\begin{pmatrix}0\\1\end{pmatrix}(1\ 0) + \sqrt{\frac{9}{16}}\begin{pmatrix}0\\1\end{pmatrix}(0\ 1),$$
$$= \frac{1}{4}\begin{pmatrix}1 & 0\\0 & 0\end{pmatrix} + \frac{\sqrt{3}}{4}\begin{pmatrix}0 & 1\\0 & 0\end{pmatrix} + \frac{\sqrt{3}}{4}\begin{pmatrix}0 & 0\\1 & 0\end{pmatrix} + \frac{3}{4}\begin{pmatrix}0 & 0\\0 & 1\end{pmatrix},$$
$$= \begin{pmatrix} \frac{1}{4} & \frac{\sqrt{3}}{4} \\ \frac{\sqrt{3}}{4} & \frac{3}{4} \end{pmatrix}. \quad (6.19)$$

For the mixed-state case, we have $k = 2$. And $|\psi_0\rangle = |\downarrow\rangle$, $P_0 = \frac{1}{4} = 0.25$, $|\psi_1\rangle = |\uparrow\rangle$, $P_1 = \frac{3}{4} = 0.75$. Therefore,

$$\rho = P_0 |\psi_0\rangle \langle \psi_0| + P_1 |\psi_1\rangle \langle \psi_1|,$$
$$= \frac{1}{4}|\downarrow\rangle\langle\downarrow| + \frac{3}{4}|\uparrow\rangle\langle\uparrow|,$$
$$= \frac{1}{4}\begin{pmatrix}1\\0\end{pmatrix}(1\ 0) + \frac{3}{4}\begin{pmatrix}0\\1\end{pmatrix}(0\ 1),$$

6.3 Density Matrix

$$= \frac{1}{4}\begin{pmatrix} 1 & 0 \\ 0 & 0 \end{pmatrix} + \frac{3}{4}\begin{pmatrix} 0 & 0 \\ 0 & 1 \end{pmatrix},$$

$$= \begin{pmatrix} \frac{1}{4} & 0 \\ 0 & \frac{3}{4} \end{pmatrix}. \tag{6.20}$$

Comparing the matrices in Eqs. (6.19) and (6.20), they have the same diagonal elements but different off-diagonal elements. Why? We will need to study the properties of density matrices.

6.3.3 Properties of Density Matrices

If there are m (e.g., 100) pure states in the mixed state, for a N-dimensional system, we need mN complex numbers to describe the system because we need N complex numbers to describe one pure state. Note that m can be much larger than N. With a density matrix, we only need N^2 numbers to describe the system.

Density matrices are **positive semi-definite (PSD)**. A PSD matrix is Hermitian (i.e., $\rho = \rho^\dagger$) and therefore, density matrices are Hermitian.

A PSD matrix is defined as having all its eigenvalues, $|e_i\rangle$, being ≥ 0. This is equivalent to $\langle v| \rho |v\rangle \geq 0$ for any vector $|v\rangle$ (that means the expectation value of the density matrix is ≥ 0) or $\rho = A^\dagger A$ which means that it can be decomposed as the product of a matrix A and its adjoint. In summary, the following are equivalent definitions of PSD matrices and we can apply them to density matrices.

$$|e_i\rangle \geq 0, \tag{6.21}$$

$$\langle v| \rho |v\rangle \geq 0, \tag{6.22}$$

$$\rho = A^\dagger A. \tag{6.23}$$

Density matrices also have a trace of one. That is,

$$\text{tr}(\rho) = 1. \tag{6.24}$$

A matrix is a density matrix if it is PSD with a trace of one. Indeed, the traces of the matrices in Eqs. (6.19) and (6.20) are both one.

Example 6.4 A mixed state contains 50% of $|0\rangle$ and 50% of $\frac{1}{\sqrt{2}}(|0\rangle + |1\rangle)$. Find the trace of its density matrix.

We have $k = 2$, $|\psi_0\rangle = |0\rangle$, $P_0 = 0.5$, $|\psi_1\rangle = \frac{1}{\sqrt{2}}(|0\rangle + |1\rangle)$, and $P_1 = 0.5$. Therefore,

$$\rho = 0.5 |0\rangle \langle 0| + 0.5 \frac{1}{\sqrt{2}}(|0\rangle + |1\rangle) \frac{1}{\sqrt{2}}(\langle 0| + \langle 1|),$$

80   6 Density Matrix and the Bloch Sphere

$$= 0.5 |0\rangle \langle 0| + 0.25(|0\rangle \langle 0| + |0\rangle \langle 1| + |1\rangle \langle 0| + |1\rangle \langle 1|),$$
$$= 0.75 |0\rangle \langle 0| + 0.25 |0\rangle \langle 1| + 0.25 |1\rangle \langle 0| + 0.25 |1\rangle \langle 1|,$$
$$= \begin{pmatrix} 0.75 & 0.25 \\ 0.25 & 0.25 \end{pmatrix}. \tag{6.25}$$

Therefore, the trace is $\text{tr}(\rho) = 0.75 + 0.25 = 1$, as expected from Eq. (6.24).

6.3.4 State Purity

How do we know if a density matrix corresponds to a pure state or a mixed state? We can use this to check,

$$tr(\rho^2) = 1 \quad \text{Pure State},$$
$$tr(\rho^2) < 1 \quad \text{Mixed State}. \tag{6.26}$$

We can understand it in this way. For a pure state, we can always diagonalize it (or choose a new basis) so that that state is one of the basis states. Then the density matrix must be a diagonal matrix with only one non-zero element (one) corresponding to that state. Then its square must also be diagonal with one non-zero element (also one). That is,

$$\rho = \begin{pmatrix} 0 & & & \\ & \ddots & & \\ & & 1 & \\ & & & \ddots \end{pmatrix}, \quad \rho^2 = \begin{pmatrix} 0 & & & \\ & \ddots & & \\ & & 1 & \\ & & & \ddots \end{pmatrix}. \tag{6.27}$$

Therefore, $tr(\rho^2) = 1$ for a pure state. Note that the trace of a diagonalized matrix (a transformed matrix) is the same as the trace of the matrix before diagonalization (transformation). For mixed state, we need to go through some proof by using matrix multiplication and Schwarz inequality. We will not do it here. However, we can consider a simple case where all the pure states in the mixed state are orthogonal (such as the case in Eq. (6.20)). In that case, the density matrix can be diagonalized in the basis of its pure states with the diagonal elements corresponding to the measurement probabilities. Then the square of the density matrix is still a diagonal

6.3 Density Matrix

matrix. That is,

$$\rho = \begin{pmatrix} P_1 & & \\ & \ddots & \\ & & P_{k-1} \end{pmatrix}, \rho^2 = \begin{pmatrix} P_1^2 & & \\ & \ddots & \\ & & P_{k-1}^2 \end{pmatrix}. \quad (6.28)$$

Since $tr(\rho) = P_1 + \cdots + P_{k-1} = 1$ and the probabilities are all smaller than one, the sum of their squares must be less than one. That is, $tr(\rho^2) = P_1^2 + \cdots + P_{k-1}^2 < 1$.

Example 6.5 Show that the density matrix in Eq. (6.19) represents a pure state.

Since,

$$\rho = \begin{pmatrix} \frac{1}{4} & \frac{\sqrt{3}}{4} \\ \frac{\sqrt{3}}{4} & \frac{3}{4} \end{pmatrix}, \quad (6.29)$$

we have,

$$\rho^2 = \begin{pmatrix} \frac{1}{4} & \frac{\sqrt{3}}{4} \\ \frac{\sqrt{3}}{4} & \frac{3}{4} \end{pmatrix} \begin{pmatrix} \frac{1}{4} & \frac{\sqrt{3}}{4} \\ \frac{\sqrt{3}}{4} & \frac{3}{4} \end{pmatrix},$$

$$= \begin{pmatrix} \frac{1}{16} + \frac{3}{16} & \frac{\sqrt{3}}{16} + \frac{3\sqrt{3}}{16} \\ \frac{\sqrt{3}}{16} + \frac{3\sqrt{3}}{16} & \frac{3}{16} + \frac{9}{16} \end{pmatrix},$$

$$= \begin{pmatrix} \frac{4}{16} & \frac{4\sqrt{3}}{16} \\ \frac{4\sqrt{3}}{16} & \frac{12}{16} \end{pmatrix}. \quad (6.30)$$

Therefore, $\rho^2 = \frac{4+12}{16} = 1$ and it is a pure state density matrix as expected. ∎

Example 6.6 Show the density matrix in Eq. (6.20) represents a mixed state.

Since,

$$\rho = \begin{pmatrix} \frac{1}{4} & 0 \\ 0 & \frac{3}{4} \end{pmatrix}, \quad (6.31)$$

we have,

$$\rho^2 = \begin{pmatrix} \frac{1}{4} & 0 \\ 0 & \frac{3}{4} \end{pmatrix} \begin{pmatrix} \frac{1}{4} & 0 \\ 0 & \frac{3}{4} \end{pmatrix},$$

$$= \begin{pmatrix} \frac{1}{16} & 0 \\ 0 & \frac{9}{16} \end{pmatrix}. \quad (5.32)$$

Therefore, $\rho^2 = \frac{1+9}{16} < 1$ and it is a mixed-state density matrix as expected. ∎

6.4 Expectation Value

We had briefly discussed the meaning of **expectation values** in Sect. 3.4.1 and Eq. (5.2). *The expectation value of an operator M in the state $|v\rangle$ is $\langle v| M |v\rangle$* (by changing the symbols in Eq. 3.31). $|v\rangle$ is a pure state.

Example 6.7 What is the expectation value of σ_z in state $|v\rangle = \frac{1}{\sqrt{4}} |0\rangle + \frac{\sqrt{3}}{\sqrt{4}} |1\rangle$?
The expectation value is,

$$\langle v| M |v\rangle = \langle v| \sigma_z |v\rangle,$$

$$= \begin{pmatrix} \frac{1}{\sqrt{4}} & \frac{\sqrt{3}}{\sqrt{4}} \end{pmatrix} \begin{pmatrix} 1 & 0 \\ 0 & -1 \end{pmatrix} \begin{pmatrix} \frac{1}{\sqrt{4}} \\ \frac{\sqrt{3}}{\sqrt{4}} \end{pmatrix},$$

$$= \begin{pmatrix} \frac{1}{\sqrt{4}} & \frac{\sqrt{3}}{\sqrt{4}} \end{pmatrix} \begin{pmatrix} \frac{1}{\sqrt{4}} \\ -\frac{\sqrt{3}}{\sqrt{4}} \end{pmatrix},$$

$$= \frac{1}{4} - \frac{3}{4} = -\frac{2}{4} = -0.5. \qquad (6.33)$$

∎

What if we have a mixed-state system? In principle, we can treat each pure state one by one and then take the average.

Example 6.8 What is the expectation value of σ_z in a mixed state of $|\psi_0\rangle = |0\rangle$, $P_0 = \frac{1}{4} = 0.25$, $|\psi_1\rangle = |1\rangle$, $P_1 = \frac{3}{4} = 0.75$ (same as the state in Eq. (6.20))?
Firstly, we find the expectation value of $|0\rangle$, that is,

$$\langle 0| \sigma_z |0\rangle = \begin{pmatrix} 1 & 0 \end{pmatrix} \begin{pmatrix} 1 & 0 \\ 0 & -1 \end{pmatrix} \begin{pmatrix} 1 \\ 0 \end{pmatrix},$$

$$= 1. \qquad (6.34)$$

Similarly, the expectation value of $|1\rangle$ is -1. Therefore, the expectation value of the mixed state is $0.25 \times 1 + 0.75 \times -1 = -0.5$. ∎

Based on the meaning of expectation value and the experience we gained in Example 6.8, if we have a mixed state with k pure states, $|i\rangle$, and each has a proportion of p_i, the expectation value of operator M is,

$$\langle M \rangle = \sum_{i=1}^{k} p_i \langle i|M|i\rangle. \qquad (6.35)$$

6.5 Density Matrix, Expectation Value, and Bloch Sphere

While we are not proving it here (see Problem 6.7), it can be shown that,

$$\langle M \rangle = \sum_{i=1}^{k} p_i \langle i|M|i \rangle,$$
$$= tr(M\rho),$$
$$= tr(\rho M),$$
$$= tr(\rho^\dagger M),$$
$$= \langle \rho | M \rangle, \quad (6.36)$$

where ρ is the density matrix of the state. This shows that the expectation value of M in a state with density matrix ρ is the *inner product M and ρ*.

Example 6.9 Redo Example 6.8 using Eq. (6.36).

We get the density matrix from Eq. (6.20). Therefore, the expectation value is,

$$tr(M\rho) = tr(\sigma_z \rho),$$
$$= tr\left(\begin{pmatrix} 1 & 0 \\ 0 & -1 \end{pmatrix} \begin{pmatrix} \frac{1}{4} & 0 \\ 0 & \frac{3}{4} \end{pmatrix}\right),$$
$$= tr\begin{pmatrix} \frac{1}{4} & 0 \\ 0 & -\frac{3}{4} \end{pmatrix},$$
$$= -0.5. \quad (6.37)$$

This is the same as the result in Example 6.8. ∎

6.5 Density Matrix, Expectation Value, and Bloch Sphere

Now we will discuss the relationship between a 1-qubit state density matrix, the expectation value of Pauli matrices, and their visualization on the Bloch sphere.

6.5.1 Pure State Expectation Values and Projections

In Section 27.3 in [1], we showed that, for a pure state, the expectation value of σ_x, σ_y, and σ_z in a state $|\psi\rangle$ is the projection of the state from the Bloch sphere surface to $\hat{x}$, $\hat{y}$, and $\hat{z}$, respectively (Fig. 6.2), where we have,

$$\langle \psi | \sigma_x | \psi \rangle = \sin\theta \cos\phi. \quad (6.38)$$

$$\langle \psi | \sigma_y | \psi \rangle = \sin\theta \sin\phi. \tag{6.39}$$

$$\langle \psi | \sigma_z | \psi \rangle = \cos\theta. \tag{6.40}$$

In general, the expectation value of a linear combination of the Pauli matrices, $n_x \sigma_x + n_y \sigma_y + n_z \sigma_z$ is,

$$\begin{aligned}
\langle \psi | (n_x \sigma_x + n_y \sigma_y + n_z \sigma_z) | \psi \rangle &= \langle \psi | \hat{n} \cdot \vec{\sigma} | \psi \rangle, \\
&= n_x \sin\theta \cos\phi + n_y \sin\theta \sin\phi + n_z \cos\theta, \\
&= \begin{pmatrix} n_x & n_y & n_z \end{pmatrix} \begin{pmatrix} \sin\theta \cos\phi \\ \sin\theta \sin\phi \\ \cos\theta \end{pmatrix}, \\
&= \hat{n} \cdot \vec{\psi}, \tag{6.41}
\end{aligned}$$

where $\hat{n} \cdot \vec{\psi}$ is just the projection of $\vec{\psi}$ on $\hat{n}$ or their inner product. In line 1, we used the definition of the Pauli vector, $\vec{\sigma}$, in Eq. (6.4). We have used Eq. (6.38) to Eq. (6.40) in line 2 and the linear property of *bra-ket* operation. We also defined a unit directional vector, $\hat{n}$, with

$$\hat{n} = \begin{pmatrix} n_x \\ n_y \\ n_z \end{pmatrix}, \tag{6.42}$$

and vector $\vec{\psi}$, which is the real space vector in the real 3D space pointing at the location where $|\psi\rangle$ resides on the embedded Bloch sphere (Fig. 6.2). That is,

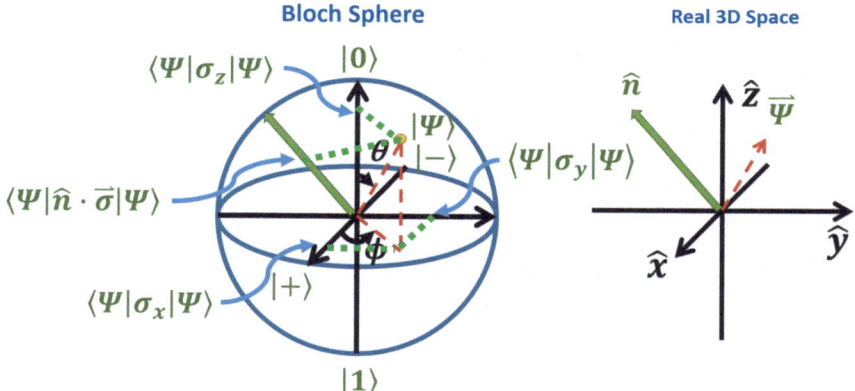

Fig. 6.2 Left: Bloch sphere showing that the expectation values of the Pauli matrices are just the projections of the state on the axes. It also shows that the expectation value of $\hat{n} \cdot \vec{\sigma}$ is the projection on $\hat{n}$. Right: The real 3D space coordinate system with $\vec{\psi}$ and $\hat{n}$ shown

6.5 Density Matrix, Expectation Value, and Bloch Sphere

$$\vec{\psi} = \begin{pmatrix} \sin\theta\cos\phi \\ \sin\theta\sin\phi \\ \cos\theta \end{pmatrix}. \tag{6.43}$$

It should be noted that the radius of the Bloch sphere is one and the lengths of $\hat{n}$ and $\vec{\psi}$ are also one. In summary, *the expectation value of $\hat{n}\cdot\vec{\sigma}$ in a state $|\psi\rangle$ is just the projection of $\vec{\psi}$ on $\hat{n}$* in our real 3D space (Fig. 6.2).

6.5.2 Density Matrix of a 1-Qubit System

As discussed earlier, a density matrix is Hermitian. A 2×2 Hermitian matrix only has 4 DoFs (Eq. (6.1)). Therefore, it can be decomposed as a linear combination of four orthogonal basis vectors (note that, here, vector means the vector in a linear space which is a 2×2 Hermitian matrix; see Sect. 6.2) with four real coefficients. Therefore, it belongs to the real vector space of 2×2 Hermitian matrices. As a result, it can be decomposed in the form of Eq. (6.6),

$$\begin{aligned} \rho &= a_0' I + \vec{a}' \cdot \vec{\sigma}, \\ &= a_0' I + a_1' \sigma_x + a_2' \sigma_y + a_3' \sigma_z. \end{aligned} \tag{6.44}$$

We now introduce new variables,

$$a_0 = 2a_0', \tag{6.45}$$

$$\vec{a} = 2\vec{a}', \tag{6.46}$$

where $\vec{a}$ is the **Bloch vector**. Then,

$$\rho = \frac{a_0 I + \vec{a}\cdot\vec{\sigma}}{2}. \tag{6.47}$$

We can further simplify it. Based on Eq. (6.12),

$$\begin{aligned} a_0 &= \mathrm{tr}(I\rho), \\ &= \mathrm{tr}(\rho), \\ &= 1, \end{aligned} \tag{6.48}$$

where we have used Eq. (6.24) in the last line for a density matrix. Therefore,

$$\rho = \frac{I + \vec{a}\cdot\vec{\sigma}}{2}. \tag{6.49}$$

Let us study the meaning of the Bloch vector, $\vec{a}$. We will first look at the matrix components of ρ^2. ρ in its matrix form is,

$$\rho = \frac{1}{2}\left(\begin{pmatrix} 1 & 0 \\ 0 & 1 \end{pmatrix} + a_x \begin{pmatrix} 0 & 1 \\ 1 & 0 \end{pmatrix} + a_y \begin{pmatrix} 0 & -i \\ i & 0 \end{pmatrix} + a_z \begin{pmatrix} 1 & 0 \\ 0 & -1 \end{pmatrix}\right),$$

$$= \frac{1}{2}\left(\begin{pmatrix} 1 & 0 \\ 0 & 1 \end{pmatrix} + \begin{pmatrix} 0 & a_x \\ a_x & 0 \end{pmatrix} + \begin{pmatrix} 0 & -ia_y \\ ia_y & 0 \end{pmatrix} + \begin{pmatrix} a_z & 0 \\ 0 & -a_z \end{pmatrix}\right),$$

$$= \frac{1}{2}\begin{pmatrix} 1+a_z & a_x - ia_y \\ a_x + ia_y & 1 - a_z \end{pmatrix}. \tag{6.50}$$

Then,

$$\rho^2 = \frac{1}{2}\begin{pmatrix} 1+a_z & a_x - ia_y \\ a_x + ia_y & 1 - a_z \end{pmatrix} \frac{1}{2}\begin{pmatrix} 1+a_z & a_x - ia_y \\ a_x + ia_y & 1 - a_z \end{pmatrix},$$

$$= \frac{1}{4}\begin{pmatrix} a_x^2 + a_y^2 + a_z^2 + 2a_z + 1 & 2(a_x - ia_y) \\ 2(a_x + ia_y) & a_x^2 + a_y^2 + a_z^2 - 2a_z + 1 \end{pmatrix}. \tag{6.51}$$

Therefore,

$$tr(\rho^2) = \frac{1}{4}(a_x^2 + a_y^2 + a_z^2 + 2a_z + 1 + a_x^2 + a_y^2 + a_z^2 - 2a_z + 1),$$

$$= \frac{a_x^2 + a_y^2 + a_z^2 + 1}{2},$$

$$= \frac{|\vec{a}|^2 + 1}{2}. \tag{6.52}$$

Based on Eq. (6.26), $tr(\rho^2) = 1$ *for a pure state; therefore,* $|\vec{a}| = 1$. *If it is a mixed state,* $tr(\rho^2) < 1$ *and therefore,* $|\vec{a}| < 1$.

To further understand the meaning of $\vec{a}$, let us consider the expectation value of $\hat{n}_a \cdot \vec{\sigma}$ in the state which gives the density matrix $\rho = \frac{I + \vec{a} \cdot \vec{\sigma}}{2}$. Since $\vec{a}$ can be smaller than one, we normalize it to $\hat{n}_a = \frac{\vec{a}}{|\vec{a}|}$ in the observable (Fig. 6.3).

This has the same form as the one in Eq. (6.41), except that now $\hat{n}$ is $\hat{n}_a$, which is the unit vector along the direction of $\vec{a}$. We will use the expectation value equation in Eq. (6.36) using the given density matrix,

$$\langle \hat{n}_a \cdot \vec{\sigma} \rangle = tr\left(\hat{n}_a \cdot \vec{\sigma} \rho\right),$$

$$= tr(\hat{n}_a \cdot \vec{\sigma} \frac{I + \vec{a} \cdot \vec{\sigma}}{2}),$$

$$= \frac{1}{2|\vec{a}|} tr(\vec{a} \cdot \vec{\sigma} + \vec{a} \cdot \vec{\sigma}\vec{a} \cdot \vec{\sigma}),$$

6.5 Density Matrix, Expectation Value, and Bloch Sphere

$$= \frac{1}{2|\vec{a}|} tr(\vec{a} \cdot \vec{\sigma} \vec{a} \cdot \vec{\sigma}),$$

$$= \frac{1}{|\vec{a}|}(a_x^2 + a_y^2 + a_z^2),$$

$$= \frac{|\vec{a}|^2}{|\vec{a}|},$$

$$= |\vec{a}|. \tag{6.53}$$

From line 3 to line 4, we have used the fact that the Pauli matrices are traceless, and therefore, $tr(\vec{a} \cdot \vec{\sigma}) = 0$ (Eq. (5.16)). In line 4, the squared term, $(\vec{a} \cdot \vec{\sigma})^2$, gives rise to terms in the form of $a_l a_m \sigma_l \sigma_m$. One type is like $a_x a_x \sigma_x \sigma_x$ which has a trace of $2a_x^2$. Another type are cross-terms like $a_x a_y \sigma_x \sigma_y$ which has a trace of 0 (see Eq. (5.18)). Therefore, only $2(a_x^2 + a_y^2 + a_z^2)$ is left after taking the trace.

The result in Eq. (6.53) tells us that the expectation value of $\hat{n}_a \cdot \vec{\sigma}$ in the state $\rho = \frac{I + \vec{a} \cdot \vec{\sigma}}{2}$ is $|\vec{a}|$.

If $|\vec{a}| = 1$, we know that this density matrix corresponds to a pure state $|\Psi\rangle$ (Eq. (6.52)). We also know that the expectation value in Eq. (6.53) is just the inner product of $\vec{\Psi}$ and $\vec{a}$ (Sect. 6.5.1 and Eq. (6.41)). Since the inner product is one, it means that $\vec{\Psi} = \vec{a}$ and this density matrix corresponds to a pure state that resides on the surface of the Bloch sphere at $\vec{a}$ (left of Fig. 6.3). Therefore, the Bloch vector of a pure state is just the location of that state on the Bloch sphere.

If $|\vec{a}| < 1$, it corresponds to a mixed state (right of Fig. 6.3). Since a mixed-state density matrix is not uniquely mapped to a mixture of pure states, we cannot deduce which mixture it is. One of the possible mixtures is having a pure state at $\hat{n}_a$ with a

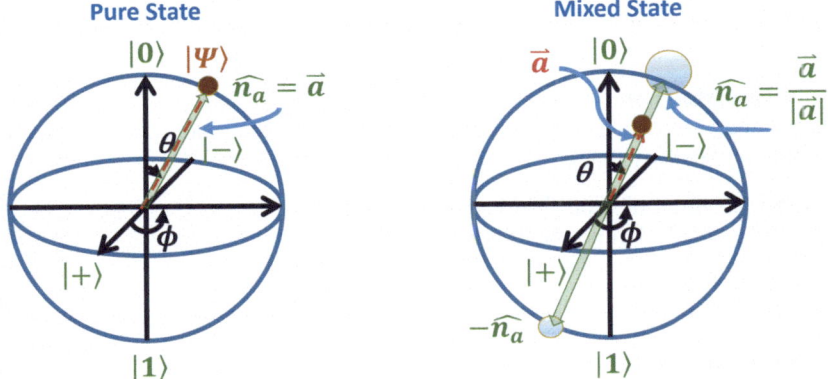

Fig. 6.3 Left: Bloch sphere with the Bloch vector $\vec{a}$ shown. Density matrix $\rho = \frac{I + \vec{a} \cdot \vec{\sigma}}{2}$ corresponds to a pure state $|\psi\rangle$ on the surface of the Bloch sphere residing at $\hat{n}_a = \vec{a}$. Therefore, $\vec{a}$ is a unit vector for a pure state. Right: When the density matrix is of a mixed state, $\vec{a}$ is smaller than 1 and resides inside the Bloch sphere

probability of P_0 and a pure state at $-\hat{n}_a$ with a probability of P_1, with $P_0 - P_1 = |\vec{a}|$ and $P_0 + P_1 = 1$ (readers are encouraged to prove this and find out what happens if $P_0 = P_1$). We may say that the mixed state is represented by a point *inside* the Bloch sphere.

6.6 Summary

In this chapter, we practice the mathematics of a more advanced linear space that has matrices as its elements. The elements are 2×2 Hermitian matrices. They can be decomposed into the linear combination of the Pauli matrices and the identity matrix. Density matrices, which are elements of this space, are convenient representations for mixed states. We study the properties of density matrices. We also study its representation on the Bloch sphere. The most important conclusion is that when the state is a mixed state, it is represented as a point inside the Bloch sphere.

Problems

6.1 Real Vector Space of Hermitian Matrices
Show it satisfies Eq. (2.1). What are the 1 and $\vec{0}$ in this space?

6.2 Hermitian Matrices
Show that a 2×2 Hermitian matrix only has 4 DOFs instead of 8 DOFs.

6.3 Inner Product of Matrices
Show that the definition of inner product in Eq. (6.13) satisfies the requirements in Eq. (2.5).

6.4 Density Matrices 1
Show that the density matrices in Eqs. (6.19), (6.20), and (6.25) satisfy the definition of a density matrix.

6.5 Inner Product of Pauli Matrices
Prove Eq. (6.17). Hints: Use Eqs. (5.8) and (5.18).

6.6 Density Matrices 2
Show that the density matrix in Eq. (6.25) is a density matrix of a mixed state.

6.7 Expectation Value
Prove Eq. (6.36).

6.8 Expectation Value 2
Prove Eq. (6.38) to Eq. (6.40).

Reference

1. Hiu-Yung Wong. *Introduction to Quantum Computing*. Springer, 2024.

Part II
Silicon Spin Qubit Architecture and Hardware

Chapter 7
Spin Qubit—Preliminary Physics

7.1 Introduction

Spin can be a very difficult and confusing concept. Mathematically, spin can only be derived in relativistic quantum mechanics (i.e., *quantum electrodynamics, QED*). The word "relativistic" means that it applies to particles at high velocity, too. In QED, *Dirac equation* (a relativistic version form of Schrödinger equation) is used and the concept of spin of an electron appears naturally. In this book, we will treat spin as a given property of a particle. Indeed, in non-relativistic quantum mechanics, we take this for granted and we have been treating spin as an *intrinsic* property of a particle. However, its relationship to angular momentum and magnetic moment and its interaction with magnetic field need to be clarified to enhance our understanding and prepare us for more advanced studies in the future.

7.1.1 Learning Outcomes

Appreciate the gyromagnetic ratio difference in classical and quantum physics; understand the concepts of magnetic moment and angular momentum; understand the interaction between the magnetic field and a spin angular momentum; be aware of the effect of the sign of the charge on the value of angular momentum and spin.

7.1.2 Teaching Videos

- Search for Ch7 in this playlist
    - https://tinyurl.com/3yhze3jn

- Other videos
  - https://youtu.be/8Q6XWXVO_6s

7.2 Magnetic Moment, Angular Momentum, and Gyromagnetic Ratio

Consider Fig. 7.1 in which a charged particle with a charge q moves along a circle with radius R, about the origin at velocity $\vec{v}$. In classical mechanics, its **angular momentum**, $\vec{L}$, is given by,

$$\begin{aligned} \vec{L} &= \vec{r} \times \vec{p}, \\ &= m\vec{r} \times \vec{v}, \end{aligned} \qquad (7.1)$$

where m, $\vec{r}$, and $\vec{p}$ are the mass, position, and linear momentum of the particle, respectively. Let us only consider the case when the particle is moving at a constant speed, $v = |\vec{v}|$. Since it is moving in a circle, then $\vec{v}$ is always perpendicular to $\vec{r}$ and $|\vec{r}| = R$. Due to the right-hand rule, which can be used to guide the direction of a cross-product, we know that the direction of $\vec{L}$ is in the $\hat{z}$ direction. Therefore, we have $m\vec{r} \times \vec{v} = m|\vec{r}|v\sin 90° \hat{z} = mvR\hat{z}$, which is constant in both magnitude and direction. Therefore,

$$\vec{L} = mvR\hat{z}. \qquad (7.2)$$

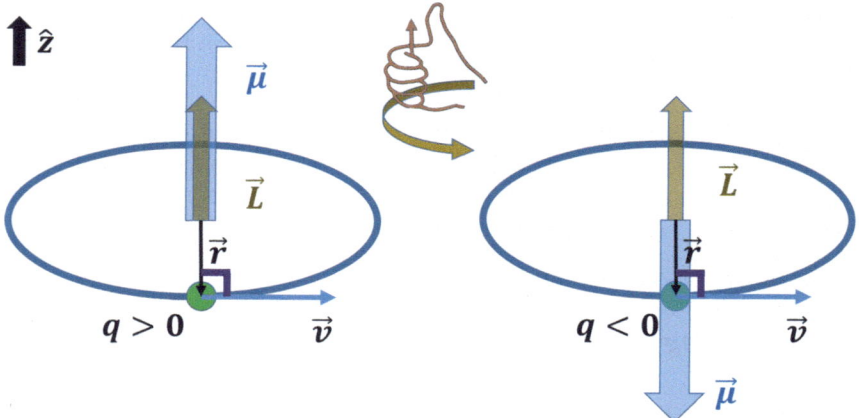

Fig. 7.1 Relationship between the angular momentum and the magnetic moment of a charged particle (Left: positive charge. Right: negative charge.). The inset shows the right-hand rule

7.2 Magnetic Moment, Angular Momentum, and Gyromagnetic Ratio

A moving charge forms a current, I. The current through a cross section is defined as the amount of charge passing through that cross section in a unit of time. Note that the cross section is perpendicular to the path of the particle. In Fig. 7.1, the cross section cuts the circle circumference. The cross section is *not* the circle drawn. Therefore,

$$I = \frac{q}{2\pi R/v},$$
$$= \frac{qv}{2\pi R}, \qquad (7.3)$$

where $2\pi R/v$ is the amount of time the particle spends to travel through the circumference of the circle.

In classical electromagnetism, it is known that a circulating current, I, along a closed path enclosing an area A creates a **magnetic moment**, $\vec{\mu}$. An **area vector**, $\vec{A}$, is defined as a vector with a magnitude A and a direction following the right-hand rule for the circulating current. The resulting magnetic moment has a magnitude of IA and a direction the same as $\vec{A}$. Therefore, in our case (left of Fig. 7.1), if the charge is positive and circulating counterclockwise, the current is also flowing counterclockwise and

$$\vec{\mu} = I\vec{A},$$
$$= IA\hat{z},$$
$$= I\pi R^2 \hat{z},$$
$$= \frac{qv}{2\pi R}\pi R^2 \hat{z},$$
$$= \frac{qvR}{2}\hat{z}, \qquad (7.4)$$

where we used the fact that $A = \pi R^2$ for a circle in line 3 and Eq. (7.3) in line 4.

Now we will derive the relationship between $\vec{\mu}$ and $\vec{L}$ by rearranging Eq. (7.2) to be $\frac{\vec{L}}{m} = vR\hat{z}$ and substitute it into Eq. (7.4) to obtain,

$$\vec{\mu} = \frac{qvR}{2}\hat{z},$$
$$= \frac{q\vec{L}}{2m},$$
$$= \gamma\vec{L}, \qquad (7.5)$$

where $\gamma = \frac{q}{2m}$ is the **gyromagnetic ratio** of the particle. Since $\gamma = \frac{q}{2m}$ and the mass of a particle must be positive, γ is positive (negative) if the charge is positive (negative). γ relates the angular momentum, $\vec{L}$, of a moving charged particle in a circle to the magnetic moment, $\vec{\mu}$, it generates. $\vec{L}$ is parallel (anti-parallel) to $\vec{\mu}$ if the charge is positive (negative).

If you are not interested in mathematics, I hope you can at least appreciate this. A moving charged particle has an angular momentum and also generates a magnetic moment. They are related through the gyromagnetic ratio in Eq. (7.5).

7.3 Spin, Spin Angular Momentum, and Spin Magnetic Moment

With the classical descriptions of angular magnetic moment and angular momentum described in the previous section, we will now discuss **spin**, **spin angular momentum**, and **spin magnetic moment**.

As discussed in Sect. 7.1, spin is an intrinsic property of an elementary particle (such as an electron and a proton). This is just like the fact that the charge is an intrinsic property of a particle. This is the safest way to understand spin. It is *wrong* to attempt to think of the spin of an elementary particle as the "spinning" of a ball. The reason is because *it is not*. If you have heard about the *color* of quarks and you accept that the color of a quark is its intrinsic property and is not the "color" we see with our eyes, then there is no difficulty in understanding that the spin is also an intrinsic property.

However, we understand Physics based on our daily lives. While an electron is not spinning, it has the properties of a spinning ball. First of all, an electron or a proton has the intrinsic property, called *spin*. In quantum mechanics, we say that it has a **spin quantum number**, S, of either $+\frac{1}{2}$ or $-\frac{1}{2}$. That is

$$S = \pm \frac{1}{2}. \tag{7.6}$$

This is given by QED. Note that it only has two possible values because *we limit ourselves in the discussion of **spin-half** particles, such as electrons and protons.* **We will only limit to the discussion of electrons** from now on. We may think of the two values corresponding to spinning clockwise and anti-clockwise, respectively. This is wrong but this is a convenient way to link it to our daily experience. But do not do that if you feel comfortable accepting that the intrinsic property of an electron can only have two possible values.

Due to the spin, it also has an associated spin angular momentum, $\vec{S}$,

$$\vec{S} = S\hbar\hat{z}, \tag{7.7}$$

which is a vector like the classical angular momentum and is arbitrarily chosen to be along the $\hat{z}$ direction (parallel or anti-parallel depending on S). Since a spinning ball has non-zero spin angular momentum, it is natural to expect an electron to have an angular momentum due to spin, even though its spin is not the classical "spin."

7.3 Spin, Spin Angular Momentum, and Spin Magnetic Moment

In Eq. (7.5), it is shown that a moving charged particle generates a magnetic moment that is correlated to its angular momentum classically. We thus expect that the spin angular momentum of an electron will also generate a spin magnetic moment, $\vec{\mu}_e$, and they are related through the gyromagnetic ratio,

$$\vec{\mu}_e = \gamma \vec{S}. \tag{7.8}$$

However, there is an important difference from the classical case. The γ of spin is about 2.002 times of the γ of the classical case for an electron. Note that an electron has a negative charge of $q = -e = -1.6 \times 10^{-19} C$. Therefore, Eq. (7.5) becomes $\gamma \approx -2.002 \frac{e}{2m}$. We now introduce a new term called the **g-factor**, g, which is about -2.002 to account for the difference. Equation (7.8) becomes

$$\vec{\mu}_e = g \frac{e}{2m} \vec{S},$$
$$= g \frac{e\hbar}{2m} \frac{\vec{S}}{\hbar},$$
$$= g \mu_B \frac{\vec{S}}{\hbar}, \tag{7.9}$$

where **Bohr magneton**, $\mu_B = \frac{e\hbar}{2m}$, is introduced. This is approximately the magnetic moment an electron particle has due to its spin (as $|\frac{g\vec{S}}{\hbar}| \approx 1$).

In summary, the gyromagnetic ratio of an electron is about $\frac{-e}{m}$, which is negative as in the classical case. This means that if an electron has a spin angular momentum in the positive direction, $\vec{S} = \frac{\hbar}{2}\hat{z}$ (or positive spin, $S = +\frac{1}{2}$), it has a spin magnetic moment, $\vec{\mu}_e$, in the negative direction and vice versa. This is shown on the right of Fig. 7.2.

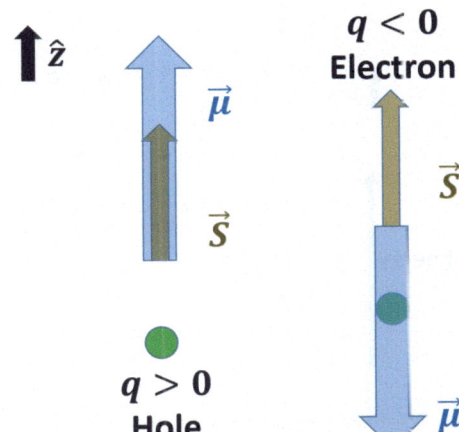

Fig. 7.2 Relationship between the spin angular momentum and the spin magnetic moment of a charged particle of a hole (left) and an electron (right)

To prepare for future discussion, a **hole** (lack of an electron) in a semiconductor is also spin-half. Its spin magnetic moment has the same direction as its spin angular momentum (left of Fig. 7.2) due to its positive charge.

7.4 Interaction Between Magnetic Moment and an External Magnetic Field

How do we know if an electron ($q = -e < 0$) or a hole ($q = e > 0$) spins up, $|\uparrow\rangle$ or $S = \frac{1}{2}$, or spins down, $|\downarrow\rangle$ or $S = -\frac{1}{2}$? We do not know until we do a measurement. While we can look at a ball to determine if it is spinning clockwise or counterclockwise, we cannot measure the spin of an elementary particle in the same way. This is because spin is an intrinsic property of an elementary particle and it is not a mechanical spin. We can try to measure its energy if different spin states have different energies.

However, if there is no external magnetic field, both states are indistinguishable (left of Fig. 7.3). In other words, an electron or a hole has the same energy in both states in the absence of a magnetic field. This is easy to understand. When there is no external magnetic field, the space is isotropic. That means every direction in the space is the same (imagine we are floating in an empty universe). So it is natural for a particle to have the same energy regardless of its spin. This is called **degeneracy** and both states are **degenerated states**.

When there is an external magnetic field, $\vec{B}$, the space is no longer isotropic for this particle. This is because the spin magnetic moment interacts with the magnetic field, resulting in different energies in different spin states. We say that the *degeneracy is lifted*. The change of energy due to the magnetic field, Δ_E, is given by

$$\Delta_E = H = -\vec{B} \cdot \vec{\mu},$$

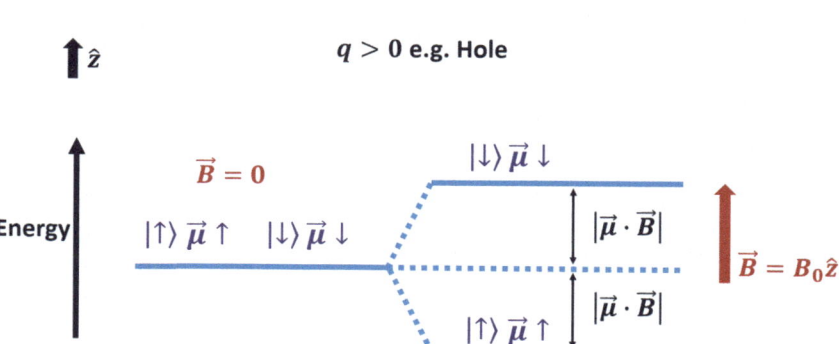

Fig. 7.3 Energy of a positively charged particle with spin under zero (left) or a finite external magnetic field (right)

7.5 Summary

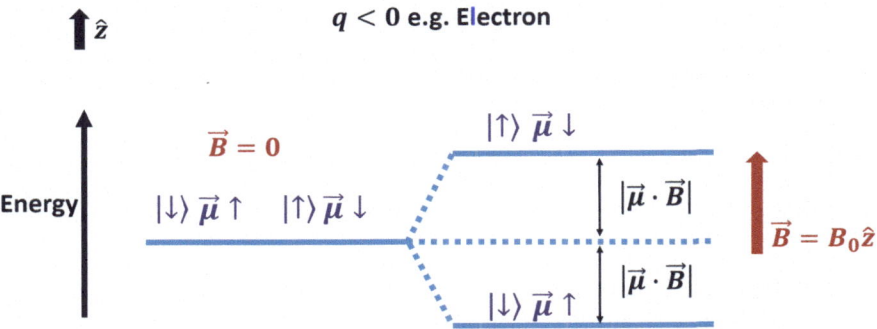

Fig. 7.4 Energy of a negatively charged particle with spin under zero (left) or a finite external magnetic field (right)

$$= -|\vec{B}||\vec{\mu}|\cos\theta, \quad (7.10)$$

where θ is the angle between $\vec{B}$ and $\vec{\mu}$. We also call the change of energy H because this is also the **interaction Hamiltonian** between the spin magnetic moment and the external magnetic field. This is a **dot/inner/scalar product** formula and the result is a scalar (energy). If $\vec{B}$ and $\vec{\mu}$ are in the same direction (parallel), $\theta = 0°$ and it has a lower energy ($H = -|\vec{B}||\vec{\mu}| < 0$). If $\vec{B}$ and $\vec{\mu}$ are in the opposite direction (anti-parallel), $\theta = 180°$ and it has a higher energy ($H = |\vec{B}||\vec{\mu}| > 0$) (right of Fig. 7.3).

In quantum computing, we use the spin quantum number (S or $|\downarrow\rangle / |\uparrow\rangle$) more often than the spin magnetic moment, $\vec{\mu}$. As discussed in the previous section and Fig. 7.2, for a positively charged particle, its spin has the same direction as the spin magnetic moment. Therefore, for a positively charged particle, when the spin is parallel (anti-parallel) to the external magnetic field, it has a lower (higher) energy as shown in Fig. 7.3. This is the case for a hole.

For a negatively charged particle such as an electron, the result is the opposite. When the spin is parallel (anti-parallel) to the external magnetic field, it has a higher (lower) energy as shown in Fig. 7.4.

7.5 Summary

Spin is an intrinsic property of elementary particles. From now on, we will only discuss electrons and holes which are spin-half particles. They have two possible spin values, $S = \pm\frac{1}{2}$. Like the classical case that a moving charged particle with an angular momentum has a magnetic moment, a charged particle with spin also has the associated spin angular momentum and spin magnetic moment. As a result, a charged particle with spin interacts with an external magnetic field resulting in a change of energy. Finally, it is important to note that a positively charged particle has a lower energy when its spin is in the same direction as (parallel to) the external magnetic field. For a negatively charged particle such as an electron, it has a higher energy when they are in parallel.

Problems

7.1 Fundamental Constants
Find the numerical values with units of e, h, $\hbar$, and μ_B.

7.2 Fundamental Constants
Calculate the γ of electron and hole. What are the masses we should use?

Chapter 8
Spin Qubit—Larmor Precession—Phase Shift Gate

8.1 Introduction

In this chapter, we use an electron spin qubit under a constant external magnetic field as an example to show how a single-qubit gate, namely, the phase shift gate, can be implemented by turning the external magnetic field on for a given time. This is *not* a practical approach because turning on and off a large DC magnetic field is difficult and cannot be done very fast. However, the example is very instructive because it clarifies many important concepts in spin qubits by using relatively simple mathematics. In the process, we will also discuss the construction of Hamiltonian, how Larmor precession can be understood on the Block sphere, and how to find the time required to implement a given phase shift gate.

8.1.1 Learning Outcomes

Be able to construct the Hamiltonian of a spin qubit under an external magnetic field; understand the meaning of Larmor precession and its relationship to a phase shift gate.

8.1.2 Teaching Videos

- Search for Ch8 in this playlist
  - https://tinyurl.com/3yhze3jn

- Other videos

  - https://youtu.be/DtdDRfFb0Zs

8.2 Construction of Single-Qubit Gate Hamiltonian Under a Constant Magnetic Field

As discussed in Sect. 7.4, a charged particle with spin has its magnetic moment, $\vec{\mu}$, interacts with an external magnetic field, $\vec{B}$, through the interaction Hamiltonian in Eq. (7.10), which is repeated here for convenience.

$$\begin{aligned} H &= -\vec{B} \cdot \vec{\mu}, \\ &= -\vec{B} \cdot \gamma \vec{S}. \end{aligned} \tag{8.1}$$

In order to study how a general spin qubit evolves under this Hamiltonian, we need to construct the Hamiltonian first. We will use *electron spin* as an example. Since an electron has a negative charge, $q = -e < 0$, where $e = 1.6 \times 10^{-19} C$, its spin has an opposite direction to its spin magnetic moment, $\vec{\mu}$, with $\gamma < 0$ (Eq. (7.8) and Fig. 7.4). Assuming the magnetic field is constant and pointing at the *negative* $\hat{z}$ direction (Fig. 8.1 which is opposite to that in Fig. 7.4), we have $\vec{B} = -B_0\hat{z}$, where $B_0 > 0$. Therefore, a spin-up state, $|\uparrow\rangle$, has a *lower* energy than without the external magnetic field by $|\vec{B}||\vec{\mu}|$ (Eq. (7.10)). This is the **ground state** and we can label it as $|g\rangle$ or $|0\rangle$. Similarly, the spin-down state, $|\downarrow\rangle$, has a *higher* energy than without the external magnetic field by $|\vec{B}||\vec{\mu}|$ and is an **excited state** ($|e\rangle$ or $|1\rangle$). Therefore,

$$\begin{aligned} |\uparrow\rangle &= |g\rangle = |0\rangle, \\ |\downarrow\rangle &= |e\rangle = |1\rangle. \end{aligned} \tag{8.2}$$

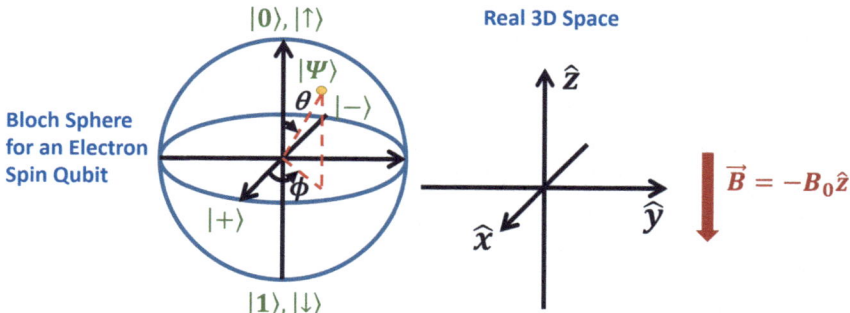

Fig. 8.1 The Bloch sphere representation of an electron spin qubit and the real 3D space coordinate system in which the direction of the external magnetic field is shown

8.2 Construction of Single-Qubit Gate Hamiltonian Under a Constant...

Before moving forward, *there are a few confusions to be clarified*. As discussed in Chaps. 5 and 6 of this book and Chapter 27 of [1], the Bloch sphere is the embedding of the abstract hyperspace in our real 3D space. It has its uses but it also creates a few confusions. Firstly, the state on top of the sphere does not have a higher energy although it appears to be on the top. Usually, it is labeled as $|0\rangle$ and is the ground state with the lowest energy just like in this case. Therefore, do not think of it as a point on the top of a ball which usually has a higher energy under gravitational force. Secondly, whether $|0\rangle$ or $|1\rangle$ has a higher energy depends on the charge of the particle and the direction of the external magnetic field. In this chapter, we assume the magnetic field is pointing downward so that the *negatively* charged electron spin has $|0\rangle$ as its ground state. It is completely legitimate if we point the magnetic field upward to have $|1\rangle$ as its ground state but it is less commonly used. Similarly, if one wants to use a similar convention for a positive charge such as a hole, it will be more convenient to apply the magnetic field upward.

Now, we know how the energy splits or how the energy degeneracy is lifted under an external magnetic field. We can construct the matrix for the Hamiltonian in Eq. (8.1). If we conduct an experiment, we will observe two possible energy values, $\lambda_0 = -|\vec{B}||\vec{\mu}|$ and $\lambda_1 = |\vec{B}||\vec{\mu}|$, for a given external magnetic field. We call them the eigenvalues of the observable operator (see also Sect. 3.4), which is just the Hamiltonian. In the experiment, we can also decide to call the corresponding eigenstates, $|\uparrow\rangle = |0\rangle$ and $|\downarrow\rangle = |1\rangle$, respectively. Using Eq. (3.9), we have,

$$H = \sum_{i=0}^{1} \lambda_i |i\rangle \langle i|,$$

$$= -|\vec{B}||\vec{\mu}| |\uparrow\rangle \langle \uparrow| + |\vec{B}||\vec{\mu}| |\downarrow\rangle \langle \downarrow|,$$

$$= -B_0 |\vec{\mu}| |0\rangle \langle 0| + B_0 |\vec{\mu}| |1\rangle \langle 1|,$$

$$= -B_0 |\vec{\mu}| \sigma_z. \tag{8.3}$$

The last step of Eq. (8.3) will be clear after the following derivation. From Eqs. (7.7) and (7.9), for an electron, $|\vec{\mu}| = |g\frac{e}{2m}\hbar S| = \frac{-ge\hbar}{4m} \approx \frac{e\hbar}{2m}$, which is just the **Bohr magneton** as expected. Note that $e > 0$. Therefore,

$$H = B_0 \frac{ge\hbar}{4m} |0\rangle \langle 0| - B_0 \frac{ge\hbar}{4m} |1\rangle \langle 1|,$$

$$= B_0 \frac{ge\hbar}{4m} \begin{pmatrix} 1 \\ 0 \end{pmatrix} (1\ 0) - B_0 \frac{ge\hbar}{4m} \begin{pmatrix} 0 \\ 1 \end{pmatrix} (0\ 1),$$

$$= B_0 \frac{ge\hbar}{4m} \begin{pmatrix} 1 & 0 \\ 0 & 0 \end{pmatrix} - B_0 \frac{ge\hbar}{4m} \begin{pmatrix} 0 & 0 \\ 0 & 1 \end{pmatrix},$$

$$= B_0 \frac{ge\hbar}{4m} \begin{pmatrix} 1 & 0 \\ 0 & -1 \end{pmatrix},$$

$$\approx -B_0 \frac{e\hbar}{2m} \begin{pmatrix} 1 & 0 \\ 0 & -1 \end{pmatrix},$$

$$= -B_0 \frac{e\hbar}{2m} \sigma_z, \quad (8.4)$$

where we have used the approximation that $g \approx -2$ from line 4 to line 5. Note also that $-B_0 \frac{e\hbar}{2m} < 0$. It can be seen that the Hamiltonian turns out to be proportional to the **Pauli spin matrix** σ_z. It can be better appreciated now why this is called the "spin" matrix and that the subscript z is due to the fact that the Hamiltonian is a result of a magnetic field in the $\hat{z}$ direction (although it is pointing in $-\hat{z}$ in this case).

8.3 Larmor Precession and Phase Shift Gate

Let us now apply the Schrödinger equation to investigate how an electron spin qubit state, $|\Psi\rangle$, will evolve under a constant external magnetic field $\vec{B} = -B_0\hat{z}$. Let $|\Psi(t)\rangle = \alpha(t)|0\rangle + \beta(t)|1\rangle$ be an arbitrary state at time t. Based on Eqs. (4.1) and (8.3),

$$i\hbar \frac{\partial |\Psi\rangle}{\partial t} = H|\psi\rangle,$$

$$i\hbar \frac{\partial |\Psi\rangle}{\partial t} = -B_0|\vec{\mu}|\sigma_z|\psi\rangle,$$

$$i\hbar \frac{\partial}{\partial t} \begin{pmatrix} \alpha(t) \\ \beta(t) \end{pmatrix} = -B_0|\vec{\mu}| \begin{pmatrix} 1 & 0 \\ 0 & -1 \end{pmatrix} \begin{pmatrix} \alpha(t) \\ \beta(t) \end{pmatrix},$$

$$\begin{pmatrix} i\hbar \frac{\partial \alpha(t)}{\partial t} \\ i\hbar \frac{\partial \beta(t)}{\partial t} \end{pmatrix} = \begin{pmatrix} -B_0|\vec{\mu}|\alpha(t) \\ B_0|\vec{\mu}|\beta(t) \end{pmatrix}. \quad (8.5)$$

By equating the vector elements on the left and right sides of the equation, we obtain two equations,

$$i\hbar \frac{\partial \alpha(t)}{\partial t} = -B_0|\vec{\mu}|\alpha(t), \quad (8.6)$$

$$i\hbar \frac{\partial \beta(t)}{\partial t} = B_0|\vec{\mu}|\beta(t). \quad (8.7)$$

8.3 Larmor Precession and Phase Shift Gate

These are two first-order differential equations for $\alpha(t)$ and $\beta(t)$, respectively. The solutions are,

$$\alpha(t) = \alpha_0 \exp\left\{\frac{-B_0|\vec{\mu}|}{i\hbar}t\right\}, \tag{8.8}$$

$$\beta(t) = \beta_0 \exp\left\{\frac{B_0|\vec{\mu}|}{i\hbar}t\right\}, \tag{8.9}$$

where α_0 and β_0 are two complex number constants. When $t = 0$, $\alpha(t) = \alpha_0$ and $\beta(t) = \beta_0$. This just represents the initial state before the magnetic field is applied. One may verify the solutions by substituting them into Eqs. (8.6) and (8.7). Apparently, the coefficients of $|\Psi\rangle$ are changing as a function of time due to the interaction Hamiltonian between the spin magnetic moment and the external magnetic field. Therefore, $|\Psi(t)\rangle$ moves on the Bloch sphere when the magnetic field is applied.

To understand how it moves, let us represent $|\Psi\rangle$ in the polar (θ) and azimuthal (ϕ) angles on the Block sphere (Eq. (5.3)). We first set $\alpha_0 = \cos\frac{\theta_0}{2}\exp\left\{-i\frac{\phi_0}{2}\right\}$ and $\beta_0 = \sin\frac{\theta_0}{2}\exp\left\{i\frac{\phi_0}{2}\right\}$, where θ_0 and ϕ_0 are the initial angles of the state on the Bloch sphere. Then from Eqs. (8.8) and (8.9),

$$|\Psi(t)\rangle = \alpha(t)|0\rangle + \beta(t)|1\rangle,$$

$$= \alpha_0 \exp\left\{\frac{-B_0|\vec{\mu}|}{i\hbar}t\right\}|0\rangle + \beta_0 \exp\left\{\frac{B_0|\vec{\mu}|}{i\hbar}t\right\}|1\rangle,$$

$$= \cos\frac{\theta_0}{2}\exp\left\{-i\frac{\phi_0}{2}\right\}\exp\left\{\frac{-B_0|\vec{\mu}|}{i\hbar}t\right\}|0\rangle$$
$$+ \sin\frac{\theta_0}{2}\exp\left\{i\frac{\phi_0}{2}\right\}\exp\left\{\frac{B_0|\vec{\mu}|}{i\hbar}t\right\}|1\rangle,$$

$$= \cos\frac{\theta_0}{2}\exp\left\{-i\frac{\phi_0 - 2B_0|\vec{\mu}|t/\hbar}{2}\right\}|0\rangle$$
$$+ \sin\frac{\theta_0}{2}\exp\left\{i\frac{\phi_0 - 2B_0|\vec{\mu}|t/\hbar}{2}\right\}|1\rangle,$$

$$= \cos\frac{\theta_0}{2}\exp\left\{-i\frac{\phi}{2}\right\}|0\rangle + \sin\frac{\theta_0}{2}\exp\left\{i\frac{\phi}{2}\right\}|1\rangle, \tag{8.10}$$

where we finally made the substitution of $\phi = \phi_0 - 2B_0|\vec{\mu}|t/\hbar$. The polar angle does not change and stays constant at θ_0 but the azimuthal angle changes at a rate of $\frac{\partial \phi}{\partial t} = -2B_0|\vec{\mu}|/\hbar$ which means the qubit state will rotate clockwise (looking from the top) as shown in the left part of Fig. 8.2 (the right part will be explained in the next sub-section). This is called the **Larmor precession** and the precession rate is called the **Larmor frequency**, $\omega_L = 2B_0|\vec{\mu}|/\hbar$ (this is an angular frequency and

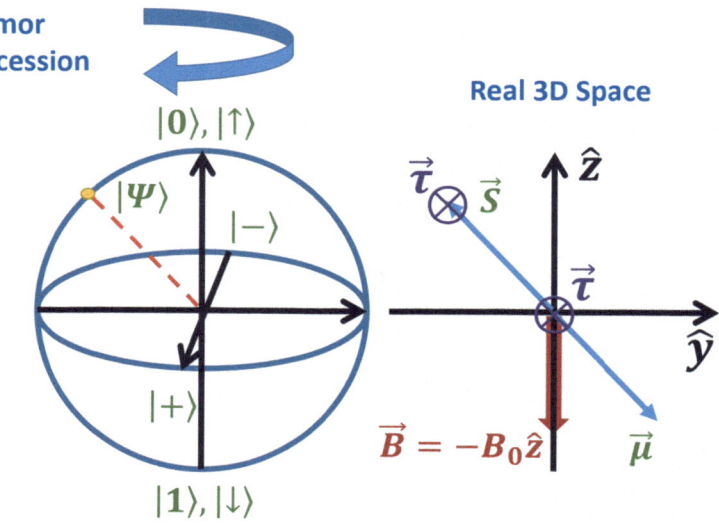

Fig. 8.2 The Bloch sphere representation of an electron spin qubit and the corresponding spin angular momentum, $\vec{S}$, and spin magnetic moment, $\vec{\mu}$, in the real 3D space. The torque, $\vec{\tau}$, generated by the interaction between $\vec{B}$ and $\vec{\mu}$ is also shown at the original. The torque is redrawn at the end of $\vec{S}$ to show how it changes $\vec{S}$ due to Eq. (8.16). The spin angular momentum will change in the direction of $\vec{\tau}$, which is pointing into the screen/paper, corresponding to the clockwise precession (seeing from the top) of the qubit state on the Bloch sphere

we also take its absolute value since the precession direction is immaterial in the definition). It can also be further expressed as

$$\begin{aligned} \omega_L &= 2B_0|\vec{\mu}|/\hbar, \\ &= 2B_0|\gamma \vec{S}|/\hbar, \\ &= \left| \frac{2ge\hbar}{4m} B_0 \right|/\hbar, \\ &\approx \frac{e}{m} B_0. \end{aligned} \qquad (8.11)$$

We can also define

$$f_L = \frac{\omega_L}{2\pi}. \qquad (8.12)$$

From the first line of Eq. (8.11), we can rewrite the eigenenergies of the system (Sect. 8.2) as

$$\lambda_0 = -|\vec{B}||\vec{\mu}|,$$

8.3 Larmor Precession and Phase Shift Gate

$$= -\hbar\omega_L/2. \tag{8.13}$$

and

$$\lambda_1 = |\vec{B}||\vec{\mu}|,$$
$$= \hbar\omega_L/2. \tag{8.14}$$

Therefore, the separation of the two energy levels ($\lambda_1 - \lambda_0$) determines the Larmor frequency.

8.3.1 Notes on Qubit Larmor Precession

This subsection may be skipped if you are not interested in going deeper. It is instructive to look deeper into the physics of Larmor precession and its relationship with the qubit on the Block sphere. The physics behind Larmor precession is due to the **torque**, $\vec{\tau}$, which is generated by the interaction between the magnetic field, $\vec{B}$, and the spin magnetic moment, $\vec{\mu}$, *changing the spin angular momentum, S*. The torque is found by,

$$\vec{\tau} = \vec{\mu} \times \vec{B}. \tag{8.15}$$

Figure 8.2 shows that for a state $|\Psi\rangle$ on the Bloch sphere embedding in the real 3D space, it has a corresponding spin angular momentum $\vec{S}$ on the $y-z$ plane. Since an electron has a negative charge, the corresponding magnetic moment, $\vec{\mu}$, is in the opposite direction but along the same line. Using the right-hand rule, a torque $\vec{\tau}$ is generated due to Eq. (8.15) which is pointing into the screen/paper. Note that this torque will modify $\vec{S}$ but *not* $\vec{\mu}$ because the torque is *the rate of change of angular momentum* (although the spin magnetic moment is the cause of the torque) through the following equation:

$$\vec{\tau} = \frac{\partial \vec{S}}{\partial t}. \tag{8.16}$$

Therefore, the angular momentum moves away from the $y - z$ plane into the paper. In the Bloch sphere, this corresponds to the state $|\Psi\rangle$ precessing clockwise when looking from the top.

8.4 Implementation of Phase Shift Gate

A phase shift gate, $U_{PS,\Phi}$, with a phase Φ has the following matrix (Eq. (4.41)):

$$U_{PS,\Phi} = \begin{pmatrix} 1 & 0 \\ 0 & e^{i\Phi} \end{pmatrix}. \tag{8.17}$$

How to implement a general phase shift gate using the physics setup we have (i.e., an electron spin qubit under a constant external magnetic field)? Based on the discussion in the previous section, the qubit will rotate about the vertical axis on the Bloch sphere at Larmor frequency *clockwise* when the external magnetic field is turned on. What is the meaning of a phase shift gate? It is a *counterclockwise* rotation about the vertical axis by an angle Φ. This can be understood through Eq. (5.4) with $\lambda = 0$, $\theta = 0$, $\alpha = 0$, and $\phi = \Phi$ as the third counterclockwise rotation (Fig. 5.2). Therefore, they have the same action except in the opposite direction. Figure 8.3 shows that the phase shift gate, $U_{PS,\Phi}$, rotates the state by an angle Φ *anti-clockwise* when looking from the top. Therefore, we need to use our setup to achieve the same effect by rotating it clockwise by $2\pi - \Phi$. As a specific example, a $U_{PS,\pi}$ is just a **Z**-gate (or σ_z) and it can be achieved by rotating the state about the vertical axis clockwise by π (Fig. 8.3).

Now let us look at the equations and find out how much time we need to turn on the external magnetic field to achieve a desirable rotation. From the second line of

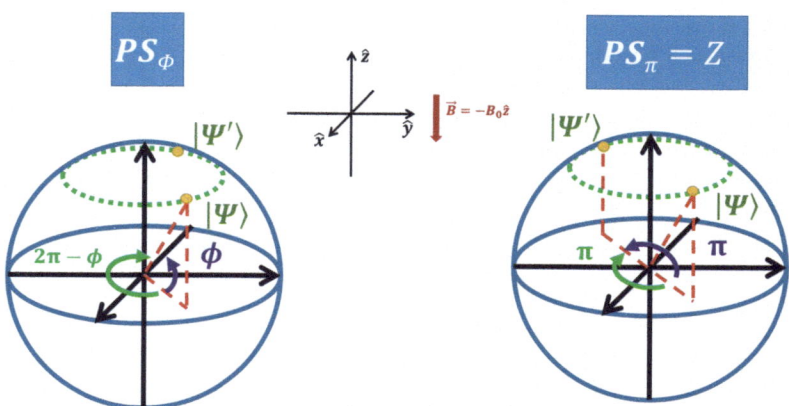

Fig. 8.3 Illustration of the action of phase shift gate, $U_{PS,\Phi}$, on a general qubit on the Bloch sphere (left: general, right: Z-gate). It can be implemented using an electron qubit under a vertical external magnetic field by rotating $2\pi - \Phi$

8.4 Implementation of Phase Shift Gate

Eq. (8.10)

$$|\Psi(t)\rangle = \alpha(t)|0\rangle + \beta(t)|1\rangle = \begin{pmatrix} \alpha(t) \\ \beta(t) \end{pmatrix},$$

$$= \alpha_0 \exp\left\{\frac{-B_0|\vec{\mu}|}{i\hbar}t\right\}|0\rangle + \beta_0 \exp\left\{\frac{B_0|\vec{\mu}|}{i\hbar}t\right\}|1\rangle,$$

$$= \begin{pmatrix} \exp\left\{\frac{-B_0|\vec{\mu}|}{i\hbar}t\right\} & 0 \\ 0 & \exp\left\{\frac{B_0|\vec{\mu}|}{i\hbar}t\right\} \end{pmatrix} \begin{pmatrix} \alpha_0 \\ \beta_0 \end{pmatrix},$$

$$= \begin{pmatrix} \exp\left\{i\frac{e}{2m}B_0 t\right\} & 0 \\ 0 & \exp\left\{-i\frac{e}{2m}B_0 t\right\} \end{pmatrix} \begin{pmatrix} \alpha_0 \\ \beta_0 \end{pmatrix},$$

$$= \exp\left\{i\frac{e}{2m}B_0 t\right\} \begin{pmatrix} 1 & 0 \\ 0 & \exp\left\{-i\frac{e}{m}B_0 t\right\} \end{pmatrix} \begin{pmatrix} \alpha_0 \\ \beta_0 \end{pmatrix}. \tag{8.18}$$

In line 1, we write the state in both the *bra-ket* notation and column vector form. Line 3 shows that the state at any time is equal to a matrix multiplying the initial state, $\begin{pmatrix} \alpha_0 \\ \beta_0 \end{pmatrix}$. Therefore, the matrix is the **quantum gate** of this physical system. We also used Eqs. (8.3) and (8.4) to go from line 3 to line 4 and the approximation of $g \approx -2$. Finally, we factorized out a global phase as this has no physical meaning (Eq. 5.2)). Our goal is now to equate the corresponding elements of the quantum gate to those of the phase shift gate,

$$\begin{pmatrix} 1 & 0 \\ 0 & \exp\left\{-i\frac{e}{m}B_0 t\right\} \end{pmatrix} = \begin{pmatrix} 1 & 0 \\ 0 & e^{i\Phi} \end{pmatrix}. \tag{8.19}$$

That is, $-\frac{e}{m}B_0 t = \Phi$. Therefore,

$$-\frac{e}{m}B_0 t = \Phi,$$

$$t = -\frac{\Phi m}{e B_0}. \tag{8.20}$$

However, this will give us a negative time. This is because, as discussed in Fig. 8.3, our physical system rotates the qubit clockwise while a phase shift gate rotates the qubit anti-clockwise. Therefore, we will rotate it by $2\pi - \Phi$ clockwise using our physical system instead. Therefore,

$$t = \frac{(2\pi - \Phi)m}{e B_0}. \tag{8.21}$$

For example, to implement a $Z - gate$ (i.e., $\Phi = \pi$), we need to turn on the magnetic field for a time, $t = \frac{\pi m}{e B_0}$.

8.5 Summary

In this chapter, we learn the physics of an electron spin qubit under an external magnetic field. We learn that it is important to set the magnetic field pointing downward if the conventional Bloch sphere notation is to be used. The qubit precesses about the vertical axis at Larmor frequency through which a single-qubit gate, namely, the phase shift gate with an arbitrary phase, can be built. This also requires precise control of the turn-on time of the magnetic field. Finally, the approximation $g \approx -2$ is used in Eq. (8.11) and after. However, it is easy to revert them to the full solutions as $\frac{e}{m}$ is just the approximated gyromagnetic ratio, γ. We may substitute all $\frac{e}{m}$ by $\gamma = \frac{-ge}{2m}$ (as $g < 0$ and $e > 0$) in the approximated equations.

8.5.1 Equations Without Approximations

Gyromagnetic ratio:

$$\gamma = \frac{ge}{2m}. \tag{8.22}$$

Interaction Hamiltonian (Eq. (8.4)):

$$H = B_0 \frac{\gamma \hbar}{2} \begin{pmatrix} 1 & 0 \\ 0 & -1 \end{pmatrix}. \tag{8.23}$$

Larmor frequency (Eq. (8.11)):

$$\omega_L = |\gamma B_0|. \tag{8.24}$$

Gate matrix corresponding to the system (Eq. (8.18)):

$$\begin{pmatrix} 1 & 0 \\ 0 & \exp\{-i|\gamma|B_0 t\} \end{pmatrix} = \begin{pmatrix} 1 & 0 \\ 0 & \exp\{-i\omega_L t\} \end{pmatrix}. \tag{8.25}$$

Turn-on time of the external magnetic field (Eq. (8.21)):

$$t = \frac{(2\pi - \Phi)}{|\gamma| B_0} = \frac{(2\pi - \Phi)}{\omega_L}. \tag{8.26}$$

Problems

8.1 Phase Shift Gates
Find the magnetic field turn-on time required to implement S and T gates.

8.2 Gate Time
Use the electron mass you found in Problem 7.2 to calculate the magnetic field strength required to implement a Z gate with a gate time of 200 ns.

8.3 Gate Time 2
In solid-state materials such as semiconductors, the effective mass of an electron is not the same as its rest mass. Assuming it is halved, how would γ change and how would the gate time in Problem 8.2 change?

Reference

1. Hiu-Yung Wong. *Introduction to Quantum Computing*. Springer, 2024.

Chapter 9
Spin Qubit—Rabi Oscillation

9.1 Introduction

In the previous chapter, we implemented a physical system to perform a phase-shift gate of arbitrary phase for an electron spin qubit. It is constructed by placing the electron spin qubit in a constant external magnetic field in the vertical direction, $\vec{B} = -B_0\hat{z}$ with $B_0 > 0$. This is a field pointing downward in the real 3D space. It causes Larmor precession of the qubit on the Bloch sphere rotating clockwise looking from the top. In this chapter, we will further apply a small oscillating magnetic field in the $\hat{x}$ direction to enable the rotation of the qubit about the $y-axis$. This is called Rabi oscillation. With this tool, we will be able to rotate a spin qubit from and to any point on the Bloch sphere and thus implement an arbitrary 1-qubit gate.

More importantly, in this process, we will clarify some mathematical skills and also introduce the concept of rotating frame and rotating wave approximation.

9.1.1 Learning Outcomes

Understand the role of the small oscillating magnetic field; be able to describe how a state moves on the Bloch sphere during Rabi oscillation; appreciate the power and limitation of the perturbation method; understand rotation wave approximation and the meaning of rotating frame.

9.1.2 Teaching Videos

- Search for Ch9 in this playlist
    - https://tinyurl.com/3yhze3jn

- Other videos
    - https://youtu.be/5u-vr6_awNc
    - https://youtu.be/XoHVvXTDyQU
    - https://youtu.be/lmFNULXkR_I
    - https://youtu.be/bEWO0bmi-M4
    - https://youtu.be/ndXeb6YcPy0
    - https://youtu.be/GLTHGGPKGuo

9.2 Spin Angular Momentum Operator

In the previous chapter, we constructed the *Hamiltonian* under a constant vertical magnetic field using the eigenenergies found in the experiment (Eq. (8.3)). When we have an oscillating magnetic field, it is difficult to use the same approach because the direction of the effective magnetic field is not constant. Therefore, we need a more formal approach and introduce the concept of **spin angular momentum operator**.

As mentioned in Chap. 7, the spin magnetic moment of an electron, $\vec{\mu}_e$, is a result of spin angular momentum, $\vec{S}$ (Eq. (7.9)). The interaction Hamiltonian between the magnetic moment and the external magnetic field, $\vec{B}$, is given in Eq. (8.1) and repeated here for convenience,

$$\begin{aligned} H &= -\vec{B} \cdot \vec{\mu}, \\ &= -\vec{B} \cdot \gamma \vec{S}. \end{aligned} \quad (9.1)$$

When deducing the eigenenergies, we implicitly set $\vec{S} = \frac{\hbar}{2}\hat{z}$ and Eq. (9.1) is just an inner product of two *real space 3D vectors*, $\vec{B}$ and $\vec{S}$. We then use the eigenenergies to construct Eq. (8.3). That means *we have treated the spin angular momentum as a vector*. In order to handle a more general case, in which *the net magnetic field direction is a variable*, we need to use another formalism by treating the spin angular momentum as an operator. We will not study the formalism and we will take it for granted. The mathematics just works out and agrees with experiments. We define the spin angular momentum operator as

$$\begin{aligned} \vec{S} &= \frac{\hbar}{2}\vec{\sigma}, \\ &= \frac{\hbar}{2}(\sigma_x \hat{x} + \sigma_y \hat{y} + \sigma_z \hat{z}), \\ &= \frac{\hbar}{2}\left[\begin{pmatrix} 0 & 1 \\ 1 & 0 \end{pmatrix} \hat{x} + \begin{pmatrix} 0 & -i \\ i & 0 \end{pmatrix} \hat{y} + \begin{pmatrix} 1 & 0 \\ 0 & -1 \end{pmatrix} \hat{z} \right], \\ &= \frac{\hbar}{2} \begin{pmatrix} \hat{z} & \hat{x} - i\hat{y} \\ \hat{x} + i\hat{y} & -\hat{z} \end{pmatrix}, \end{aligned} \quad (9.2)$$

where the Pauli vector, $\vec{\sigma}$, is used (see Chapter 7 of [1] and Eq. (6.4)). Note that $\vec{S}$ is still a vector in the linear algebra sense (it obeys the definition of vector in a vector space) but it is also an operator now.

Therefore, Eq. (9.1) becomes

$$H = -\vec{B} \cdot \gamma \vec{S}. \qquad (9.3)$$

We take the definition of spin angular momentum operator for granted but we can check if this makes sense.

Example 9.1 Find the expectation value of $\vec{S}$ in state $|0\rangle$.

We know that $|0\rangle$ is $|\uparrow\rangle$ (Fig. 8.1) and it has a spin value of $\frac{1}{2}$ and should be in the $+\hat{z}$ direction. Based on Eq. (7.7), the spin angular momentum is $\frac{\hbar}{2}\hat{z}$.

Let us find the expectation value of $\vec{S}$ in state $|0\rangle$, which is

$$\langle 0| \vec{S} |0\rangle = (1\ 0) \frac{\hbar}{2} \begin{pmatrix} \hat{z} & \hat{x} - i\hat{y} \\ \hat{x} + i\hat{y} & -\hat{z} \end{pmatrix} \begin{pmatrix} 1 \\ 0 \end{pmatrix},$$

$$= \frac{\hbar}{2} (1\ 0) \begin{pmatrix} \hat{z} \\ \hat{x} + i\hat{y} \end{pmatrix},$$

$$= \frac{\hbar}{2} \hat{z}. \qquad (9.4)$$

This is the same as what we expected with the correct magnitude and also direction. ∎

9.3 Rabi Oscillation

9.3.1 Experimental Setup and Hamiltonian

The setup for **Rabi oscillation** is shown in Fig. 9.1. It is the same as Fig. 8.1 except that a small oscillating magnetic field is applied along the $\hat{x}$ direction, with $B_1 \ll B_0$. The oscillating magnetic field oscillates at an angular frequency of ω_1. To understand how the qubit will evolve, we need to first find the *Hamiltonian* using Eq. (9.3).

Firstly, the total magnetic field, $\vec{B}$, at any time, is given by

$$\vec{B} = B_1 \cos(\omega_1 t)\hat{x} - B_0 \hat{z}. \qquad (9.5)$$

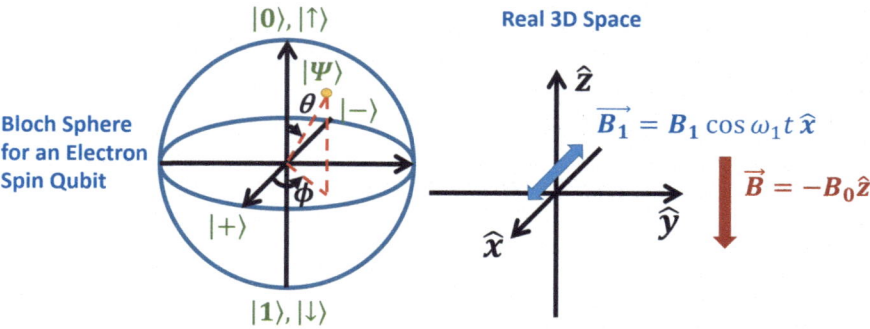

Fig. 9.1 The Bloch sphere representation of an electron spin qubit (left) and the real 3D space coordinate system in which the direction of the external constant and oscillating magnetic fields is shown

Therefore, the Hamiltonian is given by Eq. (9.3):

$$
\begin{aligned}
H &= -\vec{B} \cdot \gamma \vec{S}, \\
&\approx (\frac{e}{m}\vec{S}) \cdot \vec{B}, \\
&= \frac{e\hbar}{2m}(\sigma_x \hat{x} + \sigma_y \hat{y} + \sigma_z \hat{z}) \cdot (B_x \hat{x} + B_y \hat{y} + B_z \hat{z}), \\
&= \frac{e\hbar}{2m} \begin{pmatrix} \sigma_x & \sigma_y & \sigma_z \end{pmatrix} \begin{pmatrix} B_1 \cos(\omega_1 t) \\ 0 \\ -B_0 \end{pmatrix}, \\
&= \frac{e\hbar}{2m} B_1 \cos(\omega_1 t) \sigma_x - \frac{e\hbar}{2m} B_0 \sigma_z, \\
&= \frac{e\hbar}{2m} B_1 \cos(\omega_1 t) \sigma_x - \frac{\hbar \omega_L}{2} \sigma_z, \\
&= H_I + H_0,
\end{aligned}
\quad (9.6)
$$

where in line 2, we used the approximation of $g \approx -2$ (see Eqs. (7.8) and (7.9)). It should also be noted that $e = 1.6 \times 10^{-19} C < 0$. In line 3, we used Eq. (9.2). In line 4, Eq. (9.5) was used. In line 6, we used the definition of Larmor frequency in Eq. (8.11).

The Hamiltonian is separated into two parts, namely, $\boldsymbol{H_0} = -\frac{e\hbar}{2m} B_0 \sigma_z$ and $\boldsymbol{H_I} = \frac{e\hbar}{2m} B_1 \cos(\omega_1 t) \sigma_x$. $\boldsymbol{H_0}$ is the same as Eq. (8.4), which is the Hamiltonian due to the interaction of the vertical constant magnetic field and the spin magnetic moment. We know this causes the spin to precess about the vertical axis on the Bloch sphere. $\boldsymbol{H_I}$ is new and it is due to the transverse oscillating magnetic field and is a function of B_1 and ω_1.

9.3.2 Setup of the Schrödinger Equation

The Schrödinger equation corresponding to this system is given by

$$i\hbar \frac{\partial |\psi\rangle}{\partial t} = (\boldsymbol{H_0} + \boldsymbol{H_I})|\psi\rangle, \tag{9.7}$$

with $\boldsymbol{H_0}$ and $\boldsymbol{H_I}$ defined in Eq. (9.6). It is not trivial to solve this equation. However, as mentioned at the beginning of this section, we assume $B_1 \ll B_0$. Then the oscillating magnetic field in the $\hat{x}$ direction can be treated as a **perturbation** of the original system in Fig. 8.1 (i.e., only with a constant vertical magnetic field).

It is known that if a perturbation is added to a system, the state of the new system, $|\Psi_{perturbed}\rangle$, is a linear combination of the eigenstates ($|0\rangle$, $|1\rangle$, $\cdots$) of the unperturbed system weighted by the complex exponential of the scaled eigenenergies ($-E_0 t/\hbar$, $-E_1 t/\hbar$, $\cdots$). Interested readers can refer to time-dependent perturbation in any quantum mechanics textbook such as [2] for more details. That is,

$$|\Psi_{perturbed}\rangle = c_0 e^{-iE_0 t/\hbar}|0\rangle + c_1 e^{-iE_1 t/\hbar}|1\rangle + \cdots, \tag{9.8}$$

where $c_0, c_1, \cdots$ are complex coefficients and can be time-dependent.

In our case, the *unperturbed system* (i.e., without the oscillating field) has two eigenstates $|0\rangle$ and $|1\rangle$ with eigenenergies $-|\vec{B}||\vec{\mu}| = -\hbar\omega_L/2$ and $|\vec{B}||\vec{\mu}| = \hbar\omega_L/2$, respectively (see the discussion in Sect. 8.2 and Eqs. (8.13) and (8.14)). Therefore, the state of the new system with the perturbing oscillating magnetic field can be written as

$$\begin{aligned}|\Psi_{perturbed}\rangle &= c_0 e^{-i\frac{E_0 t}{\hbar}}|0\rangle + c_1 e^{-i\frac{E_1 t}{\hbar}}|1\rangle, \\ &= c_0 e^{-i\frac{-\hbar\omega_L t}{2\hbar}}|0\rangle + c_1 e^{-i\frac{\hbar\omega_L t}{2\hbar}}|1\rangle, \\ &= c_0 e^{i\frac{\omega_L}{2}t}|0\rangle + c_1 e^{-i\frac{\omega_L}{2}t}|1\rangle.\end{aligned} \tag{9.9}$$

Note that $|\Psi_{perturbed}\rangle$ is just the $|\psi\rangle$ in Eq. (9.7). We will now only use $|\Psi\rangle$.

We should also remember that the wavefunction needs to be normalized. Therefore,

$$\begin{aligned}|c_0|^2 + |c_1|^2 &= 1, \\ |c_0(t)|^2 + |c_1(t)|^2 &= 1,\end{aligned} \tag{9.10}$$

where we emphasize that c_0 and c_1 can be *time-dependent* in the second line.

9.3.3 Solving the Schrödinger Equation

Now, we will solve Eq. (9.7) by substituting Eq. (9.9) into it. This is a lengthy derivation. If you feel this is too long, you may skip and just trust the answer. If not, I hope you can follow closely as we will practice some very useful skills in quantum mechanics.

We first perform the substitution.

$$i\hbar \frac{\partial |\psi\rangle}{\partial t} = (H_0 + H_I) |\psi\rangle,$$

$$i\hbar \frac{\partial (c_0 e^{i\frac{\omega_L}{2}t} |0\rangle + c_1 e^{-i\frac{\omega_L}{2}t} |1\rangle)}{\partial t} = (H_0 + H_I)(c_0 e^{i\frac{\omega_L}{2}t} |0\rangle + c_1 e^{-i\frac{\omega_L}{2}t} |1\rangle). \tag{9.11}$$

Let us first simplify the *left-hand side*.

$$i\hbar \frac{\partial |\psi\rangle}{\partial t} = i\hbar \left(\dot{c}_0 e^{i\frac{\omega_L}{2}t} |0\rangle + c_0 (i\frac{\omega_L}{2}) e^{i\frac{\omega_L}{2}t} |0\rangle \right.$$
$$\left. + \dot{c}_1 e^{-i\frac{\omega_L}{2}t} |1\rangle + c_1 (-i\frac{\omega_L}{2}) e^{-i\frac{\omega_L}{2}t} |1\rangle \right), \tag{9.12}$$

where we use the chain rule in derivatives. We also use the common notation of time derivative, $\dot{c} = \frac{dc}{dt}$. We will now apply an inner product with $|0\rangle$ to Eq. (9.12). This is equivalent to applying $\langle 0|$ from the left,

$$\langle 0| i\hbar \left(\dot{c}_0 e^{i\frac{\omega_L}{2}t} |0\rangle + c_0 (i\frac{\omega_L}{2}) e^{i\frac{\omega_L}{2}t} |0\rangle \right.$$
$$\left. + \dot{c}_1 e^{-i\frac{\omega_L}{2}t} |1\rangle + c_1 (-i\frac{\omega_L}{2}) e^{-i\frac{\omega_L}{2}t} |1\rangle \right),$$
$$= i\hbar \left(\dot{c}_0 e^{i\frac{\omega_L}{2}t} \langle 0|0\rangle + c_0 (i\frac{\omega_L}{2}) e^{i\frac{\omega_L}{2}t} \langle 0|0\rangle \right.$$
$$\left. + \dot{c}_1 e^{-i\frac{\omega_L}{2}t} \langle 0|1\rangle + c_1 (-i\frac{\omega_L}{2}) e^{-i\frac{\omega_L}{2}t} \langle 0|1\rangle \right),$$
$$= i\hbar \left(\dot{c}_0 e^{i\frac{\omega_L}{2}t} + c_0 (i\frac{\omega_L}{2}) e^{i\frac{\omega_L}{2}t} \right), \tag{9.13}$$

where we have used the fact that $|0\rangle$ and $|1\rangle$ are *orthonormal* in the last step. Therefore, $\langle 0|0\rangle = 1$ and $\langle 0|1\rangle = 0$.

Now, we will simplify the right-hand side and apply $\langle 0|$ from the left. From Eq. (9.11),

$$\langle 0| (H_0 + H_I)(c_0 e^{i\frac{\omega_L}{2}t} |0\rangle + c_1 e^{-i\frac{\omega_L}{2}t} |1\rangle),$$

9.3 Rabi Oscillation

$$\begin{aligned}
&= \langle 0| \left(\boldsymbol{H_0} c_0 e^{i\frac{\omega_L}{2}t} |0\rangle + \boldsymbol{H_0} c_1 e^{-i\frac{\omega_L}{2}t} |1\rangle \right. \\
&\quad \left. + \boldsymbol{H_I} c_0 e^{i\frac{\omega_L}{2}t} |0\rangle + \boldsymbol{H_I} c_1 e^{-i\frac{\omega_L}{2}t} |1\rangle \right), \\
&= c_0 e^{i\frac{\omega_L}{2}t} \langle 0| \boldsymbol{H_0} |0\rangle + c_1 e^{-i\frac{\omega_L}{2}t} \langle 0| \boldsymbol{H_0} |1\rangle \\
&\quad + c_0 e^{i\frac{\omega_L}{2}t} \langle 0| \boldsymbol{H_I} |0\rangle + c_1 e^{-i\frac{\omega_L}{2}t} \langle 0| \boldsymbol{H_I} |1\rangle.
\end{aligned} \quad (9.14)$$

Now, we need to evaluate $\langle 0| \boldsymbol{H_0} |0\rangle$, $\langle 0| \boldsymbol{H_0} |1\rangle$, $\langle 0| \boldsymbol{H_I} |0\rangle$, and $\langle 0| \boldsymbol{H_I} |1\rangle$. Firstly, since $|0\rangle$ and $|1\rangle$ are the eigenstates of the unperturbed system, $\boldsymbol{H_0}$, we have,

$$\begin{aligned}
\langle 0| \boldsymbol{H_0} |0\rangle &= \langle 0| (-\hbar\omega_L/2 |0\rangle), \\
&= -\hbar\omega_L/2 \langle 0|0\rangle, \\
&= -\hbar\omega_L/2,
\end{aligned} \quad (9.15)$$

and

$$\begin{aligned}
\langle 0| \boldsymbol{H_0} |1\rangle &= \langle 0| \hbar\omega_L/2 |1\rangle, \\
&= \hbar\omega_L/2 \langle 0|1\rangle, \\
&= 0,
\end{aligned} \quad (9.16)$$

where we used the fact that applying the Hamiltonian to its eigenvector results in the eigenvector scaled by the corresponding eigenvalue. We could have also used the matrix form of $\boldsymbol{H_0} = -\frac{\hbar\omega_L}{2}\sigma_z$ in Eq. (9.6) to obtain the same result.

To evaluate $\langle 0| \boldsymbol{H_I} |0\rangle$ and $\langle 0| \boldsymbol{H_I} |1\rangle$, we will use the definition of $\boldsymbol{H_I} = \frac{e\hbar}{2m} B_1 \cos(\omega_1 t) \sigma_x$ in Eq. (9.6) and perform matrix multiplications.

$$\begin{aligned}
\langle 0| \boldsymbol{H_I} |0\rangle &= \langle 0| \frac{e\hbar}{2m} B_1 \cos(\omega_1 t) \sigma_x |0\rangle, \\
&= \frac{e\hbar}{2m} B_1 \cos(\omega_1 t) \langle 0| \sigma_x |0\rangle, \\
&= \frac{e\hbar}{2m} B_1 \cos(\omega_1 t) (1\ 0) \begin{pmatrix} 0 & 1 \\ 1 & 0 \end{pmatrix} \begin{pmatrix} 1 \\ 0 \end{pmatrix}, \\
&= 0,
\end{aligned} \quad (9.17)$$

and

$$\begin{aligned}
\langle 0| \boldsymbol{H_I} |1\rangle &= \langle 0| \frac{e\hbar}{2m} B_1 \cos(\omega_1 t) \sigma_x |1\rangle, \\
&= \frac{e\hbar}{2m} B_1 \cos(\omega_1 t) \langle 0| \sigma_x |1\rangle,
\end{aligned}$$

$$= \frac{e\hbar}{2m} B_1 \cos(\omega_1 t) \begin{pmatrix} 1 & 0 \end{pmatrix} \begin{pmatrix} 0 & 1 \\ 1 & 0 \end{pmatrix} \begin{pmatrix} 0 \\ 1 \end{pmatrix},$$

$$= \frac{e\hbar}{2m} B_1 \cos(\omega_1 t). \tag{9.18}$$

Therefore, only two terms are left in Eq. (9.14). By equating it to Eq. (9.13), we have

$$i\hbar \left(\dot{c}_0 e^{i\frac{\omega_L}{2}t} + c_0 (i\frac{\omega_L}{2}) e^{i\frac{\omega_L}{2}t} \right),$$

$$= c_0 e^{i\frac{\omega_L}{2}t} \langle 0| \boldsymbol{H_0} |0\rangle + c_1 e^{-i\frac{\omega_L}{2}t} \langle 0| \boldsymbol{H_I} |1\rangle ,$$

$$= c_0 e^{i\frac{\omega_L}{2}t} (-\hbar\omega_L/2) + c_1 e^{-i\frac{\omega_L}{2}t} (\frac{e\hbar}{2m} B_1 \cos(\omega_1 t)). \tag{9.19}$$

Equating line 1 and line 3 of Eq. (9.19) and recognizing $i\hbar c_0 (i\frac{\omega_L}{2}) e^{i\frac{\omega_L}{2}t}$ in line 1 and $c_0 e^{i\frac{\omega_L}{2}t}(-\hbar\omega_L/2)$ in line 3 are equal and can be canceled, we have

$$i\hbar \dot{c}_0 e^{i\frac{\omega_L}{2}t} = c_1 e^{-i\frac{\omega_L}{2}t} (\frac{e\hbar}{2m} B_1 \cos(\omega_1 t)),$$

$$i\dot{c}_0 = \frac{e}{2m} B_1 \cos(\omega_1 t) e^{-i\omega_L t} c_1. \tag{9.20}$$

What we have done so far is to perform an inner product with $|0\rangle$ so that we obtain the rate of change of c_0 as a function of c_1. Now, if we repeat the same process by applying an inner product with $|1\rangle$, we will get an equation relating the rate of change of c_1 as a function of c_0, which is

$$i\dot{c}_1 = \frac{e}{2m} B_1 \cos(\omega_1 t) e^{i\omega_L t} c_0. \tag{9.21}$$

As a reminder, $e > 0$. To solve for c_0, we can perform one more time differentiation on Eq. (9.20) and substitute Eq. (9.21) into it. However, for instructional purposes, we are not interested in the general solution here. We are interested in the case when $\omega_1 = \omega_L$.

9.4 Spin Resonance and Rotating Wave Approximation (RWA)

When $\omega_1 = \omega_L$, the system is at **electron spin resonance**. *It means that the horizontal oscillating magnetic field oscillates at the same frequency as the Larmor frequency.* By using, $\cos\theta = \frac{e^{i\theta} + e^{-i\theta}}{2}$, Eq. (9.20) becomes

9.4 Spin Resonance and Rotating Wave Approximation (RWA)

$$i\dot{c}_0 = \frac{e}{2m}B_1\cos(\omega_1 t)e^{-i\omega_L t}c_1,$$

$$= \frac{e}{4m}B_1(e^{i\omega_1 t} + e^{-i\omega_1 t})e^{-i\omega_L t}c_1,$$

$$= \frac{e}{4m}B_1(e^{i(\omega_1-\omega_L)t} + e^{-i(\omega_1+\omega_L)t})c_1,$$

$$= \frac{e}{4m}B_1(e^{i0t} + e^{-i2\omega_L t})c_1,$$

$$= \frac{e}{4m}B_1(1 + e^{-i2\omega_L t})c_1, \qquad (9.22)$$

where we have used the fact that $\omega_1 = \omega_L$ in line 4 at spin resonance. For the term $e^{-i2\omega_L t}$ which is equal to $\cos 2\omega_L t - i\sin 2\omega_L t$, it oscillates very fast compared to the time scale we are interested in. Therefore, it can be ignored. This is called the **rotating wave approximation (RWA)**. We will discuss more about the meaning of "fast" and understand it from a more intuitive point of view in the next section. For now, let us accept it and Eq. (9.22) is simplified to

$$i\dot{c}_0 = \frac{e}{4m}B_1 c_1. \qquad (9.23)$$

Similarly, by using RWA under spin resonance, Eq. (9.21) is simplified to

$$i\dot{c}_1 = \frac{e}{4m}B_1 c_0. \qquad (9.24)$$

By taking a further time derivative on Eq. (9.23) and substituting Eq. (9.24) into Eq. (9.23),

$$\frac{d(i\dot{c}_0)}{dt} = i\ddot{c}_0 = \frac{e}{4m}B_1\dot{c}_1,$$

$$= \frac{e}{4m}B_1\left(\frac{e}{4mi}B_1 c_0\right),$$

$$\ddot{c}_0 = -\left(\frac{B_1 e}{4m}\right)^2 c_0, \qquad (9.25)$$

we obtain a second-order differential equation for c_0. Defining $\omega_R' = \frac{B_1 e}{4m}$, the equation becomes $\ddot{c}_0 = -\omega_R'^2 c_0$ and the general solution is [3]

$$c_0 = A\cos\omega_R' t + B\sin\omega_R' t. \qquad (9.26)$$

To find c_1, we will use Eq. (9.23) and substitute Eq. (9.26) into it,

$$c_1 = \frac{i}{\omega_R'}\dot{c}_0 = i(-A\sin\omega_R' t + B\cos\omega_R' t),$$

$$= -iA \sin \omega'_R t + iB \cos \omega'_R t. \tag{9.27}$$

Note that c_0 and c_1 are the coefficients (ignoring the phases) of $|0\rangle$ and $|1\rangle$, respectively (Eq. (9.9)). Therefore, the squares of their magnitudes represent the probabilities of finding the electron at $|0\rangle$ and $|1\rangle$, respectively, upon a measurement. Since both of them oscillate at ω'_R, the probabilities of finding the electron at $|0\rangle$ and $|1\rangle$ thus oscillate with time.

9.5 Rabi Oscillation and Rabi Frequency

As shown in Eqs. (9.26) and (9.27), the movement of the electron state is complex. We now will inspect a special case to understand how the spin state moves on the Bloch sphere due to **Rabi oscillation**.

At time $t = 0$, by using Eqs. (9.26) and (9.27), Eq. (9.9) becomes

$$\begin{aligned} |\Psi(t=0)\rangle &= c_0 e^{i\frac{\omega_L}{2}0} |0\rangle + c_1 e^{-i\frac{\omega_L}{2}0} |1\rangle, \\ &= c_0(t=0) |0\rangle + c_1(t=0) |1\rangle, \\ &= A|0\rangle + iB|1\rangle, \end{aligned} \tag{9.28}$$

where A and iB are, thus, the coefficients of $|0\rangle$ and $|1\rangle$ at $t = 0$, respectively. However, recall that on the Bloch sphere, the state at $t = 0$ is characterized by an initial polar angle (θ_0) and an azimuthal angle (ϕ_0) (similar to Eq. (8.10)). Therefore, $A = \cos\frac{\theta_0}{2}\exp\left\{-i\frac{\phi_0}{2}\right\}$ and $iB = \sin\frac{\theta_0}{2}\exp\left\{i\frac{\phi_0}{2}\right\}$. As a result,

$$\begin{aligned} c_0 &= A\cos\omega'_R t + B\sin\omega'_R t, \\ &= \cos\frac{\theta_0}{2}\exp\left\{-i\frac{\phi_0}{2}\right\}\cos\omega'_R t - i\sin\frac{\theta_0}{2}\exp\left\{i\frac{\phi_0}{2}\right\}\sin\omega'_R t, \\ &= e^{-i\frac{\phi_0}{2}}[\cos\frac{\theta_0}{2}\cos\omega'_R t - e^{i\frac{\pi}{2}}\sin\frac{\theta_0}{2}\sin\omega'_R t e^{i\phi_0}], \\ &= e^{-i\frac{\phi_0}{2}}[\cos\frac{\theta_0}{2}\cos\omega'_R t - \sin\frac{\theta_0}{2}\sin\omega'_R t e^{i(\phi_0+\frac{\pi}{2})}], \end{aligned} \tag{9.29}$$

where we used the fact that $i = e^{i\frac{\pi}{2}}$ in line 3. Using Eq. (9.27), we obtain,

$$\begin{aligned} c_1 &= -iA\sin\omega'_R t + iB\cos\omega'_R t, \\ &= -i\cos\frac{\theta_0}{2}\exp\left\{-i\frac{\phi_0}{2}\right\}\sin\omega'_R t + \sin\frac{\theta_0}{2}\exp\left\{i\frac{\phi_0}{2}\right\}\cos\omega'_R t, \\ &= e^{i\frac{\phi_0}{2}}[\sin\frac{\theta_0}{2}\cos\omega'_R t + \cos\frac{\theta_0}{2}\sin\omega'_R t e^{-i(\phi_0+\frac{\pi}{2})}], \end{aligned} \tag{9.30}$$

9.5 Rabi Oscillation and Rabi Frequency

where we used the fact that $-i = e^{-i\frac{\pi}{2}}$. It is still difficult to visualize how the qubit evolves with an arbitrary ϕ_0. Let us set $\phi_0 = -\pi/2$ (Fig. 9.2). Then $\phi_0 + \pi/2 = 0$ and $e^{-i(\phi_0 + \frac{\pi}{2})} = 1$. Equation (9.29) and Eq. (9.30) are simplified to

$$c_0 = e^{-i\frac{\phi_0}{2}}[\cos\frac{\theta_0}{2}\cos\omega'_R t - \sin\frac{\theta_0}{2}\sin\omega'_R t],$$

$$= e^{-i\frac{\phi_0}{2}}\cos(\frac{\theta_0}{2} + \omega'_R t),$$

$$= e^{-i\frac{\phi_0}{2}}\cos\frac{\theta_0 + 2\omega'_R t}{2},$$

$$= e^{-i\frac{\phi_0}{2}}\cos\frac{\theta_0 + \omega_R t}{2}, \quad (9.31)$$

where we use the trigonometric identity $\cos(\alpha + \beta) = \cos\alpha\cos\beta - \sin\alpha\sin\beta$ in the second line. Note that we have already set $\phi_0 = \pi/2$. It is kept unsubstituted to show where this initial azimuthal angle is in the equation. Here we define **Rabi frequency**,

$$\omega_R = 2\omega'_R = \frac{B_1 e}{2m}. \quad (9.32)$$

Similarly,

$$c_1 = e^{i\frac{\phi_0}{2}}[\sin\frac{\theta_0}{2}\cos\omega'_R t + \cos\frac{\theta_0}{2}\sin\omega'_R t],$$

$$= e^{i\frac{\phi_0}{2}}\sin(\frac{\theta_0}{2} + \omega'_R t),$$

$$= e^{i\frac{\phi_0}{2}}\sin\frac{\theta_0 + 2\omega'_R t}{2},$$

$$= e^{i\frac{\phi_0}{2}}\sin\frac{\theta_0 + \omega_R t}{2}, \quad (9.33)$$

by using the trigonometric identity $\sin(\alpha + \beta) = \sin\alpha\cos\beta + \cos\alpha\sin\beta$ in the second line.

Now let us substitute Eqs. (9.31) and (9.33) into Eq. (9.9),

$$|\Psi\rangle = c_0 e^{i\frac{\omega_L}{2}t}|0\rangle + c_1 e^{-i\frac{\omega_L}{2}t}|1\rangle,$$

$$= e^{-i\frac{\phi_0}{2}}\cos\frac{\theta_0 + \omega_R t}{2}e^{i\frac{\omega_L}{2}t}|0\rangle + e^{i\frac{\phi_0}{2}}\sin\frac{\theta_0 + \omega_R t}{2}e^{-i\frac{\omega_L}{2}t}|1\rangle,$$

$$= e^{-i\frac{\phi_0 - \omega_L t}{2}}\cos\frac{\theta_0 + \omega_R t}{2}|0\rangle + e^{i\frac{\phi_0 - \omega_L t}{2}}\sin\frac{\theta_0 + \omega_R t}{2}|1\rangle. \quad (9.34)$$

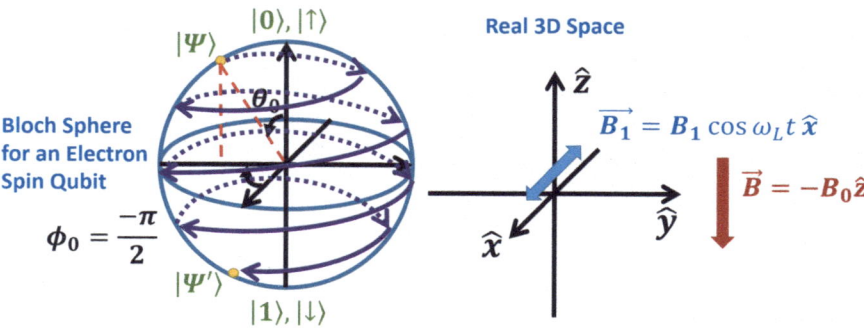

Fig. 9.2 Rabi oscillation at spin resonance when the horizontal field is oscillating at Larmor frequency ($\omega_1 = \omega_L$). The initial state $|\Psi\rangle$ has an initial azimuthal angle $\phi_0 = -\pi/2$. The left shows how the state moves on the Bloch sphere due to Larmor precession and Rabi oscillation. The right shows the setup of the experiment

Again we have already set $\phi_0 = \pi/2$ to achieve this result. Like Eq. (8.10), this equation tells us that the azimuthal angle, i.e., $\phi_0 - \omega_L t$, reduces at a rate of ω_L. This is the Larmor precession due to the vertical constant magnetic field. At the same time, the polar angle also changes as $\theta_0 + \omega_R t$ which means it changes at a rate of the Rabi frequency, ω_R (Fig. 9.2).

Example 9.2 It is instructive to understand Rabi oscillation and Rabi frequency by examining the change of probability of measuring $|0\rangle$ and $|1\rangle$ as a function of time.

Let us still consider the case when $\phi_0 = \pi/2$. We will also set $\theta_0 = 0$ which means the initial state is at the "north pole" of the Bloch sphere. Based on Eq. (9.34), the state as a function of time becomes,

$$|\Psi\rangle = e^{-i\frac{\phi_0-\omega_L t}{2}} \cos\frac{0+\omega_R t}{2}|0\rangle + e^{i\frac{\phi_0-\omega_L t}{2}} \sin\frac{0+\omega_R t}{2}|1\rangle,$$
$$= e^{-i\frac{\phi_0-\omega_L t}{2}} \cos\frac{\omega_R t}{2}|0\rangle + e^{i\frac{\phi_0-\omega_L t}{2}} \sin\frac{\omega_R t}{2}|1\rangle. \qquad (9.35)$$

The probability of finding the state at $|0\rangle$, P_0, is thus $|e^{-i\frac{\phi_0-\omega_L t}{2}} \cos\frac{\omega_R t}{2}|^2$ because it is the square of the magnitude of the coefficient of $|0\rangle$. Therefore, $P_0 = \cos^2\frac{\omega_R t}{2}$ because the exponential term has a unity length. Similarly, the probability of finding the state at $|1\rangle$, P_1, is $\sin^2\frac{\omega_R t}{2}$.

Since both P_0 and P_1 are the squares of a sinusoidal function, they have a period of π and have values between 0 and 1, which is expected as they are probabilities. Therefore, they repeat when $\frac{\omega_R t}{2} = \pi$. In other words, they have a period of $T = \frac{2\pi}{\omega_R}$. Figure 9.3 plots P_0 and P_1 as a function of time when $\omega_R = 2\pi \times 50 kHz$. It has a period of $T = \frac{2\pi}{2\pi \times 50 kHz} = 20 \mu s$. ∎

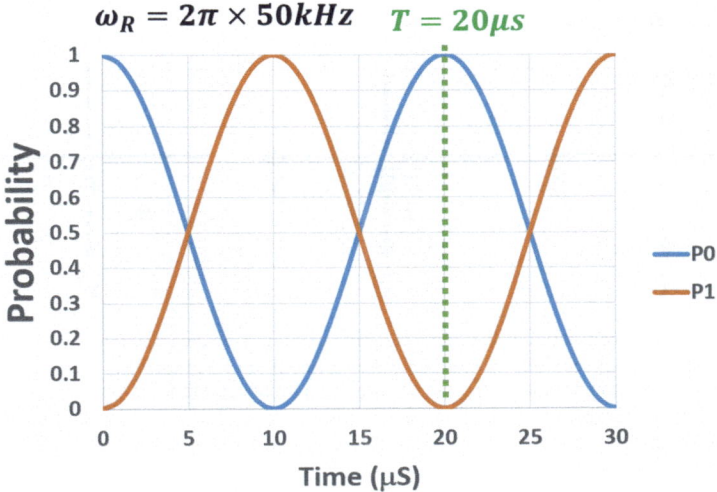

Fig. 9.3 Plots of P_0 and P_1 as a function of time when $\omega_R = 2\pi \times 50 kHz$

9.6 Intuitive View of Spin Resonance and Rotating Wave Approximation

Let us now gain more insight into the physics of the horizontal oscillating field. Since it is at spin resonance, the field is oscillating at Larmor frequency. The oscillation field is copied from Eq. (9.5) with $\omega_1 = \omega_L$ as,

$$\vec{B}_1 = B_1 \cos(\omega_1 t)\hat{x},$$
$$= \frac{B_1}{2}(\cos(\omega_1 t)\hat{x} + \sin(\omega_1 t)\hat{y} + \cos(\omega_1 t)\hat{x} - \sin(\omega_1 t)\hat{y}),$$
$$= \vec{B}_{1+} + \vec{B}_{1-}, \qquad (9.36)$$

where in line 2, the two sine terms can be canceled to restore line 1. In line 3, we made the following definitions:

$$\vec{B}_{1+} = \frac{B_1}{2}(\cos(\omega_1 t)\hat{x} + \sin(\omega_1 t)\hat{y}). \qquad (9.37)$$

$$\vec{B}_{1-} = \frac{B_1}{2}(\cos(\omega_1 t)\hat{x} - \sin(\omega_1 t)\hat{y}). \qquad (9.38)$$

Therefore, the linearly oscillating field in the $\hat{x}$ direction is decomposed into the sum of one clockwise, $\vec{B}_{1-}$, and one counterclockwise, $\vec{B}_{1+}$, rotating field with half of the original strength ($B_1/2$) at Larmor frequency (middle of Fig. 9.4).

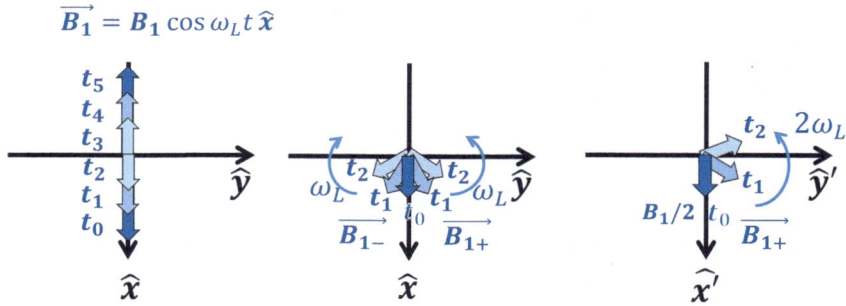

Fig. 9.4 Illustration of rotation wave approximation at spin resonance when $\omega_L = \omega_1$. Left: The change of the linearly oscillating magnetic field as a function of time with $t_0 < t_1 < t_2 < t_3 < t_4 < t_5$. Middle: Decomposition of the linearly oscillating magnetic field into two rotating fields at t_0, t_1, and t_2. Right: Transform the system to a rotating frame of Larmor precession where $\vec{B}_{1-}$ is followed and fixed. $\vec{B}_{1+}$ is then appeared to be rotating at an angular frequency of $2\omega_L$

Now, if we follow $\vec{B}_{1-}$, then $\vec{B}_{1-}$ is not moving. This is a new frame ($\hat{x}' - \hat{y}'$ at the right of Fig. 9.4) and we call it the **rotating frame** in contrast to the laboratory frame ($\hat{x} - \hat{y}$ at the middle of Fig. 9.4) where we usually stay. This is similar to the situation that if we are in an amusement park and stay on the ground (laboratory frame), we see the merry-go-round rotating. If we jump on the merry-go-round, the merry-go-round is no longer rotating to us and we are in the rotating frame. With this, we feel a constant magnetic field, $B_1/2$, pointing in the $\hat{x}'$ direction due to $\vec{B}_{1-}$. Moreover, we also feel that $\vec{B}_{1+}$ is rotating at a rate of $2\omega_L$ away from us. If for the action we are interested in, the rotation of $\vec{B}_{1+}$ is fast, then it might have rotated many cycles before we complete any meaningful action. As a result, we will not feel its effect. This is because $\vec{B}_{1+}$ has swept through all directions on the plane. Imagine that you stand still and someone pushes you from all directions within a short time; overall, there is no net effect. This is the **rotating wave approximation** in Eq. (9.22) where the effect of $e^{-i2\omega_L t}$ was ignored.

In the rotating frame, we only have a constant field, $B_1/2$, in the $\hat{x}'$ direction. This is just like the case of a constant external magnetic field and we expect that there will be Larmor precession. But this time it will precess about the $\hat{x}'$ axis. Based on Eq. (8.11), it will precess at an angular frequency of $\frac{e}{m}B_1/2$ because the constant magnetic field has a value of $B_1/2$ instead of B_0. **And this is the same as the Rabi frequency we derived in Eq. (9.32)!** Therefore, Rabi oscillation in the rotating frame is just the Larmor precession about $\hat{x}'$.

When we are in the rotating frame, we can ignore the vertical magnetic field. We can understand this from another point of view. Since $B_1 \ll B_0$, we can assume the Larmor precession due to the vertical field is not affected and the effect of the horizontal rotating field can be simply added on top of the Larmor precession. So, if we are already working in the rotating frame, $\boldsymbol{H_0}$ can be ignored because the rotating frame goes at the same angular velocity as the precession due to the vertical magnetic field.

I will leave this as an exercise. What is the Larmor precession direction in the rotating frame due to $B_1/2\hat{x}'$? Is it consistent with the direction shown in Fig. 9.2?

Lastly, let us discuss when RWA is valid. As mentioned, it is valid if the action we are interested in has a much longer time than $1/(2\omega_L)$. For example, if we are implementing a quantum gate to rotate qubits about $\hat{x}$-axis using Rabi oscillation, we want $1/\omega_R \gg 1/(2\omega_L)$. This is usually true because $1/\omega_R$ is in the μs range (Fig. 9.3) and $1/(2\omega_L)$ is much smaller.

9.7 Summary

There are a lot of equations in this chapter. However, it is worthwhile to follow to understand the details because they reinforce our understanding of some of the critical concepts. We show that by applying an oscillating magnetic field in the $\hat{x}$ direction in addition to the constant magnetic field in $-\hat{z}$, it is possible to move a state from the upper hemisphere of the Bloch sphere to the lower hemisphere. To solve the problem, we introduce the concept of the angular momentum operator. The oscillating magnetic field is usually much smaller than the constant magnetic field. This allows us to use perturbation theory to simplify the problem. We further apply the oscillating field at the same frequency as Larmor precession, resulting in spin resonance. Then we apply rotating wave approximation by ignoring the high-frequency part to arrive at the final solution to understand Rabi oscillation better. It is also very instructive to see the problem in the rotating frame to realize that Rabi oscillation in the laboratory frame is just the Larmor precession about the $\hat{x}'$-axis in the rotating frame.

Problems

9.1 Spin Angular Momentum Operator
Show that

$$\langle 1|\vec{S}|1\rangle = -\frac{\hbar}{2}\hat{z}, \tag{9.39}$$

$$\langle +|\vec{S}|+\rangle = \frac{\hbar}{2}\hat{x}. \tag{9.40}$$

9.2 Hamiltonian Elements
Redo Eqs. (9.15) and (9.16) using $\boldsymbol{H_0} = -\frac{\hbar\omega_L}{2}\boldsymbol{\sigma_Z}$.

9.3 Solving Schrödinger Equation
Derive Eq. (9.21) by following the approach for Eq. (9.20).

9.4 Rabi Oscillation 1

If $B_1 = 0.01T$, compare the Rabi oscillation period to the gate time in Problem 8.2.

9.5 Rabi Oscillation 2

Discuss if the Larmor precession in the rotating frame has the same direction as Rabi oscillation in the laboratory frame.

References

1. Hiu-Yung Wong. *Introduction to Quantum Computing*. Springer, 2024.
2. David J. Griffiths. *Introduction to Quantum Mechanics*. Pearson College Div, 1994.
3. Erwin Kreyszig. *Advanced Engineering Mathematics*. John Wiley & Sons; 9th Edition, International Edition, 2006.

Chapter 10
Spin Qubit—Rabi Oscillation Under Rotating Field Using Rotating Frame

10.1 Introduction

We learned the physics and mathematics of Rabi oscillation due to a perturbating linearly oscillating magnetic field in the previous chapter (Fig. 9.2). The skills and insights we gained are important. However, the mathematics used was cumbersome. The freedom of applying a linearly oscillating magnetic field to construct a quantum gate is also limited. In this chapter, we will consider a magnetic field rotation on the $\hat{x} - \hat{y}$ plane. It gives a great degree of freedom to rotate a state on the Bloch sphere about any axis (not just the axis along the oscillating field as in the linear case). Moreover, the magnetic field magnitude need not be small.

10.1.1 Learning Outcomes

Understand the difference between applying a linearly oscillating magnetic field and applying a rotating magnetic field; be able to solve the Schrödinger equation in the rotating frame.

10.1.2 Teaching Videos

- Search for Ch10 in this playlist
  - https://tinyurl.com/3yhze3jn
- Other videos
  - https://youtu.be/cPqby7gujFc

- https://youtu.be/r9wmOw3uxzk

10.2 Experimental Setup

The experimental setup is shown in Fig. 10.1. Besides the vertical DC magnetic field, $\vec{B} = -B_0\hat{z}$ with $B_0 > 0$, it also has a rotating magnetic field on the $\hat{x} - \hat{y}$ plane. This field rotates about the origin at an angular frequency of ω_1. Note that we set it so that it is rotating in the same direction (clockwise seen from the top) as the Larmor precession. It has the following expression:

$$\vec{B}_\perp = B_\perp(\cos(\omega_1 t + \phi_B)\hat{x} - \sin(\omega_1 t + \phi_B)\hat{y}), \tag{10.1}$$

where the initial phase of the field is ϕ_B. Its amplitude is $B_\perp$ (we do *NOT* assume it to be much smaller than B_0). Later we will see both ϕ_B and $B_\perp$ play an important role in controlling qubit state rotation and, thus, determine the qubit gate it will construct.

The total magnetic field is thus given by,

$$\begin{aligned}\vec{B} &= \vec{B}_\perp - B_0\hat{z}, \\ &= B_\perp(\cos(\omega_1 t + \phi_B)\hat{x} - \sin(\omega_1 t + \phi_B)\hat{y}) - B_0\hat{z}, \\ &= \begin{pmatrix} B_\perp \cos(\omega_1 t + \phi_B) \\ -B_\perp \sin(\omega_1 t + \phi_B) \\ -B_0 \end{pmatrix},\end{aligned} \tag{10.2}$$

where we write it in a 3D vector column form in the last line.

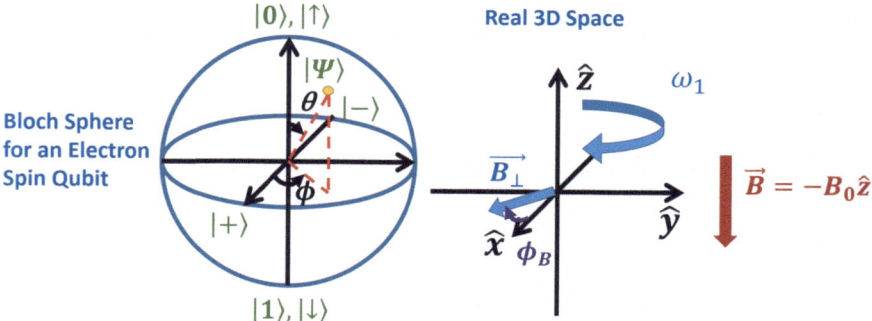

Fig. 10.1 The Bloch sphere representation of an electron spin qubit (left) and the real 3D space coordinate system in which the directions of the external constant and rotating magnetic fields are shown. The rotating field has an angular frequency ω_1 and an initial phase ϕ_B. It rotates clockwise when seeing from the top. *Note that ϕ_B is measured in the clockwise direction*

10.3 Setting Up the Hamiltonian

Therefore, the Hamiltonian is given by Eq. (9.3) as

$$
\begin{aligned}
\boldsymbol{H} &= -\vec{B} \cdot \gamma \vec{S}, \\
&\approx (\frac{e}{m}\vec{S}) \cdot \vec{B}, \\
&= \frac{e\hbar}{2m}(\sigma_x \hat{x} + \sigma_y \hat{y} + \sigma_z \hat{z}) \cdot (B_x \hat{x} + B_y \hat{y} + B_z \hat{z}), \\
&= \frac{e\hbar}{2m}\begin{pmatrix}\sigma_x & \sigma_y & \sigma_z\end{pmatrix}\begin{pmatrix} B_\perp \cos(\omega_1 t + \phi_B) \\ -B_\perp \sin(\omega_1 t + \phi_B) \\ -B_0 \end{pmatrix}, \\
&= \frac{e\hbar}{2m}B_\perp \cos(\omega_1 t + \phi_B)\sigma_x - \frac{e\hbar}{2m}B_\perp \sin(\omega_1 t + \phi_B)\sigma_y - \frac{e\hbar}{2m}B_0 \sigma_z, \\
&= \frac{\hbar \Omega_R}{2}\cos(\omega_1 t + \phi_B)\sigma_x - \frac{\hbar \Omega_R}{2}\sin(\omega_1 t + \phi_B)\sigma_y - \frac{\hbar \omega_L}{2}\sigma_z, \\
&= \boldsymbol{H_I} + \boldsymbol{H_0}. \qquad (10.3)
\end{aligned}
$$

which is similar to how we derived Eq. (9.6). In line 2, we used the approximation of $g \approx -2$ (see Eqs. (7.8) and (7.9)). It should also be noted that $e = 1.6 \times 10^{-19} C > 0$. In line 3, we used Eq. (9.2). In line 4, Eq. (10.2) is used. In line 6, we used the definition of Larmor frequency in Eq. (8.11). Again, the Hamiltonian is separated into two parts. The first part is $\boldsymbol{H_0} = \frac{-e\hbar}{2m}B_0 \sigma_z$. This is the same as Eq. (8.4) due to the interaction between the vertical constant magnetic field and the spin magnetic moment. But now $\boldsymbol{H_I} = \frac{e\hbar}{2m}B_\perp \cos(\omega_1 t + \phi_B)\sigma_x - \frac{e\hbar}{2m}B_\perp \sin(\omega_1 t + \phi_B)\sigma_y$ is due to the rotating field. We also defined a new quantity, Ω_R, which is the **Rabi frequency** in this particular experimental setup, as

$$
\begin{aligned}
\Omega_R &= \frac{e}{m}B_\perp, \\
&= -\gamma B_\perp, \qquad (10.4)
\end{aligned}
$$

where in the second line we reintroduce the *gyromagnetic ratio*, γ, to get an exact solution as $\frac{e}{m}$ is just an approximation of γ when $g \approx -2$. Note that again $e > 0$ and $\gamma < 0$ for an electron spin qubit.

It is also instructive to compare Ω_R to ω_R due to the linearly oscillating magnetic field in Eq. (9.32). Both of them are proportional to the amplitude of the field and the gyromagnetic ratio. But ω_R has an extra $\frac{1}{2}$ as it is linearly oscillating.

Now we will introduce two quantities, σ^+ and σ^-.

$$\sigma^+ = \sigma_x + i\sigma_y,$$
$$\sigma^- = \sigma_x - i\sigma_y. \qquad (10.5)$$

They are also *proportional* to the **raising** and **lowering operators** of σ_z (see Problem 10.1). Let us inspect their matrix form.

$$\sigma^+ = \sigma_x + i\sigma_y,$$
$$= \begin{pmatrix} 0 & 1 \\ 1 & 0 \end{pmatrix} + i \begin{pmatrix} 0 & -i \\ i & 0 \end{pmatrix} = \begin{pmatrix} 0 & 2 \\ 0 & 0 \end{pmatrix}. \qquad (10.6)$$

$$\sigma^- = \sigma_x - i\sigma_y,$$
$$= \begin{pmatrix} 0 & 1 \\ 1 & 0 \end{pmatrix} - i \begin{pmatrix} 0 & -i \\ i & 0 \end{pmatrix} = \begin{pmatrix} 0 & 0 \\ 2 & 0 \end{pmatrix}. \qquad (10.7)$$

Using Eq. (10.5), we can represent σ_x and σ_y in terms of σ^+ and σ^-.

$$\sigma_x = \frac{\sigma^+ + \sigma^-}{2},$$
$$\sigma_y = \frac{\sigma^+ - \sigma^-}{2i}. \qquad (10.8)$$

We now will substitute Eq. (10.8) into Eq. (10.3), and the interaction Hamiltonian becomes,

$$H_I = \frac{\hbar\Omega_R}{2}\cos(\omega_1 t + \phi_B)\sigma_x - \frac{\hbar\Omega_R}{2}\sin(\omega_1 t + \phi_B)\sigma_y,$$
$$= \frac{\hbar\Omega_R}{4}[\cos(\omega_1 t + \phi_B)(\sigma^+ + \sigma^-) - \sin(\omega_1 t + \phi_B)(\sigma^+ - \sigma^-)/i],$$
$$= \frac{\hbar\Omega_R}{4}[\cos(\omega_1 t + \phi_B)(\sigma^+ + \sigma^-) + i\sin(\omega_1 t + \phi_B)(\sigma^+ - \sigma^-)],$$
$$= \frac{\hbar\Omega_R}{4}[(\cos(\omega_1 t + \phi_B) + i\sin(\omega_1 t + \phi_B))\sigma^+ + (\cos(\omega_1 t + \phi_B)$$
$$-i\sin(\omega_1 t + \phi_B))\sigma^-],$$
$$= \frac{\hbar\Omega_R}{4}\left[\exp\{i(\omega_1 t + \phi_B)\}\sigma^+ + \exp\{-i(\omega_1 t + \phi_B)\}\sigma^-\right], \qquad (10.9)$$

where Eq. (10.8) was used in line 2. In line 5, we used the identities $e^{i\theta} = \cos\theta + i\sin\theta$ and $e^{-i\theta} = \cos\theta - i\sin\theta$.

The total Hamiltonian becomes,

$$H = H_I + H_0,$$
$$= \frac{\hbar \Omega_R}{4}[\exp\{i(\omega_1 t + \phi_B)\}\sigma^+ + \exp\{-i(\omega_1 t + \phi_B)\}\sigma^-] - \frac{\hbar \omega_L}{2}\sigma_z. \qquad (10.10)$$

10.4 Transforming to Rotating Frame

The Schrödinger equation with the time-dependent Hamiltonian in Eq. (10.10) is difficult to solve. Unlike the treatment in the previous chapter, we do NOT assume H_I to be a perturbation and we cannot use the perturbation theory (or Eq. (9.8)). To solve this, we will work on the **rotating frame** which rotates together with the rotating magnetic field $\vec{B}_\perp$ at a frequency of ω_1. Note that this is **NOT** rotating at Larmor precession as in the case in Fig. 9.4.

The Schrödinger equation to be solved in the laboratory frame is,

$$i\hbar \frac{\partial |\psi\rangle}{\partial t} = (H_0 + H_I)|\psi\rangle = H|\psi\rangle, \qquad (10.11)$$

with H given in Eq. (10.10). To solve it in the rotating frame, let us first find the unitary transformation, U, that will rotate a state at the same rate and same direction as how $\vec{B}_\perp$ is rotating. This means that in this frame, we will see $\vec{B}_\perp$ as a constant magnetic field. We can reuse the knowledge from Larmor precession even if we are rotating at ω_1. We know that Larmor precession is due to H_0, and the corresponding rotation matrix is given by (see Eq. (4.9)),

$$U_L = e^{-i\frac{H_0 t}{\hbar}},$$
$$= e^{-i\frac{-\hbar \omega_L \sigma_z t}{2\hbar}},$$
$$= e^{i\frac{\omega_L t}{2}\sigma_z}. \qquad (10.12)$$

Note that the Larmor precession and the rotating field are both rotating clockwise in this case. Therefore, the unitary matrix corresponding to *rotating a state at* ω_1 must have the same form but with ω_1 instead of ω_L in the equation. Therefore,

$$U = e^{i\frac{\omega_1 t}{2}\sigma_z}. \qquad (10.13)$$

If we are in the rotating frame, we should see the stationary wavefunction *in the laboratory frame*, $|\psi\rangle$, rotating in the opposite direction. This is equivalent to applying $U^\dagger$ to $|\psi\rangle$ in the laboratory frame if we want to describe it in the rotating frame. Therefore, in the rotating frame, a state $|\psi\rangle$ will become $U^\dagger |\psi\rangle$. We define $U_{RF} = U^\dagger$,

$$U_{RF} = U^\dagger = e^{-i\frac{\omega_1 t}{2}\sigma_z}. \tag{10.14}$$

The state $|\psi\rangle$ in the laboratory frame becomes $|\psi\rangle_{RF}$ through the following equation,

$$|\psi\rangle_{RF} = U_{RF} |\psi\rangle. \tag{10.15}$$

We can also find the inverse of U_{RF} as,

$$U_{RF}^\dagger = e^{i\frac{\omega_1 t}{2}\sigma_z}. \tag{10.16}$$

Therefore, **in the rotating frame**, the Schrödinger equation becomes,

$$i\hbar \frac{\partial |\psi\rangle_{RF}}{\partial t} = H_{RF} |\psi\rangle_{RF}, \tag{10.17}$$

where H_{RF} is the total Hamiltonian in the rotating frame. We need to find H_{RF} but it is not straightforward. We cannot just apply $U_{RF} H U_{RF}^\dagger$ to get H_{RF} because of the differential term on the left (as $U_{RF} \frac{\partial |\psi\rangle}{\partial t} \neq \frac{\partial (U_{RF}|\psi\rangle)}{\partial t}$). We will show how to solve it in the next section.

10.5 Solving the Schrödinger Equation

Now, we need to find H_{RF}. In this section, we will go through the details of the mathematics. After this section, readers are advised to read Sect. 10.7 in which canned equations can be used to speed up the process. We substitute Eq. (10.15) into the left-hand side of Eq. (10.17),

$$\begin{aligned} i\hbar \frac{\partial |\psi\rangle_{RF}}{\partial t} &= i\hbar \frac{\partial}{\partial t} (U_{RF} |\psi\rangle), \\ &= i\hbar \frac{\partial U_{RF}}{\partial t} |\psi\rangle + i\hbar U_{RF} \frac{\partial |\psi\rangle}{\partial t}, \\ &= i\hbar \frac{\partial U_{RF}}{\partial t} |\psi\rangle + U_{RF} H |\psi\rangle, \end{aligned} \tag{10.18}$$

10.5 Solving the Schrödinger Equation

where we have use the *chain rule* in line 2 and Eq. (10.11) in line 3. Since $|\psi\rangle = U_{RF}^\dagger |\psi\rangle_{RF}$ (inverse of Eq. (10.15)), we continue to change Eq. (10.18) to,

$$i\hbar \frac{\partial |\psi\rangle_{RF}}{\partial t} = i\hbar \frac{\partial U_{RF}}{\partial t} U_{RF}^\dagger |\psi\rangle_{RF} + U_{RF} H U_{RF}^\dagger |\psi\rangle_{RF},$$

$$= \left(i\hbar \frac{\partial U_{RF}}{\partial t} U_{RF}^\dagger + U_{RF} H U_{RF}^\dagger \right) |\psi\rangle_{RF},$$

$$= \boldsymbol{H}_{RF} |\psi\rangle_{RF}. \tag{10.19}$$

By using the trick above, we have successfully found that,

$$\boldsymbol{H}_{RF} = i\hbar \frac{\partial U_{RF}}{\partial t} U_{RF}^\dagger + U_{RF} H U_{RF}^\dagger. \tag{10.20}$$

Now, we will further expand $\boldsymbol{H}_{RF}$ to terms that can help us solve the Schrödinger equation. Let us evaluate the first term on the right-hand side by substituting Eqs. (10.14) and (10.16) into Eq. (10.20).

$$i\hbar \frac{\partial U_{RF}}{\partial t} U_{RF}^\dagger = i\hbar \frac{\partial}{\partial t} \left(e^{-i \frac{\omega_1 t}{2} \sigma_z} \right) e^{i \frac{\omega_1 t}{2} \sigma_z},$$

$$= i\hbar \left(-i \frac{\omega_1}{2} \sigma_z e^{-i \frac{\omega_1 t}{2} \sigma_z} \right) e^{i \frac{\omega_1 t}{2} \sigma_z},$$

$$= \frac{\hbar \omega_1}{2} \sigma_z, \tag{10.21}$$

where we performed derivative easily as σ_z is diagonal. Now consider the *second term* on the right-hand side of Eq. (10.20). We substitute Eq. (10.10) and obtain,

$$U_{RF} H U_{RF}^\dagger = U_{RF} \left(\frac{\hbar \Omega_R}{4} [\exp\{i(\omega_1 t + \phi_B)\} \sigma^+ \right.$$

$$+ \exp\{-i(\omega_1 t + \phi_B)\} \sigma^-]$$

$$\left. - \frac{\hbar \omega_L}{2} \sigma_z \right) U_{RF}^\dagger,$$

$$= \frac{\hbar \Omega_R}{4} \exp\{i(\omega_1 t + \phi_B)\} U_{RF} \sigma^+ U_{RF}^\dagger$$

$$+ \frac{\hbar \Omega_R}{4} \exp\{-i(\omega_1 t + \phi_B)\} U_{RF} \sigma^- U_{RF}^\dagger$$

$$- \frac{\hbar \omega_L}{2} U_{RF} \sigma_z U_{RF}^\dagger. \tag{10.22}$$

This is pretty daunting. But we can solve it term by term. Firstly, let us work on the last term. This one is easy if we recognize that U_{RF} and $U_{RF}^\dagger$ are the exponentiations of σ_z (Eqs. (10.14) and (10.16)) and, thus, it is diagonal (see Sect. 4.3.1 about the exponentiation of a diagonal matrix). Therefore, they commute with σ_z, i.e., $\sigma_z U_{RF}^\dagger = U_{RF}^\dagger \sigma_z$ and,

$$-\frac{\hbar\omega_L}{2} U_{RF}\sigma_z U_{RF}^\dagger = -\frac{\hbar\omega_L}{2} e^{-i\frac{\omega_1 t}{2}\sigma_z} \sigma_z e^{i\frac{\omega_1 t}{2}\sigma_z},$$

$$= -\frac{\hbar\omega_L}{2}\left(e^{-i\frac{\omega_1 t}{2}\sigma_z} e^{i\frac{\omega_1 t}{2}\sigma_z}\right)\sigma_z,$$

$$= -\frac{\hbar\omega_L}{2}\sigma_z. \qquad (10.23)$$

To find the first and second terms of Eq. (10.22), we need to find $U_{RF}\sigma^+ U_{RF}^\dagger$ and $U_{RF}\sigma^- U_{RF}^\dagger$. It turns out that

$$U_{RF}\sigma^+ U_{RF}^\dagger = e^{-i\omega_1 t}\sigma^+,$$
$$U_{RF}\sigma^- U_{RF}^\dagger = e^{i\omega_1 t}\sigma^-. \qquad (10.24)$$

Let us prove the first one and try the second one in a problem.

Example 10.1 Show that $U_{RF}\sigma^+ U_{RF}^\dagger = e^{-i\omega_1 t}\sigma^+$.

It is easy to show if we use their matrix form. Firstly, based on Eqs. (10.14) and (10.16),

$$U_{RF} = e^{-i\frac{\omega_1 t}{2}\sigma_z} = \begin{pmatrix} e^{-i\frac{\omega_1}{2}t} & 0 \\ 0 & e^{i\frac{\omega_1}{2}t} \end{pmatrix},$$

$$U_{RF}^\dagger = e^{i\frac{\omega_1 t}{2}\sigma_z} = \begin{pmatrix} e^{i\frac{\omega_1}{2}t} & 0 \\ 0 & e^{-i\frac{\omega_1}{2}t} \end{pmatrix}.$$

With Eq. (10.6), the left-hand side is,

$$U_{RF}\sigma^+ U_{RF}^\dagger = \begin{pmatrix} e^{-i\frac{\omega_1}{2}t} & 0 \\ 0 & e^{i\frac{\omega_1}{2}t} \end{pmatrix}\begin{pmatrix} 0 & 2 \\ 0 & 0 \end{pmatrix}\begin{pmatrix} e^{i\frac{\omega_1}{2}t} & 0 \\ 0 & e^{-i\frac{\omega_1}{2}t} \end{pmatrix},$$

$$= \begin{pmatrix} e^{-i\frac{\omega_1}{2}t} & 0 \\ 0 & e^{i\frac{\omega_1}{2}t} \end{pmatrix}\begin{pmatrix} 0 & 2e^{-i\frac{\omega_1}{2}t} \\ 0 & 0 \end{pmatrix},$$

$$= \begin{pmatrix} 0 & 2e^{-i\omega_1 t} \\ 0 & 0 \end{pmatrix},$$

$$= e^{-i\omega_1 t}\begin{pmatrix} 0 & 2 \\ 0 & 0 \end{pmatrix},$$

10.5 Solving the Schrödinger Equation

$$= e^{-i\omega_1 t}\sigma^+.$$

∎

Therefore, by substituting with Eqs. (10.21) and (10.22), Eq. (10.20) becomes,

$$H_{RF} = i\hbar \frac{\partial U_{RF}}{\partial t} U_{RF}^\dagger + U_{RF} H U_{RF}^\dagger,$$

$$= \frac{\hbar\omega_1}{2}\sigma_z$$

$$+ \frac{\hbar\Omega_R}{4} \exp\{i(\omega_1 t + \phi_B)\} U_{RF}\sigma^+ U_{RF}^\dagger$$

$$+ \frac{\hbar\Omega_R}{4} \exp\{-i(\omega_1 t + \phi_B)\} U_{RF}\sigma^- U_{RF}^\dagger$$

$$- \frac{\hbar\omega_L}{2} U_{RF}\sigma_z U_{RF}^\dagger. \tag{10.25}$$

We then further substitute with Eqs. (10.23) and (10.24), then

$$H_{RF} = \frac{\hbar\omega_1}{2}\sigma_z$$

$$+ \frac{\hbar\Omega_R}{4} \exp\{i(\omega_1 t + \phi_B)\} e^{-i\omega_1 t}\sigma^+$$

$$+ \frac{\hbar\Omega_R}{4} \exp\{-i(\omega_1 t + \phi_B)\} e^{i\omega_1 t}\sigma^-$$

$$- \frac{\hbar\omega_L}{2}\sigma_z$$

$$= -\frac{\hbar(\omega_L - \omega_1)}{2}\sigma_z + \frac{\hbar\Omega_R}{4}\left(e^{i\phi_B}\sigma^+ + e^{-i\phi_B}\sigma^-\right)$$

$$= -\frac{\hbar\Delta}{2}\sigma_z + \frac{\hbar\Omega_R}{4}\left(e^{i\phi_B}\sigma^+ + e^{-i\phi_B}\sigma^-\right), \tag{10.26}$$

where we have introduced the definition of **detuning**, $\Delta = \omega_L - \omega_1$, which is the *difference between the Larmor frequency and the rotating frame frequency*. In some literature, it is defined as $\omega_1 - \omega_L$. Now, let us now replace σ^+ and σ^- by σ_x and σ_y using Eq. (10.5).

$$H_{RF} = -\frac{\hbar\Delta}{2}\sigma_z + \frac{\hbar\Omega_R}{4}\left(e^{i\phi_B}(\sigma_x + i\sigma_y) + e^{-i\phi_B}(\sigma_x - i\sigma_y)\right),$$

$$= -\frac{\hbar\Delta}{2}\sigma_z + \frac{\hbar\Omega_R}{4}\left(2\cos\phi_B\sigma_x - 2\sin\phi_B\sigma_y\right),$$

$$= -\frac{\hbar\Delta}{2}\sigma_z + \frac{\hbar\Omega_R}{2}\left(\cos\phi_B\sigma_x - \sin\phi_B\sigma_y\right),$$

$$= \frac{\hbar}{2}\begin{pmatrix}\sigma_x & \sigma_y & \sigma_z\end{pmatrix} \cdot \begin{pmatrix}\Omega_R\cos\phi_B \\ -\Omega_R\sin\phi_B \\ -\Delta\end{pmatrix},$$

$$= \frac{\hbar}{2}\vec{\sigma}\cdot\vec{\Omega}_R = \frac{\hbar}{2}\vec{\Omega}_R\cdot\vec{\sigma}, \tag{10.27}$$

where in line 2, we used the identities $e^{i\theta} = \cos\theta + i\sin\theta$ and $e^{-i\theta} = \cos\theta - i\sin\theta$. In line 5, we defined the **angular frequency vector**, $\vec{\Omega}_R$, as

$$\vec{\Omega}_R = \begin{pmatrix}\Omega_R\cos\phi_B \\ -\Omega_R\sin\phi_B \\ -\Delta\end{pmatrix}. \tag{10.28}$$

We have gone through a lot of derivations and we probably forgot what we are doing. Our goal is to solve the Schrödinger equation in the rotating frame, which is Eq. (10.17). We finally express H_{RF} in the rotating frame as in Eq. (10.27). Note that now H_{RF} in Eq. (10.27) is **time independent** and we can solve it easily. The Schrödinger equation now becomes

$$i\hbar\frac{\partial|\psi\rangle_{RF}}{\partial t} = \frac{\hbar}{2}\vec{\Omega}_R\cdot\vec{\sigma}\,|\psi\rangle_{RF}. \tag{10.29}$$

While H_{RF} is a constant, it is not diagonal. We can solve it using its matrix form using the method in Sect. 4.3.2. However, we will not solve it here. We will use our intuition.

10.6 Intuitive Understanding of the Solution

Let us recall the Hamiltonian of a constant magnetic field pointing downward is given in Eq. (8.4). By using Eq. (8.11), it becomes

$$H = -B_0\frac{e\hbar}{2m}\sigma_z,$$

$$= -\frac{\hbar\omega_L}{2}\sigma_z,$$

$$= \frac{\hbar}{2}\begin{pmatrix}\sigma_x & \sigma_y & \sigma_z\end{pmatrix} \cdot \begin{pmatrix}0 \\ 0 \\ -\omega_L\end{pmatrix}. \tag{10.30}$$

10.6 Intuitive Understanding of the Solution

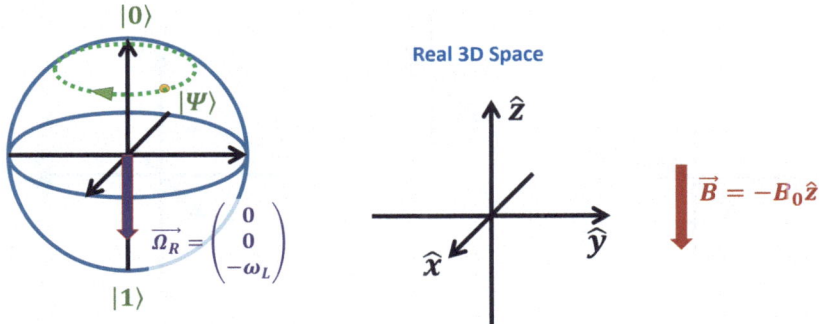

Fig. 10.2 Redraw of Fig. 8.1 with the angular velocity vector and the precession of state $|\Psi\rangle$. Left: The Bloch sphere representation of an electron spin qubit. Right: The real 3D space coordinate system in which the direction of the external magnetic field is shown. Note that it is still in the laboratory frame

We see that it is just a special case of Eq. (10.27) with $\vec{\Omega}_R = \begin{pmatrix} 0 \\ 0 \\ -\omega_L \end{pmatrix}$ and $\Delta = \omega_L$ (i.e., it is in the laboratory frame). Figure 10.2 shows the direction of $\vec{\Omega}_R$. Based on this figure, we see that Larmor precession can be understood as the rotation about $\vec{\Omega}_R$ (using right-hand-rule). Moreover, *the magnitude of $\vec{\Omega}_R$ determines the precession frequency.*

We therefore can guess that precession will occur about $\vec{\Omega}_R$ for Eq. (10.29). At the left of Fig. 10.3, it shows the precession about a general angular velocity vector. The right of the figure shows the precession about $\vec{\Omega}_R = \begin{pmatrix} \Omega_R \cos 0 \\ -\Omega_R \sin 0 \\ 0 \end{pmatrix} = \begin{pmatrix} \Omega_R \\ 0 \\ 0 \end{pmatrix}$. This is the case when $\Phi_B = 0$. And it is easy to understand because we are in the rotating frame at ω_1. If $\Phi_B = 0$, then the rotating magnetic field always points at $\hat{x}'$.

Therefore, if we set an appropriate initial phase for the rotating frame, Φ_B, an appropriate Ω_R, and an appropriate ω_1 (thus the detuning), we can control the direction and the magnitude of $\vec{\Omega}_R$. This then determines the rotation frequency and the rotation direction of a quantum state! The rotation frequency, or the **generalized Rabi frequency**, Ω'_R, is

$$\Omega'_R = |\vec{\Omega}_R| = \sqrt{\Omega_R^2 + \Delta^2}. \tag{10.31}$$

And we need to remind ourselves that we are working in the rotating frame to get the pictures in Fig. 10.3.

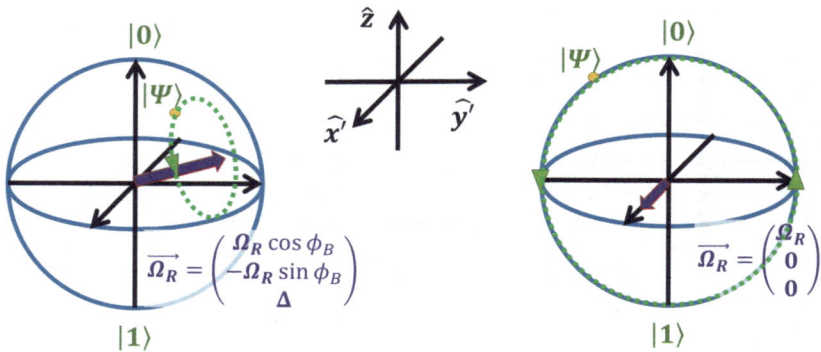

Fig. 10.3 Illustration of precession about a general angular velocity vector, $\vec{\Omega}_R$ (left), and about a special vector pointing at the $\hat{x}$ direction (right). Note that this is in the **rotating frame**

10.7 Another Method to Perform Rotating Frame Transformation

In Eq. (10.20), we showed the relationship between H_{RF} and H. We see that it has two terms. The first term is $i\hbar \frac{\partial U_{RF}}{\partial t} U_{RF}^\dagger$. This term has nothing to do with H and it is always $\frac{\hbar \omega_1}{2}\sigma_z$ (Eq. (10.21)). Together with the natural precession, they form the final detuning term.

For the second term, it is the transformation of H. If H is a linear combination of σ_x, σ_y, and σ_z, one may obtain their transformation easily by using the following identities:

$$U_{RF}\sigma_x U_{RF}^\dagger = \cos\omega_1 t\, \sigma_x + \sin\omega_1 t\, \sigma_y, \tag{10.32}$$

$$U_{RF}\sigma_y U_{RF}^\dagger = \cos\omega_1 t\, \sigma_y - \sin\omega_1 t\, \sigma_x, \tag{10.33}$$

$$U_{RF}\sigma_z U_{RF}^\dagger = \sigma_z, \tag{10.34}$$

$$U_{RF}\sigma^+ U_{RF}^\dagger = e^{-i\omega_1 t}\sigma^+, \tag{10.35}$$

$$U_{RF}\sigma^- U_{RF}^\dagger = e^{i\omega_1 t}\sigma^-. \tag{10.36}$$

For example, we may just apply them to Eq. (10.3) to obtain the results. Of course, we need to make sure to include $i\hbar \frac{\partial U_{RF}}{\partial t} U_{RF}^\dagger = \frac{\hbar\omega_1}{2}\sigma_z$.

Example 10.2 Derive Eq. (10.27) from Eq. (10.3) using Eq. (10.20) and Eqs. (10.32)–(10.34).

From Eq. (10.3), we have

$$H = \frac{\hbar\Omega_R}{2}\cos(\omega_1 t + \phi_B)\sigma_x - \frac{\hbar\Omega_R}{2}\sin(\omega_1 t + \phi_B)\sigma_y - \frac{\hbar\omega_L}{2}\sigma_z. \tag{10.37}$$

10.7 Another Method to Perform Rotating Frame Transformation

Therefore,

$$H_{RF} = i\hbar \frac{\partial U_{RF}}{\partial t} U_{RF}^\dagger + U_{RF} H U_{RF}^\dagger,$$

$$= \frac{\hbar\omega_1}{2}\sigma_z + U_{RF}\left[\frac{\hbar\Omega_R}{2}\cos(\omega_1 t + \phi_B)\sigma_x\right.$$

$$\left. - \frac{\hbar\Omega_R}{2}\sin(\omega_1 t + \phi_B)\sigma_y - \frac{\hbar\omega_L}{2}\sigma_z\right] U_{RF}^\dagger,$$

$$= -\frac{\hbar\Delta}{2}\sigma_z + \frac{\hbar\Omega_R}{2}\cos(\omega_1 t + \phi_B) U_{RF}\sigma_x U_{RF}^\dagger$$

$$- \frac{\hbar\Omega_R}{2}\sin(\omega_1 t + \phi_B) U_{RF}\sigma_y U_{RF}^\dagger, \qquad (10.38)$$

where we have used Eq. (10.34) and combined the σ_z terms as a detuning term with $\Delta = \omega_L - \omega_1$. We then use the trigonometric identities and Eqs. (10.20) and (10.32). For example,

$$\cos(\omega_1 t + \phi_B) U_{RF}\sigma_x U_{RF}^\dagger,$$

$$= \cos(\omega_1 t + \phi_B)\left(\cos\omega_1 t\,\sigma_x + \sin\omega_1 t\,\sigma_y\right),$$

$$= \cos(\omega_1 t + \phi_B)\cos\omega_1 t\,\sigma_x + \cos(\omega_1 t + \phi_B)\sin\omega_1 t\,\sigma_y. \qquad (10.39)$$

Similarly,

$$\sin(\omega_1 t + \phi_B) U_{RF}\sigma_y U_{RF}^\dagger,$$

$$= \sin(\omega_1 t + \phi_B)\left(\cos\omega_1 t\,\sigma_y - \sin\omega_1 t\,\sigma_x\right),$$

$$= \sin(\omega_1 t + \phi_B)\cos\omega_1 t\,\sigma_y - \sin(\omega_1 t + \phi_B)\sin\omega_1 t\,\sigma_x. \qquad (10.40)$$

Therefore, substituting Eqs. (10.39) and (10.40) into Eq. (10.38) and using trigonometric identities, we obtain

$$H_{RF} = -\frac{\hbar\Delta}{2}\sigma_z + \frac{\hbar\Omega_R}{2}\cos(\omega_1 t + \phi_B) U_{RF}\sigma_x U_{RF}^\dagger$$

$$- \frac{\hbar\Omega_R}{2}\sin(\omega_1 t + \phi_B) U_{RF}\sigma_y U_{RF}^\dagger,$$

$$= -\frac{\hbar\Delta}{2}\sigma_z + \frac{\hbar\Omega_R}{2}\left(\cos\phi_B\sigma_x - \sin\phi_B\sigma_y\right). \qquad (10.41)$$

This is the same as Eq. (10.27). ∎

10.8 Summary

In this chapter, we study the change of an electron spin qubit under a constant vertical magnetic field and a rotating magnetic field on the $\hat{x} - \hat{y}$ plane. We work on the rotating frame of the rotating field to simplify the problem. Under the rotating frame, the Hamiltonian becomes time-independent. The effective external magnetic field becomes a constant magnetic field pointing in a certain direction with a certain magnitude determined by the phase, frequency, and magnitude of the rotating magnetic field and the magnitude of the DC field. This then determines the generalized Rabi frequency of a qubit state and its rotation direction. The picture we see in this chapter will be reused in other types of qubits such as the superconducting qubits.

Problems

10.1 Raising and Lowering Operators
Sometimes σ^+ and σ^- are defined as,

$$\sigma^+ = \frac{\sigma_x + i\sigma_y}{2},$$

$$\sigma^- = \frac{\sigma_x - i\sigma_y}{2}, \qquad (10.42)$$

and they are called the raising and lowering operators of σ_z.

Find their matrix representations. Then apply them to $|0\rangle = \begin{pmatrix} 1 \\ 0 \end{pmatrix}$ and $|1\rangle = \begin{pmatrix} 0 \\ 1 \end{pmatrix}$. What do you see? You find that $|0\rangle = \sigma^+ |1\rangle$. So what does it raise? It does not raise $|0\rangle$ to $|1\rangle$. Instead, it raises the lower eigenvalue eigenvector of σ_z (i.e., $|1\rangle$ with eigenvalue of -1) to the higher eigenvalue eigenvector of σ_z (i.e., $|0\rangle$ with eigenvalue of 1). We need to pay attention to the ambiguities.

10.2 Rotating Frame Equations
Prove Eqs. (10.32)–(10.34).

10.3 Rotating Frame
We skipped some steps in Example 10.2. Please show all steps.

Chapter 11
Electron Spin Qubit in Semiconductor—Implementation, Initialization, and Readout

11.1 Introduction

We have learned how to rotate an electron spin state using external magnetic fields. We have been assuming that the electron floats in space. In reality, it will interact with matters and we need to decide how to set the electron in place. We will use an electron in silicon and illustrate how to achieve the five **DiVincenzo's criteria** mentioned in Sect. 1.3. In this chapter, we will discuss silicon spin qubit initialization and readout (measurement). In the next chapter, we will discuss the implementation of 1-qubit and 2-qubit gates.

11.1.1 Learning Outcomes

Understand why silicon qubit has the potential to enable large-scale integration of qubits in quantum computers; be able to describe how to implement quantum dots in a transistor; be able to describe why silicon qubit is a good qubit; be able to describe how to initialize and measure silicon qubits.

11.1.2 Teaching Videos

- Search for Ch11 in this playlist
  - https://tinyurl.com/3yhze3jn
- Other videos
  - https://youtu.be/qwye4V0QXAc

- https://youtu.be/oL9f-vRcVzc
- https://youtu.be/0JVw4xICVl0

11.2 Why Silicon?

At the moment of writing, the decoherence time of a qubit in most available architecture is not long enough. Although they are long enough for many gate operations, there are still finite errors due to interactions with the environment and due to errors from gate operations (see Chap. 25). As a result, quantum computing is unreliable because most quantum computing algorithms rely on interference and interference fails even with tiny errors (e.g., [1]). The qubit we have been discussing is called the **physical qubit**. In order to achieve **fault-tolerant quantum computing**, we need to use special algorithms and multiple physical qubits to create an error-free **logical qubit** (e.g., [2]). However, it is expected that each error-free logical qubit will consist of 100–1000 physical qubits.

At the same time, many quantum algorithms are expected to outperform classical ones only when the problem size is large enough. For example, it is expected that Shor's algorithm is only useful for a problem that needs more than 1000 *logical* qubits. Therefore, a quantum computer with 100000 to 1 million *physical* qubits is necessary to achieve practical **quantum supremacy**.

In Figs. 1.1 and 1.3, we showed that a quantum computer is just a group of passive qubits controlled by classical electronics. If there are 1 million qubits, how are we going to control them? In the superconducting case, without a careful architecture design, this seems to be impossible (Fig. 1.3). Therefore, building large-scale quantum computers using semiconductor platforms is a natural choice. This is especially true for the silicon platform as it has been proved that it can integrate more than 10 billion transistors in a few cm^2 and we can harness decades of engineering experience in the silicon semiconductor industry.

Since electron spin qubits in **quantum dots** built on semiconductor platforms are small, it is possible to integrate millions of qubits on a silicon chip. Moreover, it is also possible to integrate the control electronics on the same chip due to the mature technology. That means the control electronics (see Chap. 23) such as low-noise amplifiers (LNAs), analog-to-digital converters (ADCs), digital-to-analog converters (DACs), filters, and even signal generators and processing units can be integrated on the same platform. Readers are encouraged to take a look at Fig. 1 in [3] to see one of the possible integration schemes.

It should also be noted that when the control electronics are near the qubits and cooled to $4.2K$ or below, the **throughput** will be increased and the noise will be reduced.

Ideally, if electron spin qubit can operate at 4.2K, it can be integrated on the same chip as the control electronics and achieve minimal latency. Figure 11.1 shows an ideal integrated platform of silicon qubits and their control electronics.

11.3 Spin Qubits in Quantum Dots

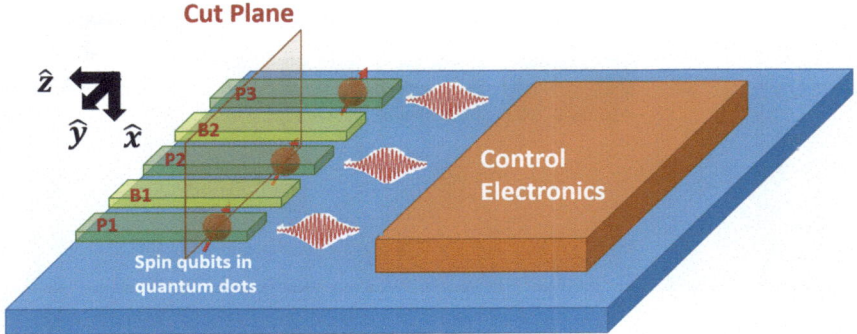

Fig. 11.1 In an *ideal* and highly integrated case, silicon spin qubits are integrated on the same silicon chip with the control electronics operating at 4.2K. This has minimal latency. The plunger gates (*P*1, *P*2, *P*3) and barrier gates (*B*1, *B*2) are shown. The plunger gates are used to deplete the electrons underneath to allow single-electron control and the barrier gates are used to form lateral barriers for the quantum dots. Only $\hat{y}$-direction confinement is shown here. *Note that this is a simplified schematic. It is necessary to confine the charge also in the $\hat{z}$ direction to form a quantum dot.* The cut plane is shown in Fig. 11.2

11.3 Spin Qubits in Quantum Dots

A common way to hold an electron carrying the spin qubit is to put it in a *quantum dot*. A quantum dot is named so because it is a tiny space in which quantum confinement becomes significant. The energy of the electron becomes discretized like in an atom (which is a quantum dot also and electrons can only occupy discrete orbitals). More importantly, due to its small size, it has a very small capacitance. In order to add an additional electron to it, a large **charging energy** is required. This is because the energy of a capacitor is given by,

$$E = \frac{1}{2}CV^2 = \frac{Q^2}{2C}, \qquad (11.1)$$

where C, V, and Q are the capacitance of the dot, the voltage across the dot, and the charge in the dot, respectively. With a small C for the same Q, to add one more charge, the energy of the whole system, E, increases a lot, and, thus, one can add or remove an individual electron to or from the dot *precisely* through an electrical method. If the energy is not enough to add an electron into the quantum dot, it is called **Coulomb blockade** which means the charge inside the dot is blocking the addition of another charge due to the large charging energy.

Of course, E needs to be compared to the thermal energy, kT, where k is the Boltzmann constant. We still need to cool the system to cryogenic temperature so that E is large compared to kT to observe the quantum effects.

Figure 11.2 shows the cross section along a qubit array (e.g., [4]). It is similar to a typical metal-oxide-semiconductor (MOS) transistor or capacitor. It has gates on

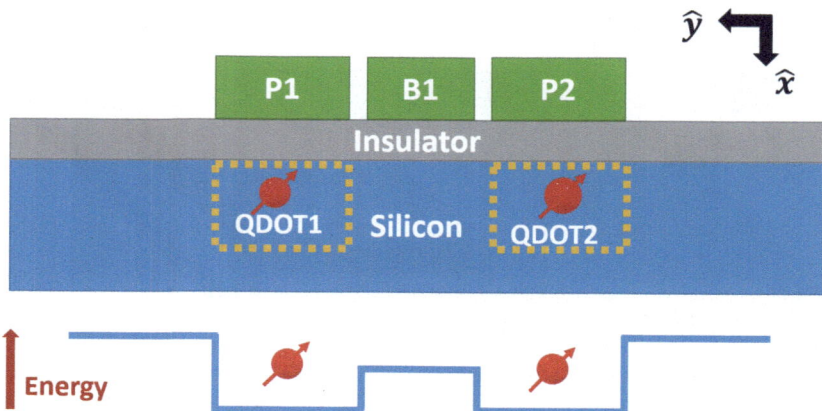

Fig. 11.2 Cut plane of Fig. 11.1. Top: Only two plunger gates and one barrier gate are shown for clarity. Note the implementation of the vertical, $\hat{x}$-direction, confinement is not shown. Bottom: The electron energy profile

top formed by either metal or heavily doped polycrystalline silicon. It is capacitively coupled to the substrate (e.g., silicon) through an insulator (e.g., oxide in silicon technology). The plunger gates, $P1$ and $P2$, are used to deplete the electron underneath and form a confinement potential in the vertical direction ($\hat{x}$). The barrier gate ($B1$) is used to control the potential between the plunger gates to allow electrons to transfer or tunnel to another quantum dot by changing the barrier height. This is a 2D cross section. In reality, the confinement should occur in the $\hat{z}$ direction also to form a quantum dot. The drawing is simplified and did not consider this. Readers can refer to [4] to see a realistic layout. The quantum dots are formed under the plunger gates and the electrons carrying the qubits reside there.

Besides silicon platforms, spin qubit has also been realized in other materials such as AlGaAs/GaAs heterostructure, where AlGaAs is used as the insulator as it has a wide bandgap and GaAs is the substrate (e.g., [5]). The advantage of using the AlGaAs/GaAs system is that it has a very perfect interface that can enhance the decoherence time. However, it is not as scalable as the silicon platform.

11.4 Silicon Spin Qubit

Silicon spin qubit is commonly referred to as spin qubit implemented on a silicon platform. Most of the time, it refers to the electron or hole spin qubits. Usually, *it has nothing to do with the silicon atom itself.* In this chapter, we only look at electron spin qubits. Hole spin qubits can provide some physics that electron spin qubits do not have [6]. For example, it has a strong **spin-orbit (SO) coupling** which allows it to be manipulated by electrostatic (instead of magnetic field) directly. Of course,

11.4 Silicon Spin Qubit

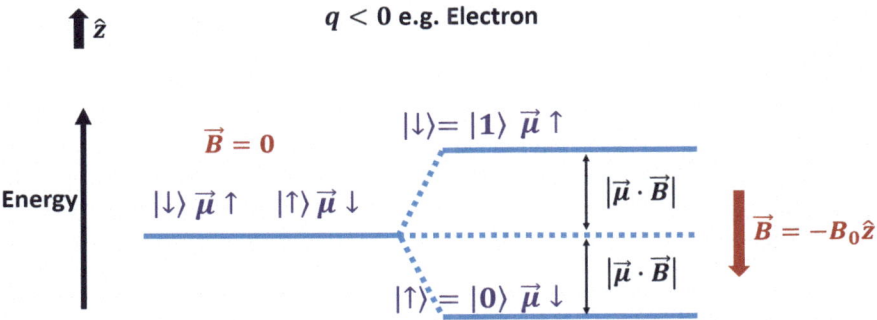

Fig. 11.3 Energy of a negatively charged particle with spin under zero (left) or a finite external magnetic field (right). This is similar to Fig. 7.4 but with the external magnetic field pointing in an opposite direction

it also means that it is more susceptible to decoherence time degradation due to the charges in the environment which are abundant.

Firstly, we need to encode the $|0\rangle$ and $|1\rangle$ states. This is to fulfill one of DiVincenzo's criteria: "A scalable physical system with well-characterized qubit". As discussed in the previous section, the silicon platform is probably the most scalable one. As for "well-characterized qubit," we just need to apply a vertical magnetic field to create an energy splitting as shown in Fig. 11.3. This is also called the **Zeeman splitting**. The physics has been discussed in the previous chapters. Here we review the important concepts and equations.

When there is no external magnetic field, the $|0\rangle$ and $|1\rangle$ states have the same energy (**degenerated**). When an external DC magnetic field is applied, the one with a magnetic moment parallel (anti-parallel) to the magnetic field will acquire a lower (higher) energy (Eq. (7.10)). Since an electron has a negative charge, it means the electron with spin parallel (anti-parallel) to the magnetic field will acquire a *higher (lower)* energy (Eq. (7.9)). In Fig. 11.3, as the magnetic field points down, the spin-down (spin-up) electron has a higher (lower) energy. The energy differences between the two states are given by

$$\begin{aligned} \Delta &= 2|\vec{B} \cdot \vec{\mu}|, \\ &= 2|B_0 \gamma S|, \\ &= 2|B_0 \gamma \hbar/2|, \\ &= \hbar \omega_L, \end{aligned} \quad (11.2)$$

where we have used Eqs. (7.8), (7.10), and (8.11). It is important to recall that the gyromagnetic ratio depends on the g-factor and also the effective electron mass, m^*. In silicon, the effective mass can change substantially. We need to make sure we use the correct mass in the calculation. Therefore,

$$\gamma = \frac{ge}{2m^*}. \tag{11.3}$$

m^* is usually written as a multiplication of a constant and the electron rest mass in the empty space, m_0. For example, an effective mass, $m^* = 0.2m_0$ means that the electron has an effective mass of $0.2 \times 9.1 \times 10^{-31}$ kg $= 1.82 \times 10^{-31}$ kg.

The g-factor may also change due to the environment. For examples, one may refer to [7].

Example 11.1 If $B_0 = 1T$, assuming the electron mass is still m_0 and $g = -2$, find the energy difference between $|0\rangle$ and $|1\rangle$.

Using Eq. (11.2),

$$\Delta = 2|B_0 \gamma \frac{\hbar}{2}|,$$
$$= 2|B_0 \frac{ge}{2m} \frac{\hbar}{2}|,$$
$$= 2 \times 1T \times \frac{2 \times 1.6 \times 10^{-19} C}{2 \times 9.1 \times 10^{-31} \text{ kg}} \times \frac{6.625 \times 10^{-34} \text{ J} \cdot \text{s}}{2 \times 2 \times 3.14},$$
$$= 1.855 \times 10^{-23} J,$$
$$= 116 \mu eV.$$

Note that we have converted the energy in $joule$ (J) to $electron - volt$ (eV) by using the definition (i.e., by dividing by $1.6 \times 10^{-19} C$). As a reminder, $1 \mu eV = 10^{-3}$ meV $= 10^{-6}$ eV. ∎

Is this a qubit with *well-separated energy levels*? That depends on the operating temperature.

Example 11.2 What is the thermal energy at 10 mK? How does it compare to the qubit energy in the previous example?

The thermal energy is given by kT, where k is the Boltzmann constant which is 1.380649×10^{-23} J/K and T is the temperature. Therefore, the thermal energy is,

$$kT = 1.380649 \times 10^{-23} \text{ J/K} \times 10 \times 10^{-3} \text{ K},$$
$$= 1.38 \times 10^{-25} J,$$
$$= 0.86 \mu eV. \tag{11.4}$$

The qubit energy separation is at least 100 times larger than the thermal energy. Therefore, the qubit states are well-defined. ∎

Besides having two well-separated states, *we also need to make sure there are no other states where the qubit can reside.* Fortunately, an electron is a half-spin particle. It can only reside at either $|0\rangle$ or $|1\rangle$ states. Therefore, the electron spin forms a **well-characterized two-dimensional Hilbert space** which is very suitable for quantum computing.

11.5 Qubit Initialization

DiVincenzo's criteria also require that an architecture should have "the ability to initialize the state of the qubits to a simple fiducial state." For silicon spin qubit, the ground state is $|0\rangle = |\uparrow\rangle$ if it is set up as in Fig. 11.3. To initialize a qubit, it means to make sure the qubit is at its ground state, $|0\rangle = |\uparrow\rangle$.

We will discuss three methods to initialize a qubit.

11.5.1 Thermalization

Thermalization is a natural and common way in all quantum computing architectures to perform qubit initialization. What it means is that we will let the qubit sit for a long time (e.g., 10 times of T_1, which is one of the decoherence times of a qubit and will be discussed in Chap. 25). If the qubit is already at its ground state, it will stay at the ground state. But if the qubit is at an excited state ($|1\rangle$ or $|\downarrow\rangle$ in Fig. 11.3), due to the interaction with the environment, it will lose its energy eventually and flip to the ground state.

Here we see the dilemma. We want to have a long decoherence time as requested by the requirement "long relevant decoherence times" in DiVincenzo's criteria. We also want to have a short T_1 time so that it can decay fast during initialization. After each unit of T_1, only $1 - e^{-1}$ of the population (or probability) will decay to $|0\rangle$ from $|1\rangle$. To make sure the qubit has decayed to $|0\rangle$, we need to wait for $10T_1$. A silicon qubit may have a T_1 as much as 100 ms. That means we need to wait for $1s$ in each initialization. This is considered long and can slow down quantum computing algorithms. Therefore, this is not a desirable initialization method.

11.5.2 State Filtering Through Tunneling

We may also try to filter and get rid of electrons in state $|1\rangle$ through electrical operations to achieve the initialization purpose. This is mentioned in [8] for another purpose but we can hijack it to perform initialization. **I have not seen people using this for initialization. But I think this is instructive and we can further discuss**

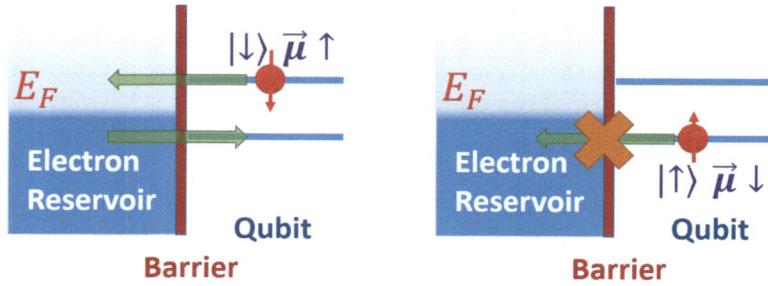

Fig. 11.4 Using state filtering to achieve initialization. The electron carrying the qubit is allowed to tunnel into an electron reservoir. The potential of the qubit is set up such that the excited state is above the Fermi level (E_F) of the reservoir and the ground state is below. On the left, if the electron is at its excited state, it can tunnel into the reservoir as there are empty states. And the electron from the reservoir can then tunnel into the ground state. On the right, since the electron has a lower energy than the Fermi level, the corresponding state is filled and it cannot tunnel

its pros and cons. It also prepares us for the spin-to-charge conversion for readout in the following section.

Figure 11.4 shows the scheme. We can make the qubit next to an electron reservoir such as a metal. It is known that in metal, there is a special energy level called the **Fermi level**, E_F. Above E_F, there is no electron, and below E_F, the states are filled with electrons. What I am saying is a simplification because the states are filled by following the so-called Fermi-Dirac distribution (see also Eq. (16.1) and the discussions). However, the picture is fairly accurate at very low temperatures, which is the case for most quantum computers (see Example 11.2).

The electron carrying the qubit is separated from the metal by an insulating barrier. Therefore, electrons can jump between the reservoir and the quantum dot (where the qubit resides) through tunneling. In principle, we can use an electrostatic method to control the thickness and height of the barrier so that it is only thin enough for significant tunneling when we want to perform qubit initialization. At the same time, we will also use an electrostatic method to bias the metal and the qubit so that the Fermi level is between the excited state level and the ground state level.

If the qubit is at its excited state, the electron will be able to tunnel into the reservoir. The electron below the Fermi level can also tunnel into the quantum dot and occupy the ground state (thus completing the initialization). However, if it is at its ground state, it will not be able to do so as the corresponding energy state in the reservoir is occupied. As a result, the qubit in the quantum dot is initialized to the ground state. Of course, we need to make sure the tunneling time is short enough so that the initialization process is fast enough.

11.5 Qubit Initialization

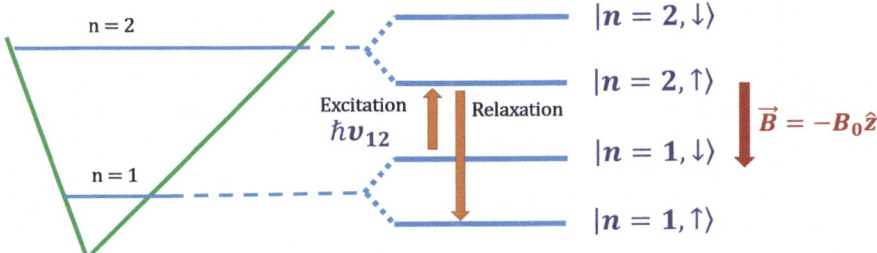

Fig. 11.5 Initialization scheme based on [9]. Left: Two orbital states formed due to an asymmetric quantum well. Right: Both orbital states split into two levels due to Zeeman splitting

11.5.3 Initialization Through Higher Orbital State

In [9], an initialization scheme through the interaction between two orbital states is used. The scheme is shown in Fig. 11.5. Due to the confinement of the potential well, there are multiple levels (just like the orbital levels of an electron in an atom). These energy levels are *not* due to the external magnetic fields. We will only discuss and use the lowest two levels, i.e., $n = 1$ and $n = 2$. Note that in the previous discussions, we have been ignoring $n = 2$. The qubit is encoded using the spin of the electron at $n = 1$. Therefore, $|0\rangle$ is $|\uparrow\rangle$ at $n = 1$ and $|1\rangle$ is $|\downarrow\rangle$ at $n = 1$. We label them as $|0\rangle = |n = 1, \uparrow\rangle$ and $|1\rangle = |n = 1, \downarrow\rangle$. We will similarly label the splitting in $n = 2$.

Note that $n = 2$ is *not* a part of the Hilbert space or the computing space. However, we will use $n = 2$ states to help us perform initialization.

During initialization, we will apply a microwave pulse with an energy equal to the energy difference between $|n = 1, \downarrow\rangle$ and $|n = 2, \uparrow\rangle$, which is $\hbar v_{12}$. Note that this is not the energy between the two orbitals before splitting. We will derive the equation in the chapter-end problem. If the qubit is $|0\rangle = |n = 1, \uparrow\rangle$, then the microwave energy is not enough to bring it to the second orbital. Note that only a quantum of energy can be absorbed (i.e., $\hbar v_{12}$). So the qubit will stay at its ground state. However, if the qubit is in its excited state, $|1\rangle = |n = 1, \downarrow\rangle$, it will absorb one photon and become $|n = 2, \uparrow\rangle$. Note that the spin direction is also flipped during the transition. This is possible because of the spin-orbit coupling which we will not discuss and please take this for granted [9]. When it is at $n = 2$, it decays rapidly to $n = 1$ because of the short lifetime (about $80ns$). The lifetime is much shorter than spin-relaxation (i.e., T_1 for relaxing $|n = 1, \downarrow\rangle$ to $|n = 1, \uparrow\rangle$) which is $> 1ms$. Therefore, we can realize very rapid initialization. For this scheme, we may say that we use an external microwave pulse to achieve spin flipping. Although it is brought to a higher energy level ($n = 2$), it will then relax naturally very fast to $n = 1$.

11.6 Qubit Readout

Readout of a qubit is equivalent to the measurement of a qubit. It is an essential part of quantum computing. It is used to read the final result or perform error correction by reading the ancillary qubits during computation. In a silicon spin qubit, the spin state is usually mapped to a charge state as there is no easy mechanism to detect the spin of the electron. This is called **spin-to-charge conversion**. We will use the example in [8] to explain one of the possible approaches.

Figure 11.6 shows the layout of a spin-to-charge conversion circuit. The electron carrying the spin qubit resides in a **quantum dot** (QDOT). The QDOT is capacitively coupled to an electron reservoir and a **quantum point contact transistor** (QPC transistor). Capacitive coupling means that they are separated by a layer of insulator and they interact with each other through the capacitor formed by the insulator. We already illustrated the meaning of electron reservoir and its interaction with a quantum dot in Sect. 11.5.2.

QPC is a type of transistor that only has a constricted region for the carrier to transport from the source to the drain [10]. It allows quantization transport which is *very sensitive* to the environmental charge. By capacitively coupling the QDOT to QPC, the current through QPC depends strongly on the charge state of the QDOT. When there is one electron in the QDOT, the current will be reduced substantially in the QPC. This is because a negative charge will increase the barrier for the carrier in the QPC to transport. Likewise, when there is no electron in the QDOT, QPC will have a larger current.

To perform the spin-to-charge conversion, we will use a scheme to let the electron stay in the QDOT if it is spin-up and empty it when it is spin-down. As a result, a larger QPC current means that it is spin-down, and a smaller current means that it is spin-up.

Figure 11.7 illustrates the scheme. Like what is discussed in Sect. 11.5.2, we will bias the QDOT so that the Fermi level of the reservoir is between the spin down ($|1\rangle$) and spin up ($|0\rangle$) energy levels. Since it is at a very low temperature (e.g.,

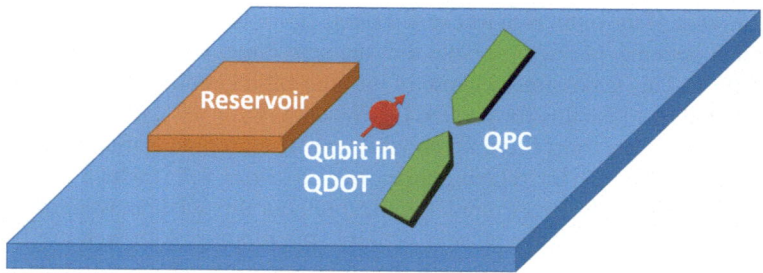

Fig. 11.6 Layout of spin-to-charge conversion for silicon spin qubit readout. The electron carrying the spin resides in the quantum dot (QDOT) which is capacitively coupled to a reservoir and a quantum point contact transistor (QPC)

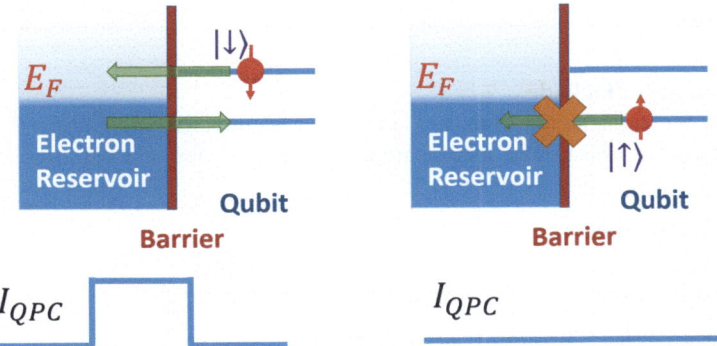

Fig. 11.7 Spin-to-charge conversion scheme for silicon spin qubit readout. Left: If the electron is in its excited state, during measurement, it will tunnel to the reservoir. The charge in the QDOT becomes less negative and thus the current in the QPC transistor, I_{QPC}, increases. However, after some time, an electron will tunnel from the reservoir to the QDOT and occupy the ground state. I_{QPC} will decrease. Right: If the electron is in its ground state, no tunneling will occur and thus I_{QPC} does not change during the measurement

10 mK), the states in the reservoir above (below) the Fermi level are empty (filled). If the qubit is spin-down ($|1\rangle$), it will then tunnel into the reservoir. As a result, the current in the QPC transistor, I_{QPC}, increases. After a certain time, an electron can tunnel from the reservoir to the QDOT and occupy the ground state and thus reduce I_{QPC} to the original level. Therefore, one will measure an I_{QPC} "pulse" if it is spin-down.

On the other hand, if the qubit is spin-up ($|0\rangle$), no tunneling will occur. There is no empty state below the Fermi level in the reservoir for the QDOT electron to tunnel to and there is no electron in the reservoir above the Fermi level to tunnel into the QDOT. As a result, I_{QPC} is constant at a low level.

By measuring I_{QPC}, one can then determine if the electron is spin-up (no pulse) or spin-down (with a pulse).

11.7 Summary

In this chapter, we have studied how an electron spin qubit is implemented on a silicon platform. Silicon spin qubit is one of the most promising platforms in terms of integration and scalability. We use a DC magnetic field to create the Hilbert space for electron spin qubit through Zeeman splitting. We study various qubit initialization schemes. Finally, we study how to use spin-to-charge conversion to read/measure a silicon spin qubit.

Problems

11.1 Coulomb Blockade
Sweeping the capacitance from $1\,aF$ to $1\,pF$, plot the charging energy if there is only one electron. How does it compare to the thermal energies at 10 mK, 100 mK, and 1 K?

11.2 Zeeman Splitting
With $|\vec{B}| = 1.4\,T$, it is found that the Zeeman splitting is 1.4 GHz. What is the corresponding energy splitting? What is the corresponding $\frac{g}{m^*}$?

11.3 Initialization Through Higher Orbitals
Continuing from Problem 11.2, if before the splitting the energy difference between $n = 2$ and $n = 1$ is 31 GHz, what is the value of ν_{12}?

References

1. Anika Zaman, Hector Jose Morrell, and Hiu Yung Wong. A step-by-step hhl algorithm walkthrough to enhance understanding of critical quantum computing concepts. *IEEE Access*, 11:77117–77131, 2023.
2. R. Acharya, et al. Suppressing quantum errors by scaling a surface code logical qubit. *Nature*, 614(7949):676–681, Feb 2023.
3. Yatao Peng, Andrea Ruffino, and Edoardo Charbon. A cryogenic broadband sub-1-db nf cmos low noise amplifier for quantum applications. *IEEE Journal of Solid-State Circuits*, 56(7):2040–2053, 2021.
4. M. Veldhorst, J. C. C. Hwang, C. H. Yang, A. W. Leenstra, B. de Ronde, J. P. Dehollain, J. T. Muhonen, F. E. Hudson, K. M. Itoh, A. Morello, and A. S. Dzurak. An addressable quantum dot qubit with fault-tolerant control-fidelity. *Nature Nanotechnology*, 9(12):981–985, Dec 2014.
5. F. H. L. Koppens, C. Buizert, K. J. Tielrooij, I. T. Vink, K. C. Nowack, T. Meunier, L. P. Kouwenhoven, and L. M. K. Vandersypen. Driven coherent oscillations of a single electron spin in a quantum dot. *Nature*, 442(7104):766–771, Aug 2006.
6. Yinan Fang, Pericles Philippopoulos, Dimitrie Culcer, W. A. Coish, and Stefano Chesi. Recent advances in hole-spin qubits, 2023.
7. Ryan M. Jock, N. Tobias Jacobson, Patrick Harvey-Collard, Andrew M. Mounce, Vanita Srinivasa, Dan R. Ward, John Anderson, Ron Manginell, Joel R. Wendt, Martin Rudolph, Tammy Pluym, John King Gamble, Andrew D. Baczewski, Wayne M. Witzel, and Malcolm S. Carroll. A silicon metal-oxide-semiconductor electron spin-orbit qubit. *Nature Communications*, 9(1):1768, May 2018.
8. J. M. Elzerman, R. Hanson, L. H. Willems van Beveren, B. Witkamp, L. M. K. Vandersypen, and L. P. Kouwenhoven. Single-shot read-out of an individual electron spin in a quantum dot. *Nature*, 430(6998):431–435, Jul 2004.
9. Mark Friesen, Charles Tahan, Robert Joynt, and M. A. Eriksson. Spin readout and initialization in a semiconductor quantum dot. *Phys. Rev. Lett.*, 92:037901, Jan 2004.
10. Henk van Houten and Carlo Beenakker. Quantum point contacts. *Physics Today*, 49(7):22–27, July 1996.

Chapter 12
Electron Spin Qubit in Semiconductor—1-Qubit and 2-Qubit Gates

12.1 Introduction

In Chap. 11, we showed how to implement a qubit using electron spin on a silicon substrate. We also demonstrated how to initialize and measure a qubit. In this chapter, we will study how to perform a universal 1-qubit gate and a 2-qubit entanglement gate to fulfill the last two DiVincenzo's criteria (Sect. 1.3). In Chap. 10, we showed that by applying a vertical DC magnetic field and a rotating horizontal magnetic field and then *working in the rotating frame*, we would be able to rotate any state on the Bloch sphere about any vector. This allows us to build a universal 1-qubit gate (Section 27.4 in [1]). However, in the literature, many silicon qubits are still implemented with the setup in Chap. 9 which means that the qubit is placed in a vertical DC magnetic field and a perturbating and linearly oscillating horizontal magnetic field. This is what we will use in this chapter. We will first summarize an experimental paper on how it implements a 1-qubit gate. Then we will discuss the implementation of a 2-qubit entanglement gate with an example.

12.1.1 *Learning Outcomes*

Be able to describe how a 1-qubit gate can be implemented for silicon spin qubits; understand how a CNOT-gate can be implemented by using a native entanglement gate of silicon spin qubits and other 1-qubit gates.

12.1.2 Teaching Videos

- Search for Ch12 in this playlist
    - https://tinyurl.com/3yhze3jn
- Other videos
    - https://youtu.be/0JVw4xICVl0
    - https://youtu.be/_CpQ-Uy0Kgo

12.2 1-Qubit Gate Implementation

We will use an example in [2] (with some variations) to demonstrate how to implement electron spin qubit in silicon and the 1-qubit gate. The setup is illustrated in Fig. 12.1.

Firstly, the Hilbert space is created by applying a DC magnetic field pointing to the right. This direction is named the $-\hat{z}$ direction. Therefore, *spin-up means that the spin is pointing to the left, and spin-down means that the spin is pointing to the right.* This is nothing special because the name of direction is completely a human definition.

The system is kept at 50 mK and $B_0 = 1.4T$. This is a spin qubit. To enhance the decoherence time, we need to avoid the spin interacting with the external environment. The electron spin can interact with the nuclear spin in the silicon easily and lose its coherence. Unfortunately, while 95.33% of natural silicon atoms have a zero nuclear spin (92.23% ^{28}Si **isotope** and 3.1% ^{30}Si isotope), 4.67% of them have spins (^{29}Si isotope). This is because ^{29}Si has 14 protons and 15 neutrons and their spins are not nullified due odd number of nucleons. The electron spin will interact strongly with the nucleus of ^{29}Si. Therefore, a highly purified silicon substrate is

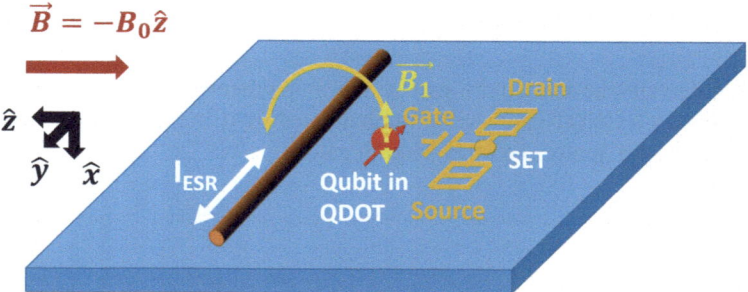

Fig. 12.1 Illustration of how an electron spin qubit is implemented in [2]. The magnetic field due to I_{ESR} is shown and it points perpendicularly to the substrate at the qubit

used, in which ^{29}Si is reduced to $800\,ppm$ (part-per-million). That means, among 1 million silicon atoms in the substrate, only 800 will be ^{29}Si.

For **readout**, it also uses spin-to-charge conversion as in Sect. 11.6. However, instead of using a quantum point contact (QPC) transistor, it uses a **single electron transistor** (SET) which has its current also very sensitive to the charge occupation in the quantum dot.

To implement a 1-qubit gate, it uses the theory we learned in Sect. 9.3. That is by applying a perturbing oscillating magnetic field in the $\hat{x}$ direction. But note that, the $\hat{x}$ direction in Fig. 12.1 is perpendicular to the silicon substrate. To achieve the oscillating magnetic field, an AC electric current, I_{ESR}, is passed through a wire next to the qubit. ESR stands for **electron spin resonance** which means that its frequency is the same as the Larmor frequency due to $\vec{B}$ (see Sect. 9.4). The current will generate a circular magnetic field about the wire due to the **Biot-Savart law** (see also Example 24.2). Although it is circular, at the point where it touches the qubit, it is pointing at the tangential direction of the circle and thus it is pointing in the $\hat{x}$ direction and perpendicular to the Si substrate. Since I_{ESR} is an AC current, the magnetic field at the qubit oscillates up and down and has the equation form given in the first term of Eq. (9.5). And as discussed in Sect. 9.6, in the rotating frame, any state will rotate about the $\hat{x}'$ due to Rabi oscillation. Combined with Larmor precession, it can be used to implement any 1-qubit gate.

12.3 2-Qubit Gate for Spin Implementation

A 2-qubit gate is usually much more difficult to implement than a 1-qubit gate. We also need to recall that, to form a universal set of quantum gates, what we need is not an arbitrary 2-qubit gate but an **entanglement gate**, which can be used to create entanglements (see Sect. 4.6). Entanglement is one of the reasons that makes quantum computing superior to classical computing.

A 2-qubit gate requires the interaction between two qubits and the external excitation under an appropriate Hamiltonian. It is not difficult to appreciate that different quantum computer architectures have different "native" entanglement gates because the physics is fairly different in different quantum computing architectures. Here, a **native entanglement gate** is a gate that can be implemented easily using the physics relevant to a given architecture and its controls.

One of the native entanglement gates for an electron spin qubit is a special form of the controlled phase shift gate. We will call it U_{Cphase} [3] to distinguish it from $U_{CPS,\pi}$ in Section 16.4 in [1],

$$U_{Cphase} = \begin{pmatrix} 1 & 0 & 0 & 0 \\ 0 & e^{i\phi} & 0 & 0 \\ 0 & 0 & e^{i(\pi-\phi)} & 0 \\ 0 & 0 & 0 & 1 \end{pmatrix}, \qquad (12.1)$$

where ϕ is the phase. We will discuss how to implement it soon. We will first show how it can be used to create an entanglement.

12.3.1 Creation of Entanglement

We have learned that we can use $CNOT$ gate, U_{XOR} to make two qubits entangled (Section 15.4 in [1] and Sect. 4.6). This is realized with the help of a Hadamard gate, H. This circuit is shown in Fig. 12.2.

The circuit implements the following operations:

$$\begin{aligned} U_{XOR}(H \otimes I)(|0\rangle \otimes |0\rangle) &= U_{XOR}(\frac{1}{\sqrt{2}}(|0\rangle + |1\rangle) \otimes |0\rangle), \\ &= U_{XOR}\frac{1}{\sqrt{2}}(|00\rangle + |10\rangle), \\ &= \frac{1}{\sqrt{2}}(|00\rangle + |11\rangle), \end{aligned} \quad (12.2)$$

where we used the definitions of H and U_{XOR} in line 1 and line 3, respectively. We can use **controlled phase shift gate** with a phase shift of π ($U_{CPS,\pi}$) (see Section 16.4 in [1]) to implement the U_{XOR} gate *by combining it with other 1-qubit gates*.

Example 12.1 Show that $U_{XOR} = (I \otimes H)U_{CPS,\pi}(I \otimes H)$.

Firstly, the matrix form of U_{XOR} is

$$U_{XOR} = \begin{pmatrix} 1 & 0 & 0 & 0 \\ 0 & 1 & 0 & 0 \\ 0 & 0 & 0 & 1 \\ 0 & 0 & 1 & 0 \end{pmatrix}, \quad (12.3)$$

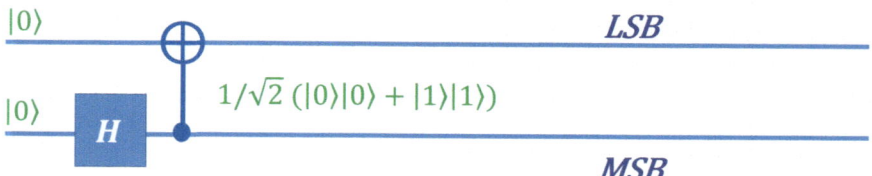

Fig. 12.2 An entanglement circuit implemented using a U_{XOR} gate and an H gate. Note that the least significant bit (LSB) and the most significant bit (MSB) are on the top and bottom, respectively

12.3 2-Qubit Gate for Spin Implementation

and the matrix form of $U_{CPS,\pi}$ is

$$U_{CPS,\pi} = \begin{pmatrix} 1 & 0 & 0 & 0 \\ 0 & 1 & 0 & 0 \\ 0 & 0 & 1 & 0 \\ 0 & 0 & 0 & e^{i\pi} \end{pmatrix},$$

$$= \begin{pmatrix} 1 & 0 & 0 & 0 \\ 0 & 1 & 0 & 0 \\ 0 & 0 & 1 & 0 \\ 0 & 0 & 0 & -1 \end{pmatrix}, \qquad (12.4)$$

as $e^{i\pi} = \cos\pi + i\sin\pi = -1$. Therefore,

$$(I \otimes H) U_{CPS,\pi} (I \otimes H)$$

$$= \begin{pmatrix} 1 & 0 \\ 0 & 1 \end{pmatrix} \otimes \frac{1}{\sqrt{2}} \begin{pmatrix} 1 & 1 \\ 1 & -1 \end{pmatrix} U_{CPS,\pi} \begin{pmatrix} 1 & 0 \\ 0 & 1 \end{pmatrix} \otimes \frac{1}{\sqrt{2}} \begin{pmatrix} 1 & 1 \\ 1 & -1 \end{pmatrix},$$

$$= \frac{1}{\sqrt{2}} \begin{pmatrix} 1 & 1 & 0 & 0 \\ 1 & -1 & 0 & 0 \\ 0 & 0 & 1 & 1 \\ 0 & 0 & 1 & -1 \end{pmatrix} \begin{pmatrix} 1 & 0 & 0 & 0 \\ 0 & 1 & 0 & 0 \\ 0 & 0 & 1 & 0 \\ 0 & 0 & 0 & -1 \end{pmatrix} \frac{1}{\sqrt{2}} \begin{pmatrix} 1 & 1 & 0 & 0 \\ 1 & -1 & 0 & 0 \\ 0 & 0 & 1 & 1 \\ 0 & 0 & 1 & -1 \end{pmatrix},$$

$$= \frac{1}{2} \begin{pmatrix} 1 & 1 & 0 & 0 \\ 1 & -1 & 0 & 0 \\ 0 & 0 & 1 & 1 \\ 0 & 0 & 1 & -1 \end{pmatrix} \begin{pmatrix} 1 & 1 & 0 & 0 \\ 1 & -1 & 0 & 0 \\ 0 & 0 & 1 & 1 \\ 0 & 0 & -1 & 1 \end{pmatrix},$$

$$= \frac{1}{2} \begin{pmatrix} 2 & 0 & 0 & 0 \\ 0 & 2 & 0 & 0 \\ 0 & 0 & 0 & 2 \\ 0 & 0 & 2 & 0 \end{pmatrix},$$

$$= \begin{pmatrix} 1 & 0 & 0 & 0 \\ 0 & 1 & 0 & 0 \\ 0 & 0 & 0 & 1 \\ 0 & 0 & 1 & 0 \end{pmatrix},$$

$$= U_{XOR}. \qquad (12.5)$$

■

Now, we will show that we can implement $U_{CPS,\pi}$ using U_{Cphase} and *other 1 qubit gates*.

Example 12.2 Show that $U_{CPS,\pi} = U_{Cphase}(I \otimes U_{PS,-\phi})(U_{PS,\phi-\pi} \otimes I)$. Recall that the matrix form of the phase shift gate, $U_{PS,\phi}$, is

$$\begin{pmatrix} 1 & 0 \\ 0 & e^{i\phi} \end{pmatrix}. \tag{12.6}$$

Therefore,

$$I \otimes U_{PS,-\phi} = \begin{pmatrix} 1 & 0 \\ 0 & 1 \end{pmatrix} \otimes \begin{pmatrix} 1 & 0 \\ 0 & e^{-i\phi} \end{pmatrix} = \begin{pmatrix} 1 & 0 & 0 & 0 \\ 0 & e^{-i\phi} & 0 & 0 \\ 0 & 0 & 1 & 0 \\ 0 & 0 & 0 & e^{-i\phi} \end{pmatrix}, \tag{12.7}$$

and

$$U_{PS,\phi-\pi} \otimes I = \begin{pmatrix} 1 & 0 \\ 0 & e^{i(\phi-\pi)} \end{pmatrix} \otimes \begin{pmatrix} 1 & 0 \\ 0 & 1 \end{pmatrix} = \begin{pmatrix} 1 & 0 & 0 & 0 \\ 0 & 1 & 0 & 0 \\ 0 & 0 & e^{i(\phi-\pi)} & 0 \\ 0 & 0 & 0 & e^{i(\phi-\pi)} \end{pmatrix}. \tag{12.8}$$

Using Eqs. (12.1), (12.7), and (12.8), we have

$$U_{Cphase}(I \otimes U_{PS,-\phi})(U_{PS,\phi-\pi} \otimes I),$$

$$= \begin{pmatrix} 1 & 0 & 0 & 0 \\ 0 & e^{i\phi} & 0 & 0 \\ 0 & 0 & e^{i(\pi-\phi)} & 0 \\ 0 & 0 & 0 & 1 \end{pmatrix} \begin{pmatrix} 1 & 0 & 0 & 0 \\ 0 & e^{-i\phi} & 0 & 0 \\ 0 & 0 & 1 & 0 \\ 0 & 0 & 0 & e^{-i\phi} \end{pmatrix} \begin{pmatrix} 1 & 0 & 0 & 0 \\ 0 & 1 & 0 & 0 \\ 0 & 0 & e^{i(\phi-\pi)} & 0 \\ 0 & 0 & 0 & e^{i(\phi-\pi)} \end{pmatrix},$$

$$= \begin{pmatrix} 1 & 0 & 0 & 0 \\ 0 & e^{i\phi} & 0 & 0 \\ 0 & 0 & e^{i(\pi-\phi)} & 0 \\ 0 & 0 & 0 & 1 \end{pmatrix} \begin{pmatrix} 1 & 0 & 0 & 0 \\ 0 & e^{-i\phi} & 0 & 0 \\ 0 & 0 & e^{i(\phi-\pi)} & 0 \\ 0 & 0 & 0 & e^{-i\pi} \end{pmatrix},$$

$$= \begin{pmatrix} 1 & 0 & 0 & 0 \\ 0 & 1 & 0 & 0 \\ 0 & 0 & 1 & 0 \\ 0 & 0 & 0 & e^{-i\pi} \end{pmatrix},$$

$$= \begin{pmatrix} 1 & 0 & 0 & 0 \\ 0 & 1 & 0 & 0 \\ 0 & 0 & 1 & 0 \\ 0 & 0 & 0 & e^{i\pi} \end{pmatrix} = \begin{pmatrix} 1 & 0 & 0 & 0 \\ 0 & 1 & 0 & 0 \\ 0 & 0 & 1 & 0 \\ 0 & 0 & 0 & -1 \end{pmatrix},$$

$$= U_{CPS,\pi}, \tag{12.9}$$

12.3 2-Qubit Gate for Spin Implementation

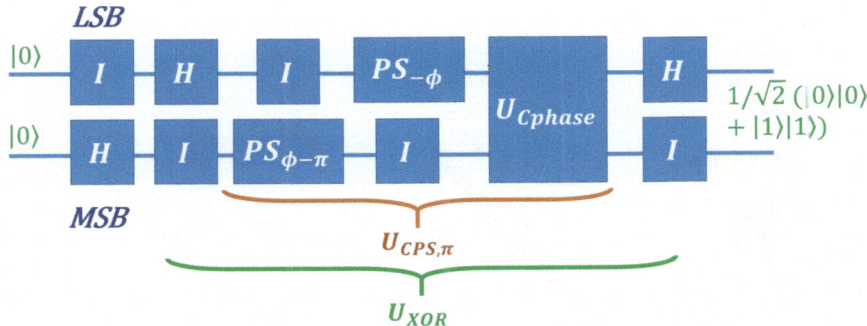

Fig. 12.3 The equivalent circuit of Fig. 12.2 implemented by 1-qubit gates and a U_{Cphase} gate

where at the end, we used the fact that $e^{-i\pi} = e^{i\pi}$. ∎

Therefore, we have successfully implemented a U_{XOR} gate for entanglement creation using the native 2-qubit gate for electron spin. Figure 12.3 shows the full entanglement circuit using 1-qubit gates and a U_{Cphase} gate.

12.3.2 Physical Implementation of U_{Cphase} Gate

12.3.2.1 Setup and Individual 1-Qubit Operations

We now will try to implement the U_{Cphase} gate in Eq. (12.1). We will follow the approaches in [3] and [4], however, with pretty significant modifications and simplifications. If there is only one thing that you can remember from this subsection, I hope you can appreciate the fact that *we need to use the physics of a system to realize an entanglement gate and we need to adjust the condition so that it can achieve the purpose most effectively*. The process of using the right physics and adjusting the conditions is called **Hamiltonian engineering**. Again, Hamiltonian refers to the total energy of the system. So the fancy word "Hamiltonian engineering" is nothing but setting up the system at the right energy at the right time to let it evolve in a desirable way.

Figure 12.4 shows the schematic of two qubits on a silicon wafer. The barrier gate is not shown for clarity. Same as the 1-qubit example in Sect. 12.2, $T = 50$ mK and $B_0 = 1.4T$ and highly purified Si is used (with only 800 ppm of ^{29}Si).

Before working on a 2-qubit gate, we need to make sure we can perform 1-qubit operations on *individual* qubit. To do so, we need to have different *electron spin resonance frequency*, which is the Larmor frequency, ω_L, for each qubit. Otherwise, we will be applying the same 1-qubit gate to both qubits at the same time. Equation (8.11) tells us that ω_L depends on B_0, the g-factor, g, and the effective mass of the electron, m^*, which is repeated here for convenience.

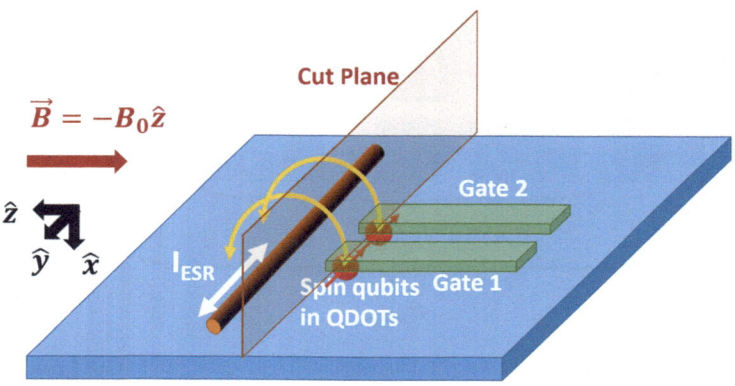

Fig. 12.4 Schematic showing a 2-qubit system. The barrier gate is not used

$$\omega_L = \left| \frac{2ge\hbar}{4m^*} B_0 \right| / \hbar. \qquad (12.10)$$

B_0 can be set to be different at each qubit by putting a micro-magnet with a different strength next to each qubit. However, this is not very scalable when millions of qubits need to be integrated. Instead of using micro-magnet, with some engineering, it is also possible to create a non-constant DC magnetic field (but still constant and pointing in $-\hat{z}$-direction) from qubit to qubit. m^* is likely to be the same because the crystal properties would not change much across the chip. Another approach is to change g which can be affected by the electric field it experiences through **Stark effect** [2]. This is very convenient because it means that we can just apply different plunger gate voltages (in combination with other electrodes) to achieve different ω_L for each qubit. We can then change the frequency of I_{ESR} so that we can address individual qubits. This is called **spectral selectivity** as we are selecting qubits based on frequency.

For example, qubit 1 might have $\omega_L = 0.34 MHz$ and qubit 1 might have $\omega_L = 0.36 MHz$. By setting I_{ESR} to oscillate at $0.34 MHz$, only qubit 1 will perform Rabi oscillation as it is at its spin resonance. Qubit 2 will not as it is off-resonant.

12.3.2.2 2-Qubit Physics

The physics and the mathematics of implementing the 2-qubit U_{Cphase} gate is pretty complicated and lengthy. We will not go through all the derivation here. We will only highlight the critical concepts so we can appreciate what the enabler is.

Figure 12.5 shows the cross section of the 2-qubit array. The gates are biased at different voltages. As a result, they have different ω_L as discussed earlier due to *Stark effect*.

12.3 2-Qubit Gate for Spin Implementation

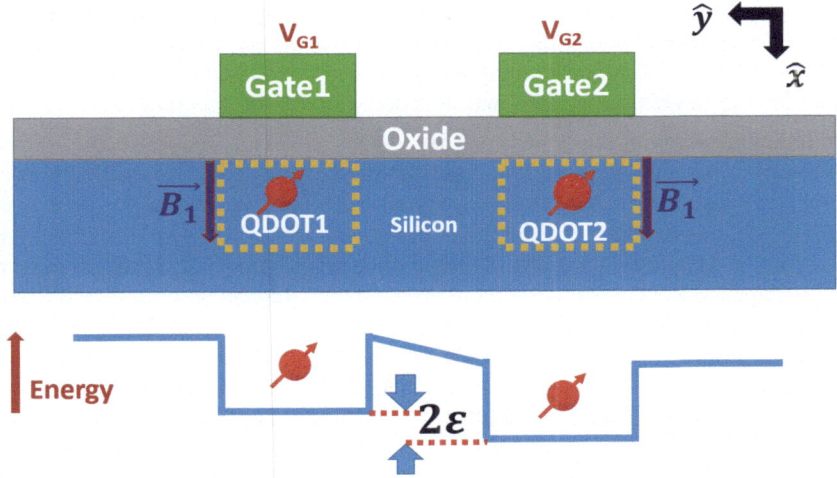

Fig. 12.5 Cut plane in Fig. 12.4 and the corresponding electron potential energy along the quantum dots. V_{G2} is more positive than V_{G1}.

Since this is a 2-qubit system, it has four basis states. One of the choices of the basis states is $|\uparrow\uparrow\rangle$, $|\uparrow\downarrow\rangle$, $|\downarrow\uparrow\rangle$, and $|\downarrow\downarrow\rangle$. When we write them in $bra-ket$ notation, the first one refers to the electron in the left quantum dot and the second one refers to the electron in the right quantum dot. For each qubit, according to Eqs. (8.4) and (8.11), its Hamiltonian is,

$$H_i = B_0 \frac{ge\hbar}{4m}\begin{pmatrix} 1 & 0 \\ 0 & -1 \end{pmatrix},$$

$$= -\frac{\hbar\omega_{Li}}{2}\begin{pmatrix} 1 & 0 \\ 0 & -1 \end{pmatrix},$$

$$= \begin{pmatrix} -\frac{\hbar\omega_{Li}}{2} & 0 \\ 0 & \frac{\hbar\omega_{Li}}{2} \end{pmatrix}, \tag{12.11}$$

where $i = 1$ and $i = 2$ for the qubit in QDOT1 and QDOT2, respectively.

Therefore, the Hamiltonian of the system, which can be obtained by the tensor product of the individual Hamiltonians, is

$$H = \begin{pmatrix} -E_z & 0 & 0 & 0 \\ 0 & \frac{-dE_z}{2} & 0 & 0 \\ 0 & 0 & \frac{dE_z}{2} & 0 \\ 0 & 0 & 0 & E_z \end{pmatrix}, \tag{12.12}$$

with $E_z = \hbar \frac{\omega_{L1}+\omega_{L2}}{2}$, which is the average of their Zeeman energies and $dE_z = \hbar\omega_{L1} - \hbar\omega_{L2}$, which is the difference of their Zeeman energies. For example, the third term along the diagonal corresponds to the energy of the state, $|\downarrow\uparrow\rangle$. Since the energy of $|\downarrow\rangle$ in QDOT1 is $\frac{\hbar\omega_{L1}}{2}$ and the energy of $|\uparrow\rangle$ in the QDOT2 is $-\frac{\hbar\omega_{L2}}{2}$, then the total energy of the system is $\frac{\hbar\omega_{L1}}{2} - \frac{\hbar\omega_{L2}}{2} = \frac{dE_z}{2}$.

However, this is *NOT* enough to construct the U_{Cphase} gate. We need to borrow another two states *outside of the computational Hilbert space*. The two states are that both electrons can be in the left quantum dot (QDOT1), $|S(2, 0)\rangle$, or in the right quantum dot (QDOT2), $|S(0, 2)\rangle$. S refers to the singlet state but we do not need to fully understand it as we do not plan to derive the mathematics rigorously. We will just accept that these are another two possible states that the two-electron system can have. In physics, when we say they might be at a certain state, it means that it is accessible for the given energy we have. There are other states (triplet states) that have a higher energy and the probability of going into those states is low.

Since the system has six states now, we need a larger matrix to describe the Hamiltonian. Again, it has more than four because it has extra degrees of freedom, namely, where to put the electrons. The Hamiltonian is,

$$H = \begin{pmatrix} -E_z & 0 & 0 & 0 & 0 & 0 \\ 0 & \frac{-dE_z}{2} & 0 & 0 & t & t \\ 0 & 0 & \frac{dE_z}{2} & 0 & -t & -t \\ 0 & 0 & 0 & E_z & 0 & 0 \\ 0 & t & -t & 0 & U-\epsilon & 0 \\ 0 & t & -t & 0 & 0 & U+\epsilon \end{pmatrix} \begin{matrix} |\uparrow\uparrow\rangle \\ |\uparrow\downarrow\rangle \\ |\downarrow\uparrow\rangle \\ |\downarrow\downarrow\rangle \\ S(0,2) \\ S(2,0) \end{matrix},$$ (12.13)

where U is the energy cost to move the two electrons to the same quantum dot (see the **charging energy** and **Coulomb blockade** discussed in Sect. 11.3) and ϵ is the energy difference between the two dots as shown in Fig. 12.5. The fifth and the sixth terms along the diagonal are the energy of $|S(0, 2)\rangle$ and $|S(2, 0)\rangle$. For example, to put both electrons in QDOT2 ($|S(0, 2)\rangle$), we need an energy of U. However, it has a lower energy due to the tilted potential by ϵ. Therefore, its energy is $U - \epsilon$.

The state associated with each row is also given. We see that there are off-diagonal terms connecting the anti-parallel spin states ($|\uparrow\downarrow\rangle$ and $|\downarrow\uparrow\rangle$) to $|S(0, 2)\rangle$ and $|S(2, 0)\rangle$. This is due to **tunneling** and the value t is related to the tunneling rate. This is easy to understand as to have a state with both electrons in QDOT1 or QDOT2, tunneling through the barrier is required. Therefore, t is called the interdot tunnel coupling. Tunneling enables the two qubits to interact and the tunneling rate is controlled by the barrier height between the two quantum dots through the barrier gate voltage (not drawn in Fig. 12.5). But why there is no coupling between the parallel spin states ($|\uparrow\uparrow\rangle$ and $|\downarrow\downarrow\rangle$) to $|S(0, 2)\rangle$ and $|S(2, 0)\rangle$? This is because $|S(0, 2)\rangle$ and $|S(2, 0)\rangle$ are singlet states in which the spins are anti-parallel. Therefore, parallel spin states of $|\uparrow\uparrow\rangle$ and $|\downarrow\downarrow\rangle$ cannot transit to $|S(0, 2)\rangle$ and $|S(2, 0)\rangle$ easily as spin flipping is needed.

Now, if we change ϵ from zero, the energies of the anti-parallel states will change due to the coupling to $|S(0, 2)\rangle$ and $|S(2, 0)\rangle$. However, the energies of the parallel states will NOT change as they are not coupled to $|S(0, 2)\rangle$ and $|S(2, 0)\rangle$. Therefore, **additional phase shift will be added to the anti-parallel spin states, $|\uparrow\downarrow\rangle$ and $|\downarrow\uparrow\rangle$** but not the parallel state after solving the Schrödinger equation (using the Schrieffer-Wolff transformation) and the gate has the following form:

$$\begin{pmatrix} 1 & 0 & 0 & 0 \\ 0 & e^{i\phi_1} & 0 & 0 \\ 0 & 0 & e^{i\phi_2} & 0 \\ 0 & 0 & 0 & 1 \end{pmatrix}, \quad (12.14)$$

where ϕ_1 and ϕ_2 are phases. By applying large enough dE_z, we can set $\phi_1 = \pi - \phi_2$ and thus create the U_{Cphase} gate in Eq. (12.1).

12.4 Summary

In this chapter, we have discussed how to implement Rabi oscillation on electron spin qubits integrated on a silicon chip. An external DC magnetic field (can be due to micro-magnet) is applied for Zeeman splitting. A wire conducting AC electric current is used to generate an AC perturbation field for Rabi oscillation, We also discuss how to implement a 2-qubit entanglement gate (U_{Cphase}) through the coupling of the anti-parallel spin state to $|S(0, 2)\rangle$ and $|S(2, 0)\rangle$. U_{Cphase} is just one of the possible entanglement gates but it highlights the importance of Hamiltonian engineering.

Problems

12.1 2-Qubit Hamiltonian
Derive Eq. (12.12) using the tensor product of H_1 and H_2.

12.2 Oscillating Magnetic Field Generation
What is the peak current required to create 0.01T of peak oscillating magnetic field if the conducting wire is $1\mu m$ away from the qubit? Assume an effective dielectric constant of 6.1. You may use Eq. (24.17).

References

1. Hiu-Yung Wong. *Introduction to Quantum Computing*. Springer, 2024.

2. M. Veldhorst, J. C. C. Hwang, C. H. Yang, A. W. Leenstra, B. de Ronde, J. P. Dehollain, J. T. Muhonen, F. E. Hudson, K. M. Itoh, A. Morello, and A. S. Dzurak. An addressable quantum dot qubit with fault-tolerant control-fidelity. *Nature Nanotechnology*, 9(12):981–985, Dec 2014.
3. T. Meunier, V. E. Calado, and L. M. K. Vandersypen. Efficient controlled-phase gate for single-spin qubits in quantum dots. *Phys. Rev. B*, 83:121403, Mar 2011.
4. M. Veldhorst, C. H. Yang, J. C. C. Hwang, W. Huang, J. P. Dehollain, J. T. Muhonen, S. Simmons, A. Laucht, F. E. Hudson, K. M. Itoh, A. Morello, and A. S. Dzurak. A two-qubit logic gate in silicon. *Nature*, 526(7573):410–414, Oct 2015.

Part III
Superconducting Qubit Architecture and Hardware

Chapter 13
Lagrangian Mechanics and Hamiltonian Mechanics

13.1 Introduction

Most of us have learned the basics of Newtonian mechanics. Newtonian mechanics is also called *vectorial mechanics* because it studies the motion of bodies under the influence of vector quantities such as *force*. However, Newtonian mechanics is not convenient in solving certain problems. There are other frameworks in theoretical physics called analytical mechanics such as **Lagrangian mechanics** and **Hamiltonian mechanics**. They are equivalent to Newtonian mechanics and they use scalar quantities such as the kinetic energy and potential energy of a system to derive the equations of motion of the system. In many problems, they appear to be more elegant and succinct than Newtonian mechanics. More importantly, the concepts in analytical mechanics can be *generalized* to hyperspace/phase space, in which we do not live. Moreover, Hamiltonian mechanics allow us to transition from classical mechanics to quantum mechanics more "smoothly." In this chapter, we will learn the *skills* of using Lagrangian and Hamiltonian mechanics. Readers are expected to learn the rules only. Readers may refer to [1] if they are interested in having a deeper appreciation of analytical mechanics.

13.1.1 Learning Outcomes

Be able to write down the Lagrangian and Hamiltonian of a given physical system; be able to derive the equations of motion of a system based on its Lagrangian and Hamiltonian.

© The Author(s), under exclusive license to Springer Nature Switzerland AG 2025
H. Y. Wong, *Quantum Computing Architecture and Hardware for Engineers*,
https://doi.org/10.1007/978-3-031-78219-0_13

13.1.2 Teaching Videos

- Search for Ch13 in this playlist
  - https://tinyurl.com/3yhze3jn
- Other videos
  - https://youtu.be/Ydj2hintCkc
  - https://youtu.be/u2SgXmf2SvQ
  - https://youtu.be/IdSF_064ZSo

13.2 Lagrangian Mechanics

13.2.1 Generalized Coordinates and Velocities

For the purposes in the following chapters, we only consider point particles, conservative forces, and non-relativistic mechanics. Let us consider a system comprised of N particles. We know that if the coordinates of each particle and the velocity of each particle are known at a given time, the system has a well-defined state. This is because the acceleration of a particle depends on the force exerted on it. And the force is the spatial derivative of its potential, which is a function of its coordinates. Therefore, if we know their positions and velocities, we know their accelerations and, thus, can deduce their past and future states.

For N particles, in our real space, there are $3N$ independent coordinates due to the three orthogonal directions. Therefore, they are the collection of $\vec{q} = \{q_1, q_2, \cdots, q_{3N}\}$, where we write it as a $3N$-dimensional vector. Similarly, it has $3N$ independent velocities, $\vec{\dot{q}} = \{\dot{q}_1, \dot{q}_2, \cdots, \dot{q}_{3N}\}$, where

$$\dot{q}_i = \frac{dq_i}{dt}. \tag{13.1}$$

Besides using real spatial coordinates and velocities to uniquely determine the state of a system, one may also use other $3N$ quantities to determine its coordinates as long as they also give the system $3N$ degrees of freedom [2]. Such quantities are called the **generalized coordinates**. The time derivatives (Eq. (13.1)) of the generalized coordinates are called the **generalized velocities**. For the formalism we will discuss later, it is easier to think and understand using spatial coordinates and velocities but we need to keep in mind and accept the fact that they are applicable to generalized coordinates and velocities, too.

13.2.2 Lagrangian and Lagrange's Equations

We will first introduce a new quantity called **Lagrangian**, $\mathcal{L}$, which is defined as,

$$\mathcal{L} = T - V, \qquad (13.2)$$

where T and V are the **kinetic energy** and **potential energy** of the system, respectively. Naturally, $\mathcal{L}$ has the unit of energy. Since T is a function of velocities, $\vec{\dot{q}}$, and V is a function of coordinates, $\vec{q}$, $\mathcal{L}$ is a function of both velocities and coordinates. Of course, they are all functions of time, t. Therefore,

$$\begin{aligned}\mathcal{L} &= T(\vec{\dot{q}}, t) - V(\vec{q}, t), \\ &= \mathcal{L}(\vec{q}, \vec{\dot{q}}, t), \\ &= \mathcal{L}(q_1, q_2, \cdots, \dot{q}_1, \dot{q}_2, \cdots, t).\end{aligned} \qquad (13.3)$$

It is given that one can derive the **equations of motion** of the system by solving the **Lagrange's equations** [1],

$$\frac{d}{dt}\left(\frac{\partial \mathcal{L}}{\partial \dot{q}_i}\right) - \frac{\partial \mathcal{L}}{\partial q_i} = 0, \qquad (13.4)$$

for i from 1 to $3N$. Lagrange's equations contain the Lagrangian of the system with the *coordinates and velocities being the independent variables*. This forms the basics of Lagrangian mechanics. It should be noted that when working with Lagrangian mechanics, one needs to make sure to **express the Lagrangian explicitly as a function of coordinates and velocities**. Of course, this includes the *generalized* coordinates and velocities.

It should also be noted that the Lagrangian of a system is *not unique*. As long as it gives the correct equations of motion through Eq. (13.4), it is a valid Lagrangian.

In this book, we take Lagrange's equations as given like how we trust $F = ma$ when we study **Newtonian mechanics**. But Lagrange's equations can be derived from a more fundamental principle, namely, the **principle of least action** or **Hamilton's principle** [2]. The principle defines **action**, S, as,

$$S = \int_{t_1}^{t_2} \mathcal{L}(\vec{q}, \vec{\dot{q}}, t) dt. \qquad (13.5)$$

The action, S, integrates the Lagrangian of a system from time t_1 to time t_2 and mandates that the system should evolve from time t_1 to time t_2 in a way such that S is minimal based on which the Lagrange's equations in Eq. (13.4) are derived [2].

Again, for the purpose of this book, we just need to learn the skills to construct the Lagrangian and solve the Lagrange's equations of a given system. Let us look at two examples.

Example 13.1 Find the equations of motion of a free particle with mass, m. Assume that the particle is moving in the $\hat{x}$-direction with speed v at time t_0.

We already know from Newton's first law that a free particle will keep moving at a constant speed. Let us see if we will obtain the same result by using Lagrangian mechanics.

Set $\vec{q} = \{q_1 = x, q_2 = y, q_3 = z\}$. Its velocity is $\vec{\dot{q}} = \{\dot{q}_1, \dot{q}_2, \dot{q}_3\}$. As a free particle, there is no external force and thus it experiences a constant potential energy that can be set to a constant ϕ. Therefore, its kinetic energy is

$$T = \frac{1}{2}m\dot{q}_1{}^2 + \frac{1}{2}m\dot{q}_2{}^2 + \frac{1}{2}m\dot{q}_3{}^2, \tag{13.6}$$

and its potential energy is

$$V = \phi. \tag{13.7}$$

The Lagrangian of the system is

$$\mathcal{L} = T - V,$$
$$= \frac{1}{2}m\dot{q}_1{}^2 + \frac{1}{2}m\dot{q}_2{}^2 + \frac{1}{2}m\dot{q}_3{}^2 - \phi. \tag{13.8}$$

To find its equations of motion, we solve Lagrange's equations in Eq. (13.4). Note that there are three coordinates (for $i = 1$ to $i = 3$ with $q_1 = x, q_2 = y, q_3 = z$). Therefore, we have three equations,

$$\frac{d}{dt}\left(\frac{\partial \mathcal{L}}{\partial \dot{q}_1}\right) - \frac{\partial \mathcal{L}}{\partial q_1} = 0, \tag{13.9}$$

$$\frac{d}{dt}\left(\frac{\partial \mathcal{L}}{\partial \dot{q}_2}\right) - \frac{\partial \mathcal{L}}{\partial q_2} = 0, \tag{13.10}$$

$$\frac{d}{dt}\left(\frac{\partial \mathcal{L}}{\partial \dot{q}_3}\right) - \frac{\partial \mathcal{L}}{\partial q_3} = 0. \tag{13.11}$$

The equations involve partial derivatives with respect to $q_1, q_2, q_3, \dot{q}_1, \dot{q}_2$, and $\dot{q}_3$. But none of the terms in Eq. (13.8) depends on q_1, q_2, and q_3 as ϕ is a constant. $\mathcal{L}$ only depends on $\dot{q}_1, \dot{q}_2$, and $\dot{q}_3$ Therefore, the three Lagrange's equations become,

$$\frac{d}{dt}\left(\frac{\partial \mathcal{L}}{\partial \dot{q}_1}\right) - 0 = 0, \tag{13.12}$$

$$\frac{d}{dt}\left(\frac{\partial \mathcal{L}}{\partial \dot{q}_2}\right) - 0 = 0, \tag{13.13}$$

$$\frac{d}{dt}\left(\frac{\partial \mathcal{L}}{\partial \dot{q}_3}\right) - 0 = 0. \tag{13.14}$$

13.2 Lagrangian Mechanics

Let us only solve Eq. (13.12),

$$\frac{d}{dt}\left(\frac{\partial \mathcal{L}}{\partial \dot{q}_1}\right) = 0,$$

$$\frac{d}{dt}\left(\frac{\partial(\frac{1}{2}m\dot{q}_1^2 + \frac{1}{2}m\dot{q}_2^2 + \frac{1}{2}m\dot{q}_3^2 - \phi)}{\partial \dot{q}_1}\right) = 0,$$

$$\frac{d(m\dot{q}_1)}{dt} = 0,$$

$$\ddot{q}_1 = 0. \tag{13.15}$$

$\ddot{q}_1$ is the time derivative of velocity, which is the acceleration in the $\hat{x}$ direction. It means the particle will move at a constant velocity in the $\hat{x}$ direction. Since at $t = t_0$, $\dot{q}_1 = v$, then it will be moving at speed v in the $\hat{x}$ direction forever. Similarly, $\ddot{q}_2 = \ddot{q}_3 = 0$; therefore, $q_2 = q_3 = 0$ at all time.

Using Lagrangian mechanics, we obtain the same conclusion as Newton's first law. ∎

Example 13.2 Find the equations of motion of a mass, m, attached to a fixed wall through a spring with a spring constant of k (Fig. 13.1). This is the famous **simple harmonic oscillator (SHO)** problem.

This is a 1D system with 1 particle. So there is only one coordinate and one velocity component. That is, $\vec{q} = \{q_1 = x\}$ and $\vec{\dot{q}} = \{\dot{q}_1\}$. We assume at equilibrium (i.e., the spring is not stretched nor compressed), the particle is at $x = 0$. The kinetic energy, potential energy, and the Lagrangian of the system are given by

$$T = \frac{1}{2}m\dot{q}_1^2, \tag{13.16}$$

$$V = \frac{1}{2}kq_1^2, \tag{13.17}$$

$$\mathcal{L} = \frac{1}{2}m\dot{q}_1^2 - \frac{1}{2}kq_1^2. \tag{13.18}$$

Fig. 13.1 The simple harmonic oscillator considered in Example 13.2. The equilibrium position of the particle is at $x = 0$

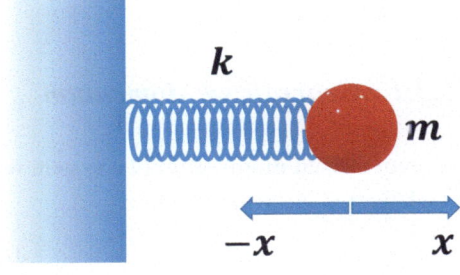

Substitute Eq. (13.18) into Lagrange's equations Eq. (13.4),

$$\frac{d}{dt}\left(\frac{\partial \mathcal{L}}{\partial \dot{q}_1}\right) - \frac{\partial \mathcal{L}}{\partial q_1} = 0,$$

$$\frac{d}{dt}\left(\frac{\partial(\frac{1}{2}m\dot{q}_1^2 - \frac{1}{2}kq_1^2)}{\partial \dot{q}_1}\right) - \frac{\partial\left(\frac{1}{2}m\dot{q}_1^2 - \frac{1}{2}kq_1^2\right)}{\partial q_1} = 0,$$

$$\frac{d(m\dot{q}_1)}{dt} + kq_1 = 0,$$

$$m\ddot{q}_1 + kq_1 = 0,$$

$$\ddot{q}_1 + \frac{k}{m}q_1 = 0,$$

$$\ddot{q}_1 + \omega_0^2 q_1 = 0, \qquad (13.19)$$

which are the equations of motion of the system and we have defined the resonance frequency of the system as $\omega_0 = \sqrt{\frac{k}{m}}$ in the last line. This is a differential equation and is exactly what we learned in Newtonian mechanics. ∎

13.3 Hamiltonian Mechanics

In the previous section, we learned the skills of using Lagrangian mechanics. Lagrangian mechanics has the generalized coordinates (q_i) and generalized velocities ($\dot{q}_i$) as its independent variables. We first construct the Lagrangian based on kinetic energy and potential energy. Then we use Lagrange's equations to find the equations of motion of the system. In Lagrange's equations, derivatives of the Lagrangian are taken with respect to independent variables, q_i and $\dot{q}_i$.

In the previous paragraph, if you replace "Lagrangian" with "Hamiltonian," "Lagrange's equation" with "Hamilton's equation," and "generalized velocities" with "generalized momentum," it will become a summary of Hamiltonian mechanics.

13.3.1 Generalized Momentum

The **generalized momentum** of a system with $3N$ degrees of freedom is defined as $\vec{p} = \{p_1, p_2, \cdots, p_{3N}\}$, with

$$p_i = \frac{\partial \mathcal{L}}{\partial \dot{q}_i}, \qquad (13.20)$$

13.4 Hamiltonian and Hamilton's Equation

for $i = 1$ to $i = 3N$. We will not spend more time discussing generalized momentum. But we will look at two examples to make sure it makes sense.

Example 13.3 Find the generalized momenta of the free particle in Example 13.1. Substitute Eq. (13.8) into Eq. (13.20),

$$p_i = \frac{\partial \mathcal{L}}{\partial \dot{q}_i},$$
$$= \frac{\partial(\frac{1}{2}m\dot{q}_1^2 + \frac{1}{2}m\dot{q}_2^2 + \frac{1}{2}m\dot{q}_3^2 - \phi)}{\partial \dot{q}_i}, \quad (13.21)$$

where $i = 1$ to $i = 3$. ϕ is a constant and does not depend on $\dot{q}_i$. Let us find the case when $i = 1$. We have

$$p_i = p_1 = \frac{\partial(\frac{1}{2}m\dot{q}_1^2 + \frac{1}{2}m\dot{q}_2^2 + \frac{1}{2}m\dot{q}_3^2 - \phi)}{\partial \dot{q}_1},$$
$$= m\dot{q}_1. \quad (13.22)$$

Similarly, $p_2 = m\dot{q}_2$ and $p_3 = m\dot{q}_3$. This is exactly the definition of linear momentum in elementary physics where momentum is the product of mass and velocity. ∎

13.4 Hamiltonian and Hamilton's Equation

Like in Lagrangian mechanics, we introduce a new quantity called **Hamiltonian**, $\mathcal{H}$, which is defined as,

$$\mathcal{H} = \sum_{i=1}^{3N} p_i \dot{q}_i - \mathcal{L}, \quad (13.23)$$

which is called the **Legendre transform**. $\mathcal{H}$ also has the unit of energy. Of course, $\mathcal{H}$ is the same as H, which is the matrix representation of the Hamiltonian of a quantum state that we have been using so far. It is given that one can derive the **equations of motion** of the system by solving the **Hamilton's equations** [1],

$$\dot{p}_i = -\frac{\partial \mathcal{H}}{\partial q_i}, \quad (13.24)$$

$$\dot{q}_i = \frac{\partial \mathcal{H}}{\partial p_i}, \quad (13.25)$$

for i from 1 to $3N$. Hamilton's equations contain the Hamiltonian of the system with the *coordinates and momenta being the independent variables*. This forms the basics of Hamiltonian mechanics. It should be noted that when working with Hamiltonian mechanics, one needs to make sure to **express the Hamiltonian explicitly as a function of generalized coordinates and generalized momenta**. That is,

$$\mathcal{H} = \mathcal{H}(q_1, q_2, \cdots, p_1, p_2, \cdots, t). \tag{13.26}$$

Let us first practice with a simple example to gain a deeper understanding of Hamiltonian.

Example 13.4 Find the equations of motion of a free particle with mass, m, using Hamilton's equations. Assume that the particle is moving in the $\hat{x}$-direction with speed v at time t_0.

When going over this example, it will be instructive to compare it with the Lagrangian mechanics approach in Example 13.1. Firstly, we need to find the Hamiltonian of the system. From Eqs. (13.23) and (13.8),

$$\mathcal{H} = \sum_{i=1}^{3N} p_i \dot{q}_i - \mathcal{L},$$

$$= p_1 \dot{q}_1 + p_2 \dot{q}_2 + p_3 \dot{q}_3 - \left(\frac{1}{2} m \dot{q}_1^2 + \frac{1}{2} m \dot{q}_2^2 + \frac{1}{2} m \dot{q}_3^2 - \phi \right). \tag{13.27}$$

We have found the momenta, q_i, in Eq. (13.22). *Our goal is to eliminate $\dot{q}_i$ using Eq. (13.22) so that $\mathcal{H}$ is expressed as a function of q_i and p_i only (see Eq. (13.26))*. We will substitute $\dot{q}_i = \frac{p_i}{m}$.

$$\mathcal{H} = p_1 \frac{p_1}{m} + p_2 \frac{p_2}{m} + p_3 \frac{p_3}{m} - \left(\frac{1}{2} m \left(\frac{p_1}{m}\right)^2 + \frac{1}{2} m \left(\frac{p_2}{m}\right)^2 + \frac{1}{2} m \left(\frac{p_3}{m}\right)^2 - \phi \right),$$

$$= \frac{1}{2} \frac{p_1^2}{m} + \frac{1}{2} \frac{p_2^2}{m} + \frac{1}{2} \frac{p_3^2}{m} + \phi. \tag{13.28}$$

This is just the *sum of the kinetic energy, T, and the potential energy, V, of the system*. This is true if the constraints in the problem do not depend explicitly on time [3]. Therefore,

$$\mathcal{H} = T(\vec{p}, t) + V(\vec{q}, t), \tag{13.29}$$

$$= \frac{1}{2} \frac{p_1^2}{m} + \frac{1}{2} \frac{p_2^2}{m} + \frac{1}{2} \frac{p_3^2}{m} + \phi. \tag{13.30}$$

Readers should compare these equations against Eqs. (13.3) and (13.8) to appreciate how the generalized momenta have replaced the generalized velocities of the system in Hamiltonian.

13.4 Hamiltonian and Hamilton's Equation

To find the equations of motion, we just need to substitute Eq. (13.30) into Hamilton's equations (Eqs. (13.24) and (13.25)). Firstly,

$$\dot{p}_i = -\frac{\partial \mathcal{H}}{\partial q_i},$$

$$= -\frac{\partial \left(\frac{1}{2}\frac{p_1^2}{m} + \frac{1}{2}\frac{p_2^2}{m} + \frac{1}{2}\frac{p_3^2}{m} + \phi \right)}{\partial q_i},$$

$$= 0. \tag{13.31}$$

This is because ϕ is a constant and $\mathcal{H}$ does not depend on q_i. Therefore, the momenta in all three spatial directions are constant. Since at $t = t_0$, $\dot{q}_1 = v$, $p_1 = mv$ at all time and $p_2 = p_3 = 0$. Now let us consider Eq. (13.25).

$$\dot{q}_i = \frac{\partial H}{\partial p_i},$$

$$= \frac{\partial (\frac{1}{2}\frac{p_1^2}{m} + \frac{1}{2}\frac{p_2^2}{m} + \frac{1}{2}\frac{p_3^2}{m} + \phi)}{\partial p_i},$$

$$= \frac{p_i}{m}, \tag{13.32}$$

with $i = 1$ to 3. They just tell us that the velocity in a certain direction is just the momentum in that direction divided by m.

Finally, recognizing $F = \dot{p}$ (i.e., force equals to the rate of change of momentum), by applying a time derivative to Eq. (13.32),

$$\ddot{q}_i = \frac{\dot{p}_i}{m},$$

$$a = \frac{F}{m}, \tag{13.33}$$

which is just Newton's second law, $F = ma$! ∎

Now, let us apply Hamilton's equations to solve the simple harmonic oscillator problem in Example 13.2.

Example 13.5 Find the equations of motion of a mass, m, attached to a fixed wall through a spring with a spring constant of k (Fig. 13.1) using Hamiltonian mechanics.

Note again that this is a 1D problem and $\vec{q} = \{q_1 = x\}$ and $\dot{\vec{q}} = \{\dot{q}_1\}$. We summarize the process in the following steps.

Step 1: Find the Momenta

Using Eq. (13.20) for the definition of momentum and Eq. (13.18) for the Lagrangian of this system, we have,

$$p_1 = \frac{\partial \mathcal{L}}{\partial \dot{q}_1},$$

$$= \frac{\partial(\frac{1}{2}m\dot{q}_1^2 - \frac{1}{2}kq_1^2)}{\partial \dot{q}_1},$$

$$= m\dot{q}_1. \tag{13.34}$$

Therefore, $\dot{q}_1 = \frac{p_1}{m}$.

Step 2: Find the Hamiltonian

We need to find the Hamiltonian in terms of q_1 and p_1. From Eqs. (13.23) and (13.18),

$$\mathcal{H} = \sum_{i=1}^{3N} p_i \dot{q}_i - \mathcal{L},$$

$$= p_1 \dot{q}_1 - \left(\frac{1}{2}m\dot{q}_1^2 - \frac{1}{2}kq_1^2\right),$$

$$= p_1 \frac{p_1}{m} - \left(\frac{1}{2}m(\frac{p_1}{m})^2 - \frac{1}{2}kq_1^2\right),$$

$$= \frac{1}{2}\frac{p_1^2}{m} + \frac{1}{2}kq_1^2. \tag{13.35}$$

Again, $\mathcal{H} = T + V$!

Step 3: Set up the Hamilton's Equations

Finally, we find the equations of motion by substituting Eq. (13.35) into Hamilton's equations. For Eq. (13.24),

$$\dot{p}_1 = -\frac{\partial \mathcal{H}}{\partial q_1},$$

$$= -\frac{\partial(\frac{1}{2}\frac{p_1^2}{m} + \frac{1}{2}kq_1^2)}{\partial q_1},$$

$$= -kq_1, \tag{13.36}$$

and for Eq. (13.25),

$$\dot{q}_1 = \frac{\partial \mathcal{H}}{\partial p_1},$$

$$= \frac{\partial(\frac{1}{2}\frac{p_1^2}{m} + \frac{1}{2}kq_1^2)}{\partial p_1},$$

$$= \frac{p_1}{m}. \tag{13.37}$$

We then take another time derivative of Eq. (13.37) and substitute Eq. (13.36) into it, we obtain

$$\ddot{q}_1 = \frac{\dot{p}_1}{m},$$

$$\ddot{q}_1 = \frac{-kq_1}{m},$$

$$\ddot{q}_1 + \frac{k}{m}q_1 = 0,$$

$$\ddot{q}_1 + \omega_0^2 q_1 = 0, \tag{13.38}$$

with $\omega_0 = \sqrt{\frac{k}{m}}$ and is the same as using Lagrangian mechanics (see Eq. (13.19)).
Therefore, Hamiltonian mechanics is equivalent to Lagrangian mechanics. ∎

13.5 Summary

We have learned the "skills" of using Lagrangian mechanics and Hamiltonian mechanics. As demonstrated, they derive the equations of motion of a system by involving scalar quantities such as kinetic energy, T, and potential energy, V. Lagrangian mechanics uses the Lagrangian ($\mathcal{L} = T - V$) of a system in Lagrange's equations while Hamiltonian mechanics uses the Hamiltonian ($\mathcal{H} = T + V$) of a system in Hamilton's equations. We note that in Lagrangian mechanics, generalized coordinates and generalized velocities are the independent variables. The Lagrangian of a system is not unique and is valid as long as it gives the correct equations of motion. The Hamiltonian of a system can be derived from Lagrangian and generalized coordinates and generalized momenta are its independent variables. Both analytical mechanics are equivalent. In the next chapter, we will show how Hamiltonian mechanics can link classical mechanics to quantum mechanics.

Problems

13.1 Equations of Motion from Lagrangian
Derive the equations of motion in the $\hat{y}$ and $\hat{z}$ direction in Example 13.1.

13.2 Generalized Coordinates, Velocities, and Momenta
A pendulum has a length of L with a mass m attached at the end. Use Lagrangian mechanics (purely classical) to find its equations of motion (no need to solve it). You can use its swing angle, θ, as the generalized coordinate and the rate of change of θ, $\dot{\theta}$, as the generalized velocity.

References

1. Herbert Goldstein, Charles Poole, and John Safko. *Classical Mechanics*. Pearson, 2001.
2. L.D. Landau and E.M. Lifshitz. *Mechanics: Volume 1 (Course of Theoretical Physics S)*. Butterworth-Heinemann, 1976.
3. Lagrange's and hamilton's equations. https://www.britannica.com/science/mechanics/Lagranges-and-Hamiltons-equations#ref612270. Accessed:2024-05-04.

Chapter 14
Quantization of Simple Harmonic Oscillator

14.1 Introduction

In the previous chapter, we studied Hamiltonian mechanics. In this chapter, we will show how to use Hamiltonian mechanics to transition from classical mechanics to quantum mechanics. We will use the *quantization* of a simple harmonic oscillator (SHO) as an example. If readers do not have a strong background in quantum mechanics, they are advised to revisit wave mechanics (e.g. [1]) and matrix mechanics (e.g. [2]) treatments of SHO. However, just like how we studied Lagrangian mechanics and Hamiltonian mechanics in the previous chapter, our goal is only to learn the skills and rules of performing quantization and promoting a variable to an operator.

14.1.1 Learning Outcomes

Be able to convert a classical Hamiltonian to a matrix in quantum mechanics; appreciate the link between the classical solution and quantum mechanical solution of an SHO; understand the meaning of generalized coordinates, generalized momenta, and conjugate variables.

14.1.2 Teaching Videos

- Search for Ch14 in this playlist
    - https://tinyurl.com/3yhze3jn

- Other videos
    - https://youtu.be/X3Rpj_hVOac
    - https://youtu.be/ALidb1_XbwY
    - https://youtu.be/bRWEz3W5aZE

14.2 SHO Hamiltonian and Promotion of Operators

In Chap. 13, we have derived the Hamiltonian of a **simple harmonic oscillator** (**SHO**) (Fig. 13.1) in Eq. (13.35). It is copied here for convenience:

$$\begin{aligned} \mathcal{H} &= \frac{1}{2}\frac{p_1^2}{m} + \frac{1}{2}kq_1^2, \\ &= \frac{1}{2}\frac{p^2}{m} + \frac{1}{2}kx^2, \end{aligned} \tag{14.1}$$

where we set $p_1 = p$ and $q_1 = x$ so that we use the variables we usually see. We should remind ourselves that generalized momenta and generalized coordinates are the dynamical variables in Hamiltonian mechanics. Also, we set $\omega_0 = \sqrt{\frac{k}{m}}$ (Eq. (13.38)) and we have:

$$\begin{aligned} \mathcal{H} &= \frac{1}{2}\frac{p^2}{m} + \frac{1}{2}kx^2, \\ &= \frac{p^2}{2m} + \frac{\omega_0^2 m}{2}x^2. \end{aligned} \tag{14.2}$$

14.2.1 Promotion of Classical Variables to Quantum Operators

To bridge classical mechanics to quantum mechanics, one can simply **promote** relevant classical variables such as p and x to **quantum operators**. By promoting variables to quantum operators, we can keep the form of classical mechanics in quantum mechanics with minimal modifications to equations. The promotions need to result in a solution that is consistent with experiment and classical mechanics. We will take it for granted. In the promotion process, for example, in SHO, all we need to do is just to rewrite p as $\hat{p}$, x as $\hat{x}$, and $\mathcal{H}$ as $\hat{\mathcal{H}}$. Note that $\hat{x}$ *is the operator of position x, not the unit vector in the x-direction*. Also, we if write the Hamiltonian in matrix form, we will use **H** which we have been using in the earlier chapters. There should be no ambiguity as they usually do not appear in the same equation.

14.2 SHO Hamiltonian and Promotion of Operators

The Hamiltonian becomes:

$$\hat{\mathcal{H}} = \frac{\hat{p}^2}{2m} + \frac{\omega_0^2 m}{2}\hat{x}^2. \qquad (14.3)$$

We see that the Hamiltonian also becomes an operator. The promotion process looks very simple and seems to be useless and arbitrary. However, this is not arbitrary. Let us understand the meaning of $\hat{x}$ and $\hat{p}$. We also want to emphasize that $\hat{p}$, $\hat{x}$, and $\hat{\mathcal{H}}$ are **operators**, while others variables such as m and ω_0 are scalar and numbers. *Operators obey the rule of the matrices in linear algebra* (see Sect. 3.2).

14.2.2 Position Operator and Eigenvector

When a particle has a definite location, x, (Fig. 14.1), its state can be represented as a **position vector**, $|x\rangle$. For example, if a particle is at location $3 cm$, then its position vector is $|x = 3cm\rangle = |3cm\rangle$. Since position is a real number, there are infinite numbers of possible states a particle can have.

$\hat{x}$ is called the **position operator**. It does not rotate a position vector but scales it with the value of the position vector. Writing in an equation, we have:

$$\hat{x}|x\rangle = x|x\rangle, \qquad (14.4)$$

which also means that the position vector $|x\rangle$ is the **eigenvector** of $\hat{x}$ with an **eigenvalue** of x (see Sect. 3.3.1). Since position is an observable, we expect $\hat{x}$ to be **Hermitian** and therefore,

$$\hat{x}^\dagger = \hat{x}. \qquad (14.5)$$

To gain a deeper understanding, it is instructive to know that $|x\rangle$ forms an orthonormal basis. Since x is continuous and, thus, $|x\rangle$ forms a **continuous basis**, it has the following **orthonormal** property:

$$\langle x|x'\rangle = \delta(x - x'), \qquad (14.6)$$

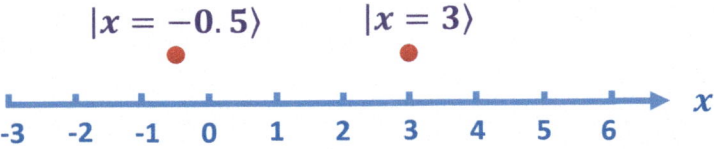

Fig. 14.1 Illustration of position eigenvectors and the locations of a particle

as opposed to that of a discrete basis

$$\langle n|n'\rangle = \delta_{n,n'}, \tag{14.7}$$

where $\delta(x - x')$ is the **Dirac delta function** and $\delta_{n,n'}$ is the **Kronecker delta function**.

It also has the following **completeness** property:

$$\int_{-\infty}^{+\infty} |x\rangle\langle x|\, dx = I, \tag{14.8}$$

as opposed to that of a discrete basis

$$\sum |n\rangle\langle n| = I. \tag{14.9}$$

14.2.3 Momentum Operator and Eigenvector

A particle can also be in a definite momentum state which means that it has a well-defined momentum, p (Fig. 14.2). Its state can be represented as a **momentum vector**, $|p\rangle$. If a particle has a definite momentum, it can be represented with a delta function in the momentum space, such as $|p = -0.5\rangle$ and $|p = 2\rangle$ where the unit is ignored, without loss of generality. In the real space, they are sinusoidal functions with $|p = 2\rangle$ having a spatial frequency four times that of $|p = -0.5\rangle$. Since momentum is a real number, there are infinite numbers of possible states a particle can have.

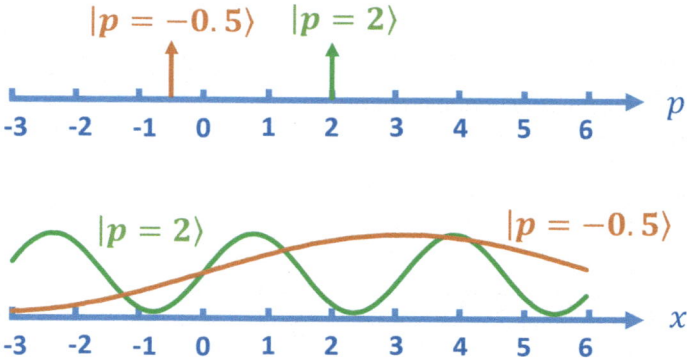

Fig. 14.2 Illustration of momentum eigenvectors and their representation in the momentum (top) and position (bottom) spaces

$\hat{p}$ is called the **momentum operator**. It does not rotate a momentum vector but scales it with the value of the momentum vector. Writing in an equation, we have:

$$\hat{p}|p\rangle = p|p\rangle, \qquad (14.10)$$

which also means that the momentum vector $|p\rangle$ is the **eigenvector** of $\hat{p}$ with an **eigenvalue** of p. Since momentum is an observable, we expect $\hat{p}$ to be **Hermitian** and,

$$\hat{p}^\dagger = \hat{p}. \qquad (14.11)$$

Similar to position vectors, momentum vectors also form an orthonormal basis:

$$\langle p|p'\rangle = \delta(p - p'), \qquad (14.12)$$

with the following completeness property

$$\int_{-\infty}^{+\infty} |p\rangle\langle p|\, dx = \mathbf{I}. \qquad (14.13)$$

14.2.4 Commutation Relation of $\hat{x}$ and $\hat{p}$

The reason x and p are promoted to operators is because they will then follow different "rules" to satisfy experimental observations in quantum mechanics such as **quantization** and **uncertainty principle**. The rules they will follow are those of the matrices in linear algebra. One of them is the **noncommutative** relationship. While $xp - px = 0$ before promotion (or in classical mechanics), in quantum mechanics, it has the following relation:

$$[\hat{x}, \hat{p}] = \hat{x}\hat{p} - \hat{p}\hat{x} = i\hbar. \qquad (14.14)$$

This is the very reason causing quantization as we will see later. *Variables of pairs of observables whose operators are not commuting with each other like the case of $\hat{x}$ and $\hat{p}$ are called the **conjugate variables***.

14.3 Annihilation, Creation, and Number Operators

With the introduction of the momentum operator, $\hat{p}$, and position operator, $\hat{x}$, through simple mathematics, we can create other operators that will make quantum mechanics very beautiful and succinct.

14.3.1 Annihilation and Creation Operators

The **annihilation operator** for SHO, $\hat{a}$, is defined as

$$\hat{a} = \sqrt{\frac{m\omega_0}{2\hbar}} \left(\hat{x} + \frac{i\hat{p}}{m\omega_0} \right), \tag{14.15}$$

where m and ω_0 are the mass of the particle and the resonant frequency of the SHO, respectively, as defined earlier. We also define a **creation operator**, $\hat{a}^\dagger$:

$$\hat{a}^\dagger = \sqrt{\frac{m\omega_0}{2\hbar}} \left(\hat{x} - \frac{i\hat{p}}{m\omega_0} \right). \tag{14.16}$$

Is the creation operator the **Hermitian** conjugate of the annihilation operator? Indeed,

$$\begin{aligned}
(\hat{a})^\dagger &= \left(\sqrt{\frac{m\omega_0}{2\hbar}} \left(\hat{x} + \frac{i\hat{p}}{m\omega_0} \right) \right)^\dagger, \\
&= \left(\sqrt{\frac{m\omega_0}{2\hbar}} \left(\hat{x}^\dagger + \frac{(-i)\hat{p}^\dagger}{m\omega_0} \right) \right), \\
&= \sqrt{\frac{m\omega_0}{2\hbar}} \left(\hat{x} - \frac{i\hat{p}}{m\omega_0} \right), \\
&= \hat{a}^\dagger,
\end{aligned} \tag{14.17}$$

where, from line 1 to line 2, we have used the fact that $i^\dagger = -i$ and, from line 2 to line 3, we have used the fact that $\hat{x}$ and $\hat{p}$ are Hermitian (Eqs. (14.5) and (14.11)).

Let us now practice the arithmetic for operators and check the commutation relation of $\hat{a}$ and $\hat{a}^\dagger$.

Example 14.1 Find $[\hat{a}, \hat{a}^\dagger]$.

This is straightforward. The most important thing is to remind ourselves that *operators are not numbers*. Therefore, usually, they do not commute with each other:

$$\begin{aligned}
[\hat{a}, \hat{a}^\dagger] &= \hat{a}\hat{a}^\dagger - \hat{a}^\dagger\hat{a}, \\
&= \frac{m\omega_0}{2\hbar}\left[\left(\hat{x} + \frac{i\hat{p}}{m\omega_0}\right)\left(\hat{x} - \frac{i\hat{p}}{m\omega_0}\right) - \left(\hat{x} - \frac{i\hat{p}}{m\omega_0}\right)\left(\hat{x} + \frac{i\hat{p}}{m\omega_0}\right) \right], \\
&= \frac{m\omega_0}{2\hbar}\Big[\hat{x}\hat{x} + \hat{x}\left(-\frac{i\hat{p}}{m\omega_0}\right) + \frac{i\hat{p}}{m\omega_0}\hat{x} + \left(\frac{i\hat{p}}{m\omega_0}\right)\left(-\frac{i\hat{p}}{m\omega_0}\right) \\
&\quad -\hat{x}\hat{x} - \hat{x}\left(\frac{i\hat{p}}{m\omega_0}\right) - \left(-\frac{i\hat{p}}{m\omega_0}\right)\hat{x} - \left(-\frac{i\hat{p}}{m\omega_0}\right)\left(\frac{i\hat{p}}{m\omega_0}\right) \Big],
\end{aligned}$$

14.3 Annihilation, Creation, and Number Operators

$$= \frac{m\omega_0}{2\hbar}\left[\frac{-i\hat{x}\hat{p}}{m\omega_0} + \frac{i\hat{p}\hat{x}}{m\omega_0} - \frac{i\hat{x}\hat{p}}{m\omega_0} + \frac{i\hat{p}\hat{x}}{m\omega_0}\right],$$

$$= \frac{m\omega_0}{2\hbar}\left[\frac{-i}{m\omega_0}(i\hbar) + \frac{-i}{m\omega_0}(i\hbar)\right],$$

$$= 1, \tag{14.18}$$

where we have used the commutation relation of $\hat{x}$ and $\hat{p}$ from the third last line to the second last line (Eq. (14.14)). It should be emphasized that the order of the terms is carefully kept during expansion as operators cannot be assumed to be commutative. ∎

Now, let us also express $\hat{x}$ and $\hat{p}$ in terms of $\hat{a}$ and $\hat{a}^\dagger$. This can be done by solving Eqs. (14.15) and (14.16). By adding the two equations, we get:

$$\hat{a} + \hat{a}^\dagger = \sqrt{\frac{m\omega_0}{2\hbar}}\left(\hat{x} + \frac{i\hat{p}}{m\omega_0}\right) + \sqrt{\frac{m\omega_0}{2\hbar}}\left(\hat{x} - \frac{i\hat{p}}{m\omega_0}\right),$$

$$\hat{a} + \hat{a}^\dagger = \sqrt{\frac{m\omega_0}{2\hbar}}\left(\hat{x} + \frac{i\hat{p}}{m\omega_0} + \hat{x} - \frac{i\hat{p}}{m\omega_0}\right),$$

$$\hat{x} = \sqrt{\frac{\hbar}{2m\omega_0}}\left(\hat{a} + \hat{a}^\dagger\right),$$

$$\hat{x} = X_{zpf}\left(\hat{a} + \hat{a}^\dagger\right). \tag{14.19}$$

Similarly, by finding the difference between the two equations, we have:

$$\hat{a} - \hat{a}^\dagger = \sqrt{\frac{m\omega_0}{2\hbar}}\left(\hat{x} + \frac{i\hat{p}}{m\omega_0}\right) - \sqrt{\frac{m\omega_0}{2\hbar}}\left(\hat{x} - \frac{i\hat{p}}{m\omega_0}\right),$$

$$\hat{a} - \hat{a}^\dagger = \sqrt{\frac{m\omega_0}{2\hbar}}\left(\hat{x} + \frac{i\hat{p}}{m\omega_0} - \hat{x} + \frac{i\hat{p}}{m\omega_0}\right),$$

$$\hat{p} = i\sqrt{\frac{m\omega_0\hbar}{2}}\left(-\hat{a} + \hat{a}^\dagger\right),$$

$$\hat{p} = iP_{zpf}(-\hat{a} + \hat{a}^\dagger). \tag{14.20}$$

In the derivations, we introduced two variables, X_{zpf} and P_{zpf} which are the **zero-point fluctuation** of the position and momentum of the SHO. We will discuss their meanings later in this chapter.

It is very often that we will work on two-level systems. Then, the creation operator and the annihilation operator can be written as 2×2 matrices.

14.3.2 Number Operator

Now, we will discuss the **number operator** after which we can appreciate the meaning of creation and annihilation operators better. We define number operator, $\hat{N}$, as

$$\hat{N} = \hat{a}^\dagger \hat{a}. \tag{14.21}$$

Let us expand it using Eqs. (14.15) and (14.16) to see if we can find out its meaning:

$$\begin{aligned}
\hat{N} &= \hat{a}^\dagger \hat{a}, \\
&= \left[\sqrt{\frac{m\omega_0}{2\hbar}}\left(\hat{x} - \frac{i\hat{p}}{m\omega_0}\right)\right]\left[\sqrt{\frac{m\omega_0}{2\hbar}}\left(\hat{x} + \frac{i\hat{p}}{m\omega_0}\right)\right], \\
&= \frac{m\omega_0}{2\hbar}\left(\hat{x}^2 - \frac{i\hat{p}\hat{x}}{m\omega_0} + \frac{i\hat{x}\hat{p}}{m\omega_0} - \frac{i^2\hat{p}^2}{m^2\omega_0^2}\right), \\
&= \frac{m\omega_0 \hat{x}^2}{2\hbar} + \frac{\hat{p}^2}{2\hbar m\omega_0} + \frac{m\omega_0}{2\hbar}\frac{i}{m\omega_0}(\hat{x}\hat{p} - \hat{p}\hat{x}), \\
&= \left[\frac{\hat{p}^2}{2\hbar m\omega_0} + \frac{m\omega_0 \hat{x}^2}{2\hbar}\right] + \frac{i}{2\hbar}[\hat{x}, \hat{p}], \\
&= \frac{1}{\hbar\omega_0}\left[\frac{\hat{p}^2}{2m} + \frac{m\omega_0^2 \hat{x}^2}{2}\right] + \frac{i}{2\hbar}(i\hbar), \\
&= \frac{1}{\hbar\omega_0}\hat{\mathcal{H}} - \frac{1}{2},
\end{aligned} \tag{14.22}$$

where we had performed expansions and were careful not to swap operators. We also used Eq. (14.14) in the last three lines. Equation (14.3) was used in the last line. By rearranging the terms, therefore,

$$\hat{\mathcal{H}} = \hbar\omega_0\left(\hat{N} + \frac{1}{2}\right). \tag{14.23}$$

Recalling that Hamiltonian is the total energy of a system (Eq. (13.35)), it means that the total energy can be regarded as the sum of some quanta of energy $\hbar\omega_0$. At this stage, you probably understand why $\hat{N}$ is called the number operator because it is related to the number of energy quanta in SHO. However, we need to remind ourselves again that this is an operator, not a real number. But we can boldly assume that the SHO can be at a definite state with a certain number, N, of $\hbar\omega_0$. This is just like a particle at a state of definite position, $|x\rangle$, or momentum, $|p\rangle$ (Figs. 14.1 and 14.2). Here, the whole SHO system is at the state of a definite number, $|N\rangle$. Let us

14.3 Annihilation, Creation, and Number Operators

assume $|N\rangle$ to be the eigenstate of $\hat{N}$ with eigenvalue N, then,

$$\hat{N}|N\rangle = N|N\rangle. \tag{14.24}$$

Now, apply $\hat{\mathcal{H}}$ to $|N\rangle$:

$$\hat{\mathcal{H}}|N\rangle = \hbar\omega_0(\hat{N} + \frac{1}{2})|N\rangle,$$

$$= \hbar\omega_0(\hat{N}|N\rangle + \frac{1}{2}|N\rangle),$$

$$= \hbar\omega_0(N|N\rangle + \frac{1}{2}|N\rangle),$$

$$= \hbar\omega_0(N + \frac{1}{2})|N\rangle, \tag{14.25}$$

from which we have shown that $|N\rangle$ is also an eigenvector of $\hat{\mathcal{H}}$ but with eigenvalue $\hbar\omega_0(N + \frac{1}{2})$. Therefore, $|N\rangle$ are called the **energy eigenstates**. We can define:

$$E_N = \hbar\omega_0\left(N + \frac{1}{2}\right). \tag{14.26}$$

Then,

$$\hat{\mathcal{H}}|N\rangle = E_N|N\rangle. \tag{14.27}$$

Therefore, our assumption makes sense. $\hat{N}$ indeed is the operator with eigenvector $|N\rangle$. This is because, based on this assumption, we deduced that by applying the Hamiltonian to $|N\rangle$, it would return the total energy of the SHO as $\hbar\omega_0(N + \frac{1}{2})$. Of course, this is not exactly what we expected as we expect that with N quanta, the total energy should be $N\hbar\omega_0$. But now, when $N = 0$, the SHO still has an energy of $E_0 = \frac{\hbar\omega_0}{2}$. It actually tells us about something we do not get in classical mechanics. This means that there is a **zero-point energy** or **vacuum energy** even when there are no energy quanta in the SHO. This is also a result of the noncommutative relation of $\hat{x}$ and $\hat{p}$ or, equivalently, the uncertainty principle.

It should also be noted that $|N\rangle$ forms an infinite discrete orthonormal and complete basis (see Eqs. (14.7) and (14.9)). Therefore,

$$\langle N|N'\rangle = \delta_{N,N'}, \tag{14.28}$$

and

$$\sum |N\rangle\langle N| = \boldsymbol{I}. \tag{14.29}$$

14.3.3 Meaning of the Creation and Annihilation Operators

After introducing the number operator, you might have guessed the meaning of the creation and annihilation operators. The annihilation operator probably annihilates (destroys) a quantum of energy in the SHO system, and the creation operator probably creates a quantum of energy in the SHO system. If our guesses are correct, then we have the following:

$$\hat{a} |N\rangle = d_- |N - 1\rangle ,$$
$$\hat{a}^\dagger |N\rangle = d_+ |N + 1\rangle , \qquad (14.30)$$

where d_- and d_+ are numbers. They have not been determined yet. These two equations only tell us that after applying $\hat{a}$ ($\hat{a}^\dagger$) to $|N\rangle$ which has N quanta of energy, the state becomes $|N - 1\rangle$ ($|N + 1\rangle$) which has $N - 1$ ($N + 1$) quanta of energy. However, we still do not know its prefactor, d_- (d_+), yet.

Now let us prove what we guess is correct and find d_- and d_+. Firstly, let us find $\left[\hat{N}, \hat{a}\right]$.

Example 14.2 Find $\left[\hat{N}, \hat{a}\right]$:

$$\left[\hat{N}, \hat{a}\right] = \hat{N}\hat{a} - \hat{a}\hat{N},$$
$$= \hat{a}^\dagger \hat{a} \hat{a} - \hat{a} \hat{a}^\dagger \hat{a}, \qquad (14.31)$$

where we have used the definition of commutator bracket in line 1 and Eq. (14.21) in line 2. Since $\hat{a}\hat{a}^\dagger - \hat{a}^\dagger \hat{a} = 1$ (Eq. (14.18)), we have $\hat{a}\hat{a}^\dagger = 1 + \hat{a}^\dagger \hat{a}$. Substituting it into Eq. (14.31)

$$\left[\hat{N}, \hat{a}\right] = \hat{a}^\dagger \hat{a} \hat{a} - (1 + \hat{a}^\dagger \hat{a})\hat{a}$$
$$= -\hat{a}. \qquad (14.32)$$

■

Therefore,

$$\hat{N}\hat{a} - \hat{a}\hat{N} = -\hat{a},$$
$$\hat{N}\hat{a} = \hat{a}\hat{N} - \hat{a}. \qquad (14.33)$$

To prove that $\hat{a}$ annihilates one quantum of energy, we can apply the number operator after $\hat{a}$ has been applied to $|N\rangle$:

$$\hat{N}(\hat{a} |N\rangle) = (\hat{N}\hat{a}) |N\rangle ,$$

14.3 Annihilation, Creation, and Number Operators

$$= (\hat{a}\hat{N} - \hat{a})|N\rangle,$$
$$= \hat{a}\hat{N}|N\rangle - \hat{a}|N\rangle,$$
$$= \hat{a}N|N\rangle - \hat{a}|N\rangle,$$
$$= \hat{a}(N-1)|N\rangle,$$
$$= (N-1)(\hat{a}|N\rangle), \qquad (14.34)$$

where in line 1, we have used the associative property. In line 2, we have used Eq. (14.33). In line 4, we used the fact that $|N\rangle$ is an eigenvector of $\hat{N}$ with eigenvalue N (Eq. (14.24)). This shows that $\hat{a}|N\rangle$ has $N-1$ quanta of energy. Therefore, $\hat{a}$ destroys one quantum of energy as we guessed in Eq. (14.30).

Now, we will find the value of d_- in Eq. (14.30).

Example 14.3 Use the fact that $|N\rangle$ forms an orthonormal basis (Eq. (14.28)) to find d_-.

Since we will use the orthonormal property of $|N\rangle$, we can try to explore its inner product after $\hat{a}$ is applied. After $\hat{a}$ is applied to $|N\rangle$, the state becomes $\hat{a}|N\rangle$. To perform an inner product, we need to find its bra version which is $\langle N|\hat{a}^\dagger$. Therefore, its inner product is $(\langle N|\hat{a}^\dagger)(\hat{a}|N\rangle)$ and,

$$\langle N|\hat{a}^\dagger a|N\rangle = \langle N-1|d_-^* d_- |N-1\rangle,$$
$$\langle N|\hat{N}|N\rangle = \langle N-1||d_-|^2|N-1\rangle,$$
$$\langle N|N|N\rangle = |d_-|^2 \langle N-1|N-1\rangle,$$
$$N\langle N|N\rangle = |d_-|^2 \langle N-1|N-1\rangle,$$
$$N = |d_-|^2,$$
$$d_- = \sqrt{N}, \qquad (14.35)$$

where in line 1, we used Eq. (14.30) and the fact that the bra version of $d_-|N-1\rangle$ is $\langle N-1|d_-^*$. In line 2, we used the definition of $\hat{N} = \hat{a}^\dagger a$ on the left-hand side. From line 4 to line 5, we recognized that $|N\rangle$ was normalized and therefore, $\langle N|N\rangle = \langle N-1|N-1\rangle = 1$. Finally, we took d_- as real and got $d_- = \sqrt{N}$. ∎

Similarly, we can prove that $d_+ = \sqrt{N+1}$ (see Problems). As a result, Eq. (14.30) becomes:

$$\hat{a}|N\rangle = \sqrt{N}|N-1\rangle, \qquad (14.36)$$
$$\hat{a}^\dagger|N\rangle = \sqrt{N+1}|N+1\rangle. \qquad (14.37)$$

14.3.4 Values of N

We had been assuming that N is an integer in our previous discussion and claimed that N represents the number of quanta of energy an SHO has. However, we have only shown that this assumption makes sense (e.g., the total energy is N times $\hbar\omega_0$). We have not proved it yet. Can it be fractional or even irrational? The answer is no. It can only be a natural number (0 or positive integer) which fits the meaning of "number of quanta." To prove this, let us consider $\langle N| \hat{N} |N\rangle$:

$$\langle N| \hat{N} |N\rangle = \langle N| N |N\rangle ,$$
$$= N \langle N|N\rangle ,$$
$$= N, \tag{14.38}$$

where we use the fact that $|N\rangle$ is normalized (Eq. (14.28)). But $\hat{N} = \hat{a}^\dagger \hat{a}$. Therefore,

$$\langle N| \hat{a}^\dagger \hat{a} |N\rangle = N \tag{14.39}$$

However, $\langle N| \hat{a}^\dagger \hat{a} |N\rangle \geq 0$ because it is just the *norm* of $\hat{a}|N\rangle$. Therefore, $N \geq 0$.

Since the annihilation operator will remove a quantum of energy from the SHO, N cannot be non-integer. For example, $\hat{a}|N=0.5\rangle = \sqrt{0.5}|N=-0.5\rangle$ and this violates the requirement that $N \geq 0$. How about applying $\hat{a}$ to $|0\rangle$? Will we get $N = -1$? Since $\hat{a}|N=0\rangle = \sqrt{0}|-1\rangle = 0$, we see that it terminates at 0 and will not go to negative numbers. Therefore, N can only take non-negative integers.

Example 14.4 Show that $a^\dagger = \begin{pmatrix} 0 & 0 \\ 1 & 0 \end{pmatrix}$ in the eigenenergy basis.

The eigenenergy basis is the same as the number basis as demonstrated in Eq. (14.27). Since a creation operator is to increase N by 1, therefore, it should change $|N=0\rangle$ to $|N=1\rangle$ and $|N=1\rangle$ to nothing (as it is out of the two-level space):

$$a^\dagger |0\rangle = \begin{pmatrix} 0 & 0 \\ 1 & 0 \end{pmatrix} \begin{pmatrix} 1 \\ 0 \end{pmatrix} ,$$
$$= \begin{pmatrix} 0 \\ 1 \end{pmatrix} = |1\rangle . \tag{14.40}$$

$$a^\dagger |1\rangle = \begin{pmatrix} 0 & 0 \\ 1 & 0 \end{pmatrix} \begin{pmatrix} 0 \\ 1 \end{pmatrix} ,$$
$$= 0. \tag{14.41}$$

Therefore, we have shown $\hat{a}^\dagger = \begin{pmatrix} 0 & 0 \\ 1 & 0 \end{pmatrix}$. Similarly, $a = \begin{pmatrix} 0 & 1 \\ 0 & 0 \end{pmatrix}$. ∎

14.4 Energy Eigenstates, Coherent State, and Linkage to Classical SHO

14.4.1 Energy Eigenstates and Wavefunction

By promoting x and p to operators and following the noncommutative relation of $\hat{x}$ and $\hat{p}$ (Eq. (14.14)), we have gone through an interesting journey to show that energy eigenstates, $|N\rangle$, take only non-negative integers and can be regarded as having N quanta of energy of $\hbar\omega_0$. We also show that the annihilation operator, $\hat{a}$, and creation operator, $\hat{a}^\dagger$, reduces and increases the number of quanta by one. We can relate $|N\rangle$ to $|0\rangle$ through $\hat{a}^\dagger$. For natural number n, since $\hat{a}^\dagger |n-1\rangle = \sqrt{n}\,|n\rangle$, therefore, $|n\rangle = \frac{\hat{a}^\dagger |n-1\rangle}{\sqrt{n}}$ (see Eq. (14.37) with $N = n - 1$). So,

$$\begin{aligned}
|N\rangle &= \frac{\hat{a}^\dagger}{\sqrt{N}} |N-1\rangle, \\
&= \frac{\hat{a}^\dagger}{\sqrt{N}} \frac{\hat{a}^\dagger}{\sqrt{N-1}} |N-2\rangle, \\
&= \frac{\hat{a}^\dagger}{\sqrt{N}} \frac{\hat{a}^\dagger}{\sqrt{N-1}} \frac{\hat{a}^\dagger}{\sqrt{N-2}} |N-3\rangle, \\
&= \frac{\hat{a}^\dagger}{\sqrt{N}} \frac{\hat{a}^\dagger}{\sqrt{N-1}} \frac{\hat{a}^\dagger}{\sqrt{N-2}} \cdots \frac{\hat{a}^\dagger}{\sqrt{1}} |0\rangle, \\
&= \frac{(\hat{a}^\dagger)^N}{\sqrt{N!}} |0\rangle.
\end{aligned} \qquad (14.42)$$

How do the energy eigenstates, $|N\rangle$, look like in real space? When we say "real space," we mean its wavefunction in our real space. We are usually familiar with the concept of wavefunction represented as $\Psi(x)$, and we know that $|\Psi(x)|^2$ is the probability density of finding a particle at location x. This is the Schrödinger wave mechanics. To do so, we will **project** $|N\rangle$ on to the position eigenvector, $|x\rangle$ (Sect. 14.2.2) by performing $\langle x|N\rangle$. That is

$$\Psi_N(x) = \langle x|N\rangle. \qquad (14.43)$$

This makes sense because the projection is just to find out how much overlap there is between $|x\rangle$ and $|N\rangle$. The square of the magnitude of the projection represents the **probability density** of finding $|N\rangle$ at position x and is consistent with the meaning of **wavefunction**. That is,

$$|\Psi_N(x)|^2 = |\langle x|N\rangle|^2. \tag{14.44}$$

Readers can also review Sect. 3.3.3 and realize that the wavefunction at x is the coefficient after applying the project operator $P_{|x\rangle}$ to $|N\rangle$. Also, note that $\Psi_N(x)$ is a number (can be imaginary) as a function of x. The *projection* process can help represent the energy eigenstate as a linear combination of the position eigenstates. That is,

$$\begin{aligned} |N\rangle &= I\,|N\rangle, \\ &= \left(\int_{-\infty}^{+\infty} |x\rangle\langle x|\,dx\right)|N\rangle, \\ &= \int_{-\infty}^{+\infty} |x\rangle\langle x|N\rangle\,dx, \\ &= \int_{-\infty}^{+\infty} |x\rangle\,\Psi_N(x)dx, \\ &= \int_{-\infty}^{+\infty} \Psi_N(x)\,|x\rangle\,dx, \end{aligned} \tag{14.45}$$

where we have used the completeness property of position eigenvector (Eq. (14.8)) from line 1 to line 2 and Eq. (14.43) from line 3 to line 4. In case it is not clear, although Eq. (14.45) is an integral, it is still a sum of $|x\rangle$ with $\Psi_N(x)$ as its coefficients. Projection to $|x\rangle$ is special just because we also live in the real space of x. We can also project $|N\rangle$ to other spaces such as the momentum space in Sect. 14.2.3. Naturally, we can project to its own space $|N\rangle$ (which means it is still itself) and call it the wavefunction in the energy eigenspace. That is why we sometimes also call a state a wavefunction.

Let us study the wavefunction of the energy eigenstate in the real space, $\Psi_N(x)$. We will derive $\Psi_0(x)$. Since, $\hat{a}\,|0\rangle = 0$ (Eq. (14.36)),

$$\langle x|\,\hat{a}\,|0\rangle = 0,$$

$$\langle x|\sqrt{\frac{m\omega_0}{2\hbar}}\left(\hat{x} + \frac{i\hat{p}}{m\omega_0}\right)|0\rangle = 0,$$

$$\langle x|\,\hat{x} + \frac{i\hat{p}}{m\omega_0}\,|0\rangle = 0,$$

$$\langle x|\,\hat{x}\,|0\rangle + \langle x|\,\frac{i\hat{p}}{m\omega_0}\,|0\rangle = 0, \tag{14.46}$$

where we used Eq. (14.15) in the second line. For the first term, it is just $\langle x|\,\hat{x}\,|0\rangle = \langle x|\,x\,|0\rangle = x\,\langle x|0\rangle = x\Psi_0(x)$ because $|x\rangle$ is the eigenstate of $\hat{x}$ with eigenvalue x. For the second term, we need to convert $\hat{p}$ to a differential operator. We will not

14.4 Energy Eigenstates, Coherent State, and Linkage to Classical SHO

prove it and will just copy Eq. (1.7.17) in [2]:

$$\langle x'| \hat{p} |\alpha\rangle = -i\hbar \frac{\partial}{\partial x'} \langle x'|\alpha\rangle, \tag{14.47}$$

where $|x'\rangle$ is a position vector and $|\alpha\rangle$ is an arbitrary state. Therefore,

$$\langle x| \frac{i\hat{p}}{m\omega_0} |0\rangle = \frac{i}{m\omega_0} \langle x| \hat{p} |0\rangle,$$

$$= \frac{i}{m\omega_0} \left(-i\hbar \frac{\partial}{\partial x} \langle x|0\rangle\right),$$

$$= \frac{\hbar}{m\omega_0} \frac{\partial}{\partial x} \Psi_0(x). \tag{14.48}$$

Therefore, Eq. (14.46) becomes:

$$x\Psi_0(x) + \frac{\hbar}{m\omega_0} \frac{\partial}{\partial x} \Psi_0(x) = 0,$$

$$\left(x + \frac{\hbar}{m\omega_0} \frac{\partial}{\partial x}\right) \Psi_0(x) = 0, \tag{14.49}$$

which is a differential equation and the solution is

$$\Psi_0(x) = \langle x|0\rangle = \frac{1}{\pi^{\frac{1}{4}} \sqrt{x_0}} \exp\left[-\frac{1}{2}\left(\frac{x}{x_0}\right)^2\right], \tag{14.50}$$

with $x_0 = \sqrt{\frac{\hbar}{m\omega_0}}$.

In Fig. 14.3, the wavefunctions of the first three energy eigenstates are plotted. This plot is very informative but can be confusing. Firstly, it plots the energy as a function of position, x. The parabolic curve represents the potential energy of the SHO for a given stretch/compression of the spring, x. It is parabolic because the potential energy is $V = \frac{1}{2}kx^2 = \frac{\omega_0^2 m}{2}x^2$ (Eqs. (14.1) and (14.2)). This is purely classical. For a given total energy, classically, the spring will not be stretched or compressed more than the boundary set by this potential energy. Beyond the boundary is the forbidden region. It also plots the eigenenergies (total energies) of the Hamiltonian. These are the energies the SHO can attain with 0, 1, and 2 quanta of $\hbar\omega_0$ for states $|0\rangle$, $|1\rangle$, and $|2\rangle$, respectively. The corresponding wavefunctions in the real space are also plotted ($\Psi_0(x) = \langle x|0\rangle$, $\Psi_1(x) = \langle x|1\rangle$, and $\Psi_2(x) = \langle x|2\rangle$). Of course, they don't share the same y-axis as the energies. They are plotted so that their zero values align with the corresponding energy eigenvalues and plotted in a way to show that wavefunctions are penetrating the forbidden region due to **quantum mechanical tunneling**.

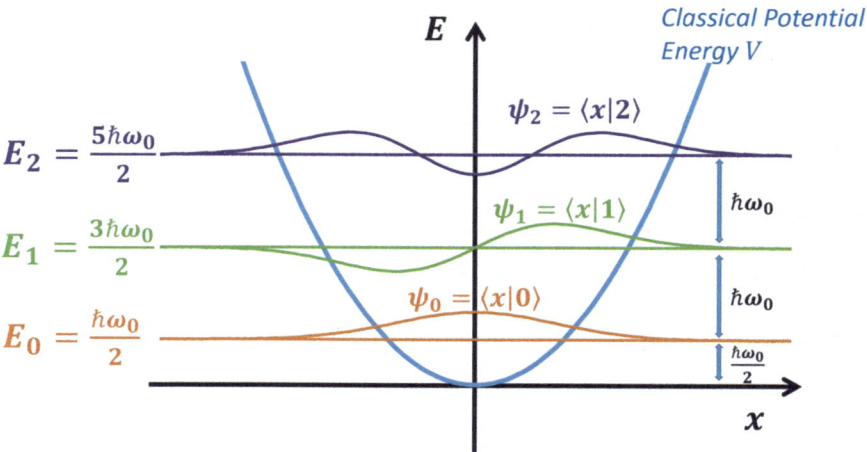

Fig. 14.3 Illustration of SHO energy landscape. The classical parabolic potential energy is plotted as a function of x (the stretch and compression of the spring). The energy eigenvalues and the real space energy eigen-wavefunctions are shown. Note that the wavefunctions do not share the y-axis with the energy functions. Moreover, the wavefunctions are shifted vertically for illustration

In Eqs. (14.19) and (14.20), we introduced the zero-point fluctuation of position and momentum. They refer to the standard deviation of the position and momentum wavefunctions of $|N=0\rangle$, respectively. For example, X_{zpf} is the standard deviation of wavefunction $\Psi_0(x)$ shown in Fig. 14.3. Similarly, if we had projected $|N=0\rangle$ to the momentum space, P_{zpf} is the standard deviation of $\langle p|0\rangle$.

14.4.2 Coherent States

The energy eigenstates are quantum mechanical states. Their wavefunctions do not change (up to a phase) as a function of time if we try to solve with time-dependent Schrödinger equation. The phase is $\exp(-i\frac{E_N t}{\hbar})$. This does not resemble the classical view we have about an SHO. In a classical SHO, we expect the particle to have the maximum speed at $x = 0$. It then moves to the right and slows down when it converts its kinetic energy to potential energy. It has zero speed when it reaches the maximum potential energy given by the parabola (Fig. 14.3), and it will move back to the left and convert its potential energy to kinetic energy. This process will be repeated resulting in oscillation in the real space. This is obviously not the case for the energy eigenstate as the maximum probability density of finding the particle is independent of time.

14.4 Energy Eigenstates, Coherent State, and Linkage to Classical SHO

The state that resembles the classical behavior is the **coherent state**, $|\alpha\rangle$. It is the eigenstate of the annihilation operator. That is,

$$\hat{a}|\alpha\rangle = \alpha|\alpha\rangle, \tag{14.51}$$

where α is its eigenvalue. Note that, unlike $|x\rangle$, $|p\rangle$, and $|N\rangle$, $|\alpha\rangle$ is not orthonormal and it is also **over complete**, which means that it has more basis vectors than required to form a complete space. $|\alpha\rangle$ is given by

$$|\alpha\rangle = e^{-\frac{1}{2}|\alpha|^2} \sum_{N=0}^{\infty} \frac{\alpha^N}{\sqrt{N!}} |N\rangle. \tag{14.52}$$

Example 14.5 Prove Eq. (14.52).
This is equivalent to showing that it satisfies Eq. (14.51):

$$\begin{aligned}
\hat{a}|\alpha\rangle &= \hat{a}\left(e^{-\frac{1}{2}|\alpha|^2} \sum_{N=0}^{\infty} \frac{\alpha^N}{\sqrt{N!}} |N\rangle\right), \\
&= e^{-\frac{1}{2}|\alpha|^2} \sum_{N=0}^{\infty} \frac{\alpha^N}{\sqrt{N!}} \hat{a}|N\rangle, \\
&= e^{-\frac{1}{2}|\alpha|^2} \sum_{N=1}^{\infty} \frac{\alpha^N}{\sqrt{N!}} \sqrt{N}|N-1\rangle, \\
&= e^{-\frac{1}{2}|\alpha|^2} \sum_{N=0}^{\infty} \frac{\alpha^{N+1}}{\sqrt{(N+1)!}} \sqrt{N+1}|N\rangle, \\
&= \alpha\left(e^{-\frac{1}{2}|\alpha|^2} \sum_{N=0}^{\infty} \frac{\alpha^N}{\sqrt{(N)!}} |N\rangle\right), \\
&= \alpha|\alpha\rangle, \tag{14.53}
\end{aligned}$$

where from line 2 to line 3, we used the fact that $\hat{a}|0\rangle = 0$ to eliminate the $N = 0$ term in the sum. In line 4, since it is summed to infinity, we started the sum from $N = 0$ again by substituting N by $N + 1$. Then, we performed simplification in line 5 and factorized α out. ■

The coherent state is a linear combination of the energy eigenstates. Each of the energy eigenstates will have a phase of $\exp(-i\frac{E_N t}{\hbar})$ to satisfy the time-dependent Schrödinger equation. Therefore,

$$|\alpha(t)\rangle = e^{-\frac{1}{2}|\alpha(0)|^2} \sum_{N=0}^{\infty} \frac{\alpha(0)^N}{\sqrt{N!}} |N\rangle e^{-i\frac{E_N t}{\hbar}},$$

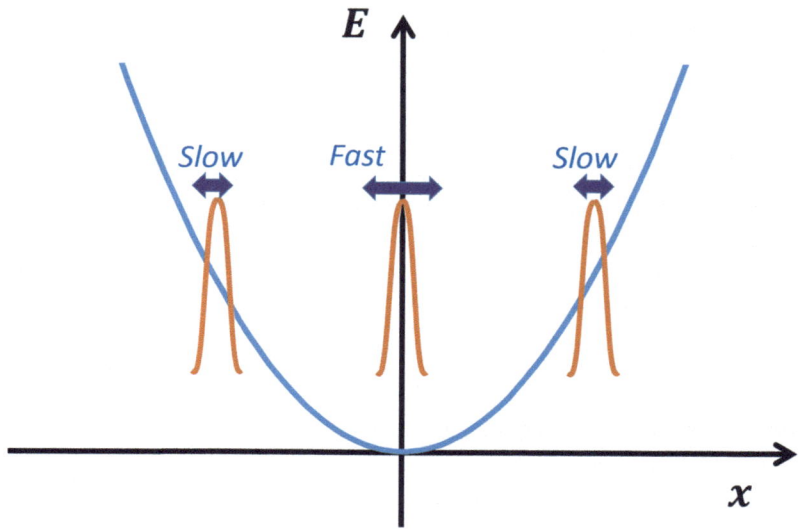

Fig. 14.4 Illustration of a coherent state oscillating in real space

$$= e^{-\frac{1}{2}|\alpha(0)|^2} \sum_{N=0}^{\infty} \frac{\alpha(0)^N}{\sqrt{N!}} |N\rangle \, e^{-i\omega_0(N+\frac{1}{2})t}. \qquad (14.54)$$

In real space, its wavefunction is thus

$$\Psi_\alpha(x) = \langle x|\alpha\rangle,$$

$$= e^{-\frac{1}{2}|\alpha(0)|^2} \sum_{N=0}^{\infty} \frac{\alpha(0)^N}{\sqrt{N!}} \langle x|N\rangle \, e^{-i\omega_0(N+\frac{1}{2})t},$$

$$= e^{-\frac{1}{2}|\alpha(0)|^2} \sum_{N=0}^{\infty} \frac{\alpha(0)^N}{\sqrt{N!}} \Psi_N(x) e^{-i\omega_0(N+\frac{1}{2})t}. \qquad (14.55)$$

Figure 14.4 shows that the wavefunction of a coherent state is a narrow wave packet and oscillates as a function of time from left to right like a classical particle.

14.5 Summary

In this chapter, we learn to promote a variable to an operator. By promoting x and p to $\hat{x}$ and $\hat{p}$, respectively, we naturally obtain results consistent with quantum mechanical experiments. Operators follow the arithmetic of matrices and usually do not commute with each other. By following the rules carefully, we define the annihilation operator, creation operator, and number operator. They have very intuitive meanings in an SHO. We obtain the energy eigenstates which are also the eigenstates of the number operator. It is found that the energy eigenstates represent the number of quanta of energy in the SHO and it still has finite vacuum energy even when $N = 0$. This is a result of the non-commutation relation of $\hat{x}$ and $\hat{p}$. While the energy eigenstates are quantum mechanical states, their linear combination with appropriate coefficients can be used to create coherence states that resemble a classical SHO. These form the basics of the following chapters when we quantize an electrical circuit.

Problems

14.1 Effect of the Creation Operator, $\hat{a}^\dagger$
Prove Eq. (14.37) by following the examples in Sect. 14.3.3 where Eq. (14.36) is proved.

14.2 Solution of SHO
By substituting Eq. (14.50) into Eq. (14.49), show that Eq. (14.50) is a solution of Eq. (14.49).

14.3 Time-Dependent Schrödinger Equation
The time-dependent Schrödinger equation is given in Eq. (4.1). Substitute Eq. (14.55) into it to show that $|\alpha\rangle$ is indeed a solution.

14.4 Matrix of Annihilation Operator
Following the approach in Example 14.4, show that $a = \begin{pmatrix} 0 & 1 \\ 0 & 0 \end{pmatrix}$.

References

1. David J. Griffiths and Darrell F. Schroeter. *Modern Quantum Mechanics*. Cambridge University Press, 2018.
2. J.J. Sakurai. *Modern Quantum Mechanics*. Addison-Wesley, 1993.

Chapter 15
Quantization of an *LC* Tank: A Bad Qubit

15.1 Introduction

We had a fairly detailed study of a mechanical simple harmonic oscillator in the previous chapter. The most important thing we have learned was to promote a pair of conjugate variables to operators ($\hat{x}$ and $\hat{p}$ in the mechanical SHO). They have a special commutation relationship resulting in quantization as we saw. The same idea is not limited to the mechanical SHO. In other problems, they have their own conjugate variables. For example, in an electrical circuit, *charge* and *flux* are one of the conjugate variable pairs. In this chapter, we will apply the procedure in the previous chapter to a simple circuit, the *LC* tank, and quantize it. This forms the foundation of a superconducting qubit. We will also appreciate why a dilution refrigerator is necessary to operate a superconducting qubit. Some of the materials in this chapter are based on [1].

15.1.1 Learning Outcomes

Appreciate the similarity between a mechanical and electrical simple harmonic oscillator; be able to quantize a simple electrical circuit; be able to calculate the energy separation in a quantized electrical circuit and compare it to thermal energy.

15.1.2 Teaching Videos

- Search for Ch15 in this playlist
    - https://tinyurl.com/3yhze3jn

- Other videos
  - https://youtu.be/Yh1n7ZBhzQM
  - https://youtu.be/L74mOLWtEMg

15.2 Classical LC Tank

An LC tank is a simple oscillating circuit (Fig. 15.1). It has a capacitor of capacitance C and an inductor of inductance L in parallel. Let us first study its classical behavior.

The directions of the voltages and currents are labeled in the figure. Using **Kirchhoff's voltage law** and **current law**, we have:

$$V_L = -V_C = V, \tag{15.1}$$

$$I_L = I_C = I. \tag{15.2}$$

Our goal is to understand how each of the variables behaves as a function of time. We will start with V, and we need to use physics to relate voltages to currents by using the **constitutive relations** of the capacitor and the inductor:

$$I_C = C\frac{dV_C}{dt}, \tag{15.3}$$

$$V_L = L\frac{dI_L}{dt}. \tag{15.4}$$

As a remark, these constitutive relations are consistent with the voltage and current directions defined in Fig. 15.1. Therefore, starting from Eq. (15.1), we have:

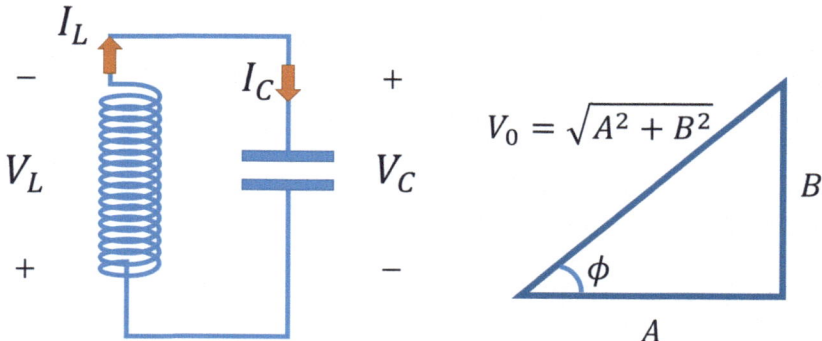

Fig. 15.1 Left: An LC tank circuit. Right: Illustration of the initial phase of the circuit (Example 15.1)

15.2 Classical LC Tank

$$V_L + V_C = 0,$$

$$L\frac{dI_L}{dt} + V_C = 0,$$

$$L\frac{dI_C}{dt} + V_C = 0,$$

$$L\frac{d(C\frac{dV_C}{dt})}{dt} + V_C = 0,$$

$$L\frac{d(C\frac{dV}{dt})}{dt} + V = 0,$$

$$LC\frac{d^2V}{dt^2} + V = 0,$$

$$\frac{d^2V}{dt^2} + \frac{1}{LC}V = 0,$$

$$\frac{d^2V}{dt^2} + \omega_0^2 V = 0, \tag{15.5}$$

where we substituted Eq. (15.4) in line 2. In line 3, we used Eq. (15.2). In line 4, we used Eq. (15.3). Finally, we used $V = V_C$ and defined $\omega_0 = \frac{1}{\sqrt{LC}}$, which is the **resonant frequency**. The ω_0 here plays the same role as the ω_0 in the mechanical SHO in Eq. (13.19).

Example 15.1 Find the solution for Eq. (15.5).

This is a homogeneous second-order differential equation with complex roots (actually purely imaginary roots). Therefore, its general solution is [2]

$$V(t) = A\sin(\omega_0 t) + B\cos(\omega_0 t). \tag{15.6}$$

Note that A and B are real because $V(t)$ is real. We set $V_0 = \sqrt{A^2 + B^2}$ and $\phi = \arctan\frac{B}{A}$. Then $A = V_0 \cos(\phi)$ and $B = V_0 \sin(\phi)$ (Fig. 15.1). Therefore,

$$V(t) = V_0 \cos(\phi)\sin(\omega_0 t) + V_0 \sin(\phi)\cos(\omega_0 t),$$
$$= V_0 \sin(\omega_0 t + \phi). \tag{15.7}$$

When $t = 0$, $V(t) = V_0 \sin(\phi)$. Therefore, ϕ is the initial phase.

We can then find I as a function of t by using Eq. (15.3).

$$I(t) = I_C,$$
$$= C\frac{dV_C}{dt},$$
$$= -C\frac{dV}{dt},$$

$$= -C\frac{d(V_0 \sin(\omega_0 t + \phi))}{dt},$$
$$= -V_0 C \omega_0 \cos(\omega_0 t + \phi),$$
$$= -I_0 \cos(\omega_0 t + \phi), \tag{15.8}$$

where we defined $I_0 = V_0 C \omega_0$. It can also be derived that

$$Z_0 = \frac{V_0}{I_0} = \sqrt{\frac{L}{C}}, \tag{15.9}$$

where Z_0 is the **charateristic impedance**. We can see that the voltage and current of the capacitor and inductor oscillate sinusoidally. ∎

15.2.1 Energy in LC Tank

But what we care about the most is the energy stored in the inductor and the capacitor. The energy stored in the inductor, E_L, is given by

$$E_L(t) = \frac{1}{2} L I_L^2,$$
$$= \frac{1}{2} L (-I_0 \cos(\omega_0 t + \phi))^2,$$
$$= \frac{1}{2} L I_0^2 \cos^2(\omega_0 t + \phi). \tag{15.10}$$

Similarly, the energy stored in the capacitor, E_C, is

$$E_C(t) = \frac{1}{2} C V_C^2,$$
$$= \frac{1}{2} C (-V_0 \sin(\omega_0 t + \phi))^2,$$
$$= \frac{1}{2} C V_0^2 \sin^2(\omega_0 t + \phi). \tag{15.11}$$

Figure 15.2 plots $E_L(t)$ and $E_C(t)$ as a function of time. It can be seen that *energy is transferred back and forth between the inductor and the capacitor* and the total energy of the system, $E_L(t) + E_C(t)$, is constant. We may just call it the **Hamiltonian** of the circuit. And it is easy to prove that it is constant:

$$\mathcal{H} = E_C(t) + E_L(t),$$

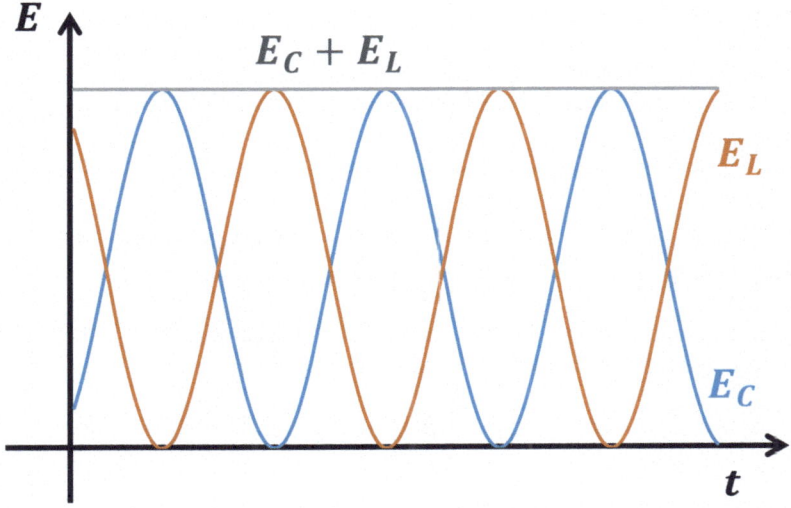

Fig. 15.2 Energy stored in the capacitor and the inductor as a function of time in an LC tank

$$\begin{aligned}
&= \frac{1}{2}CV_0^2 \sin^2(\omega_0 t + \phi) + \frac{1}{2}LI_0^2 \cos^2(\omega_0 t + \phi), \\
&= \frac{1}{2}CV_0^2 \left(\sin^2(\omega_0 t + \phi) + \frac{LI_0^2}{CV_0^2} \cos^2(\omega_0 t + \phi) \right), \\
&= \frac{1}{2}CV_0^2 \left(\sin^2(\omega_0 t + \phi) + \cos^2(\omega_0 t + \phi) \right), \\
&= \frac{1}{2}CV_0^2 = \frac{1}{2}LI_0^2,
\end{aligned} \quad (15.12)$$

which is just the maximum energy the capacitor (or inductor) stores. In lines 4 and 5, we have used Eq. (15.9).

This behavior resembles what we learned in a mechanical simple harmonic oscillator where the energy is transferred back and forth between the potential energy and the kinetic energy. We are naturally curious if we can repeat what we have done for a mechanical SHO in Chaps. 13 and 14. It *seems* to us that I_L and V_C play the roles of x and p. However, this is *not* true. The **conjugate variables** turn out to be the **magnetic flux**, Φ, stored in the inductor and the **electric charge**, Q, stored in the capacitor [1], and we take this for granted in this book. We can use the following equation to further transform Eq. (15.12):

$$\Phi = LI_L, \quad (15.13)$$

$$Q = CV_C. \quad (15.14)$$

Therefore,

$$\mathcal{H} = \frac{1}{2}\frac{Q_0^2}{C}\sin^2(\omega_0 t + \phi) + \frac{1}{2}\frac{\Phi_0^2}{L}\cos^2(\omega_0 t + \phi),$$

$$= \frac{1}{2}\frac{Q(t)^2}{C} + \frac{1}{2}\frac{\Phi(t)^2}{L}, \quad (15.15)$$

where we defined $\Phi_0 = LI_0$ and $Q_0 = CV_0$. Although we have deliberately indicated that $Q(t)$ and $\Phi(t)$ are time-dependent, $\mathcal{H}$ is a constant as shown in Eq. (15.12).

We now compare this to the Hamiltonian of a mechanical SHO in Eq. (13.35), which is copied here for convenience (with $p_1 = p$ and $q_1 = q$):

$$\mathcal{H} = \frac{1}{2}\frac{p^2}{m} + \frac{1}{2}kq^2. \quad (15.16)$$

We have the following correspondence:

$$Q \iff p, \quad (15.17)$$
$$\Phi \iff q \text{ or } x, \quad (15.18)$$
$$C \iff m, \quad (15.19)$$
$$L \iff \frac{1}{k}. \quad (15.20)$$

15.3 Generalized Coordinate, Velocity, and Momentum in the *LC* Tank

In Chap. 13, we introduced the concepts of **generalized coordinate**, **generalized velocity**, and **generalized momentum**, but we only worked on real space coordinate q, velocity $\dot{q}$, and momentum p. Here, with the LC tank system, we choose Q and Φ to be the generalized momentum and generalized coordinate of the *LC* tank, respectively. Naturally, its generalized velocity is thus the voltage, $V_L = V$, because, based on physics,

$$\Phi(t) = \int_{-\infty}^{t} V_L(\tau)d\tau, \quad (15.21)$$

$$V(t) = \frac{d\Phi(t)}{dt} = \dot{\Phi}. \quad (15.22)$$

To use the formal approach to obtain the Hamiltonian in Eq. (15.15), we should build the Lagrangian of the system first. Based on Eqs. (13.3), (13.18), and (15.17)–

15.3 Generalized Coordinate, Velocity, and Momentum in the LC Tank

(15.20)

$$\mathcal{L} = T(\vec{q}, t) - V(\vec{q}, t),$$
$$= \frac{1}{2}m\dot{q}_1^2 - \frac{1}{2}kq_1^2,$$
$$= \frac{1}{2}CV^2 - \frac{1}{2L}\Phi^2, \qquad (15.23)$$

where we deliberately wrote the Hamiltonian of a mechanical SHO in line 2 and then substituted the corresponding generalized variables in line 3 for an LC tank. We can also verify that the definitions of general velocity and momentum in Eq. (15.21) are correct by examining Eq. (13.20):

$$p_i = \frac{\partial \mathcal{L}}{\partial \dot{q}_i},$$
$$Q = \frac{\partial \mathcal{L}}{\partial \dot{\Phi}},$$
$$= \frac{\partial(\frac{1}{2}CV^2 - \frac{1}{2L}\Phi^2)}{\partial \dot{\Phi}},$$
$$= \frac{\partial(\frac{1}{2}CV^2 - \frac{1}{2L}\Phi^2)}{\partial V},$$
$$= CV, \qquad (15.24)$$

which is consistent with the capacitor equation $Q = CV$.

Then, we should apply **Legendre transform** to obtain its Hamiltonian (Eq. (13.23)):

$$\mathcal{H} = \sum_{i=1}^{3N} p_i \dot{q}_i - \mathcal{L},$$
$$= Q\dot{\Phi} - \mathcal{L}. \qquad (15.25)$$

Note that the LC tank problem is a 1-D problem with one generalized coordinate, velocity, and momentum. Readers are urged to complete Eq. (15.25) in Problem 15.1.

Since $E = \frac{1}{2}CV_C^2 + \frac{1}{2}LI_L^2$, why not use V_C and I_L as the generalized momentum and coordinate, respectively? This is because they would not give consistent and meaningful results as we have just shown for Q and Φ. For example, what is the corresponding general velocity? Try Problem 15.2.

15.4 Quantization of an LC Tank

We obtained the Hamiltonian of an *LC* tank in Eq. (15.16) which can be more formally obtained by using Eq. (15.25). Now, we can quantize the circuit. However, we have obtained exactly the same equation as a mechanical SHO if we use the mapping in Eqs. (15.17)–(15.20). If we go through what was done in Chap. 14 step-by-step, we should get the same results! **To facilitate comparisons, in the following equation sets, the first one is for the mechanical SHO in Chap. 14, and the second one is for the *LC* tank.**

Firstly, we will promote Φ and Q to operators as $\hat{\Phi}$ and $\hat{Q}$, respectively:

$$\hat{\mathcal{H}} = \frac{\hat{p}^2}{2m} + \frac{\omega_0^2 m}{2}\hat{x}^2, \quad mechanical,$$

$$\hat{\mathcal{H}} = \frac{\hat{Q}^2}{2C} + \frac{\omega_0^2 C}{2}\hat{\Phi}^2, \quad LC\text{-}tank, \qquad (15.26)$$

where the resonant frequency is defined as

$$\omega_0 = \sqrt{\frac{k}{m}}, \quad mechanical,$$

$$\omega_0 = \sqrt{\frac{1/L}{C}} = \frac{1}{\sqrt{LC}}, \quad LC\text{-}tank. \qquad (15.27)$$

Then, we need to make sure that $\hat{\Phi}$ and $\hat{Q}$ have the same commutation relation as $\hat{x}$ and $\hat{p}$. We will take this for granted:

$$[\hat{x}, \hat{p}] = \hat{x}\hat{p} - \hat{p}\hat{x} = i\hbar, \quad mechanical,$$

$$[\hat{\Phi}, \hat{Q}] = \hat{\Phi}\hat{Q} - \hat{Q}\hat{\Phi} = i\hbar, \quad LC\text{-}tank. \qquad (15.28)$$

We thus can also define the annihilation operator as

$$\hat{a} = \sqrt{\frac{m\omega_0}{2\hbar}}\left(\hat{x} + \frac{i\hat{p}}{m\omega_0}\right), \quad mechanical,$$

$$\hat{a} = \sqrt{\frac{C\omega_0}{2\hbar}}\left(\hat{\Phi} + \frac{i\hat{Q}}{C\omega_0}\right), \quad LC\text{-}tank, \qquad (15.29)$$

and the creation operator as

$$\hat{a}^\dagger = \sqrt{\frac{m\omega_0}{2\hbar}}\left(\hat{x} - \frac{i\hat{p}}{m\omega_0}\right), \quad mechanical,$$

15.4 Quantization of an LC Tank

$$\hat{a}^\dagger = \sqrt{\frac{C\omega_0}{2\hbar}} \left(\hat{\Phi} - \frac{i\hat{Q}}{C\omega_0} \right), \quad LC\text{-}tank. \tag{15.30}$$

We also have the **zero point fluctuation** for $\hat{\Phi}$ and $\hat{Q}$ in the LC tank:

$$\hat{x} = \sqrt{\frac{\hbar}{2m\omega_0}}(\hat{a} + \hat{a}^\dagger) = X_{zpf}(\hat{a} + \hat{a}^\dagger), \quad mechanical,$$

$$\hat{\Phi} = \sqrt{\frac{\hbar}{2C\omega_0}}(\hat{a} + \hat{a}^\dagger) = \Phi_{zpf}(\hat{a} + \hat{a}^\dagger), \quad LC\text{-}tank, \tag{15.31}$$

and

$$\hat{p} = i\sqrt{\frac{m\omega_0\hbar}{2}}(-\hat{a} + \hat{a}^\dagger) = iP_{zpf}(-\hat{a} + \hat{a}^\dagger), \quad mechanical,$$

$$\hat{Q} = i\sqrt{\frac{C\omega_0\hbar}{2}}(-\hat{a} + \hat{a}^\dagger) = iQ_{zpf}(-\hat{a} + \hat{a}^\dagger), \quad LC\text{-}tank. \tag{15.32}$$

The zero-point fluctuation of the generalized coordinate, Φ_{zpf}, can be further expressed as

$$\Phi_{zpf} = \sqrt{\frac{\hbar}{2C\omega_0}},$$

$$= \sqrt{\frac{\hbar}{2C\frac{1}{\sqrt{LC}}}},$$

$$= \sqrt{\frac{\hbar}{2}\sqrt{\frac{L}{C}}},$$

$$= \sqrt{\frac{\hbar}{2}Z_0}, \tag{15.33}$$

where we have used the definitions of characteristic impedance (Eq. (15.9)) and resonant frequency (Eq. (15.27)). That of the generalized momentum, Q_{zpf}, is

$$Q_{zpf} = \sqrt{\frac{C\omega_0\hbar}{2}},$$

$$= \sqrt{\frac{\hbar}{2Z_0}}, \tag{15.34}$$

where we recognized $C\omega_0 = \frac{1}{Z_0}$.

Finally, we will obtain the same equations for annihilation/creation operators commutation relation, number operator, eigenenergies, and Hamiltonian with N being the number of quanta of energy in the LC tank. That is,

$$\left[\hat{a}, \hat{a}^\dagger\right] = 1, \tag{15.35}$$

$$\hat{N} = \hat{a}^\dagger \hat{a}, \tag{15.36}$$

$$\hat{\mathcal{H}} = \hbar\omega_0 \left(\hat{N} + \frac{1}{2}\right), \tag{15.37}$$

$$\hat{N}|N\rangle = N|N\rangle, \tag{15.38}$$

$$E_N = \hbar\omega_0 \left(N + \frac{1}{2}\right). \tag{15.39}$$

What is the meaning of an energy quantum in an LC tank? It is $\hbar\omega_0$ and, thus, it is the energy of a **photon** with frequency $f_0 = \frac{\omega_0}{2\pi}$. Therefore, the energy eigenstates of an LC tank refer to the number of *photons* it contains! It seems to be a little bit strange to talk about photons in an electrical circuit. Please be reminded a photon is just **electromagnetic (EM) wave**. The photon in an LC tank is just the EM wave with a much longer wavelength than those we can see with our eyes.

Like what we did for a mechanical SHO, we can also plot the energies and wavefunctions of energy eigenstates in the *flux space* (the generalized coordinate). Figure 15.3 shows the plot and it is exactly the same as Fig. 14.3 except that the x-axis is the flux, Φ, instead of the coordinate, x. Of course, please be reminded that Φ is the *generalized coordinate*. Moreover, the wavefunction is the projection of the energy eigenstate on Φ. For example, the shape of $\langle\Phi|0\rangle$ tells the probability density of finding $|0\rangle$ for a given Φ value. We should also note that there is **quantum mechanical tunneling** that a wavefunction can have flux value in the forbidden region in the *flux space*.

Example 15.2 Let $f_0 = 1GHz$. Find the energy levels in Fig. 15.3.

The energy of the photon is

$$\hbar\omega_0 = \frac{6.626 \times 10^{-34}}{2\pi} \text{J} \cdot \text{s} \times 2\pi \times 10^9 \text{ Hz},$$

$$= 6.626 \times 10^{-25} \text{ J},$$

$$= 6.626 \times 10^{-25} \text{ J} \times \frac{1}{1.6 \times 10^{-19}} \text{ eV/J},$$

15.4 Quantization of an LC Tank

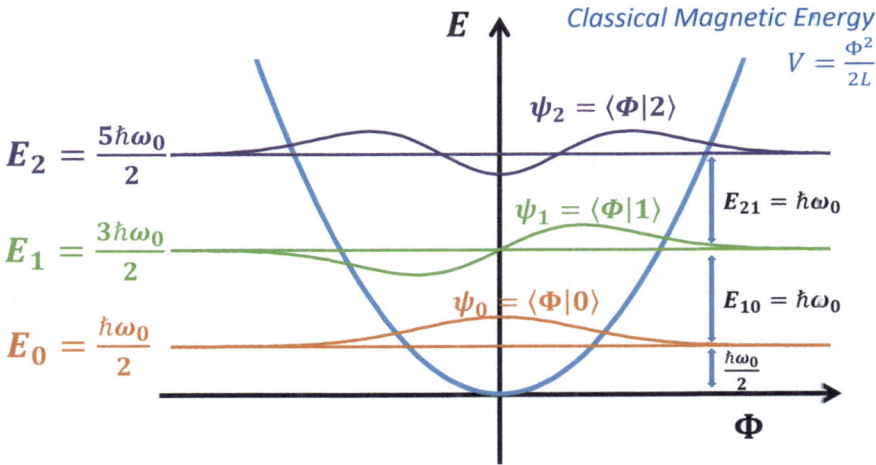

Fig. 15.3 Illustration of the energy landscape of an LC tank. This resembles that of a classical mechanical oscillator. The classical parabolic magnetic energy is plotted as a function of Φ. This is the equivalent potential energy of the system. The energy eigenvalues and the Φ-space energy eigen-wavefunctions are shown. Note that the wavefunctions do not share the y-axis with the energy functions and they are shifted vertically for illustration

$$= 4.14 \times 10^{-6}\,\text{eV},$$
$$= 4.14\,\mu\text{eV}. \tag{15.40}$$

Therefore, the spacing $\Delta E = \hbar\omega_0 = 4.14\,\mu\text{eV}$, $E_0 = 2.07\,\mu\text{eV}$, $E_1 = 6.21\,\mu\text{eV}$, and $E_2 = 10.35\,\mu\text{eV}$. ∎

Example 15.3 What is the temperature that gives a thermal energy similar to the energy spacing in Example 15.2?

The thermal energy is given by kT which is about 25 meV at 300K. Therefore, ΔE corresponds to

$$T = \frac{4.14\,\mu\text{eV}}{25\,\text{meV}} \times 300K,$$
$$= 50\,\text{mK}. \tag{15.41}$$

∎

We may also construct coherent states for the LC tank like what we did for the mechanical SHO in Sect. 14.4.2. Since C, Q, and Φ are not involved explicitly in the equation, the LC tank has the same equation as the mechanical SHO and is copied here for convenience:

$$\begin{aligned}|\alpha(t)\rangle &= e^{-\frac{1}{2}|\alpha(0)|^2} \sum_{N=0}^{\infty} \frac{\alpha(0)^N}{\sqrt{N!}} |N\rangle \, e^{-i\frac{E_N t}{\hbar}}, \\ &= e^{-\frac{1}{2}|\alpha(0)|^2} \sum_{N=0}^{\infty} \frac{\alpha(0)^N}{\sqrt{N!}} |N\rangle \, e^{-i\omega_0 (N+\frac{1}{2})t}. \end{aligned} \quad (15.42)$$

In the flux space, it oscillates like a mechanical SHO from left to right and right to left (Fig. 15.4). We cannot feel the flux space. The oscillation corresponds to the conversion of magnetic energy stored in the inductor (at left/right) to electric energy stored in the capacitor (at the center) and vice versa. This corresponds to the energy oscillations in Fig. 15.2.

15.4.1 Is an LC Tank a Good Qubit?

If we want to use an *LC* tank as a qubit, we may assign $|N=0\rangle$ as state $|0\rangle$ and $|N=1\rangle$ as state $|1\rangle$. As long as the temperature is much lower than $\hbar\omega_0$ (e.g., 5 mK which is $\frac{1}{10}$-th of 50 mK in Example 15.3), it has well-defined energy levels. However, the energy separations between adjacent levels are the same. For example,

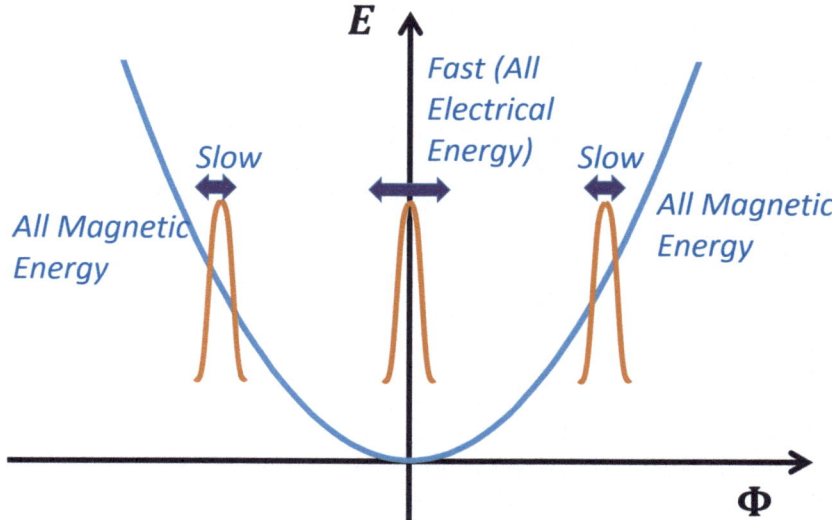

Fig. 15.4 Illustration of a coherent state oscillating in the flux, Φ, space. This is equivalent to the transfer of energy back and forth between the electric field in the capacitor ($\Phi = 0$) and the magnetic field in the inductor (maximum Φ) in the classical view of Fig. 15.2

$E_{21} = E_{10}$ in Fig. 15.3. This means that $|N = 1\rangle$ may be excited to $|N = 2\rangle$ when a pulse is applied intending to bring it to $|N = 0\rangle$. Therefore, the system might not stay in its two-level Hilbert space during qubit manipulation and it violates one of DiVincenzo's criteria (Sect. 1.3) that it needs to have a "well-characterized qubit."

How to solve this problem? We will discuss in the future chapters by replacing the inductor with a nonlinear inductor which can be implemented using **Josephson junction**. This will allow nonuniform energy spacing which is called the **anharmonicity**.

15.5 Summary

In this chapter, we apply what we have learned in a mechanical simple harmonic oscillator to an LC tank. We point out that Φ and Q are the generalized coordinate and generalized momentum for this system, respectively. By using the same procedure when treating a mechanical SHO, namely, constructing Hamiltonian, promoting conjugate variables to operators, and applying commutation relation of the operators, the LC tank is quantized. Its energy eigenstates represent the number of photons of the resonant frequency in the system. At microwave frequency, the energy separation of the eigenenergies is in the order of 0.1K. Therefore, a cryogenic environment is needed to observe the quantum behavior. Moreover, an LC tank cannot be used as a qubit because the energy separation between adjacent levels is the same. However, this chapter lays an important foundation for us to appreciate the design and constraints of a superconducting qubit to be discussed in future chapters.

Problems

15.1 Legendre Transform of LC Tank Lagrangian
Complete the derivations in Eq. (15.25) to get Eq. (15.26).

15.2 Generalize Coordinate, Velocity, and Momentum of LC Tanks
By going through the same process as in Problem 15.1, show that V_C and I_L are not suitable generalized coordinate and momentum, respectively.

15.3 High-Temperature Qubit
It is desirable to operate a qubit at higher temperatures. Find the resonant frequency required in order to operate an LC tank at 4.2K (liquid helium boiling point) and 300K as a qubit. What should be the values of L and C? Are they easy to obtain?

15.4 Quantization of an LC Tank
We used variable mapping to reuse the equations in mechanical SHO to obtain the quantization equations for an LC tank. Repeat Chap. 14 for an LC tank step-by-step.

References

1. Alexandre Blais, Arne L. Grimsmo, S. M. Girvin, and Andreas Wallraff. Circuit quantum electrodynamics. *Rev. Mod. Phys.*, 93:025005, May 2021.
2. Erwin Kreyszig. *Advanced Engineering Mathematics*. John Wiley & Sons; 9th Edition, International Edition, 2006.

Chapter 16
Superconductor and Josephson Junction

16.1 Introduction

In Chap. 15, we showed that a quantized LC tank could not be used as a qubit because the separation of adjacent eigenenergies is constant (e.g. $E_{10} = E_{21}$). To have a working qubit that has its states confined in the two-dimensional Hilbert space, we need to have non-uniform separations. This can be achieved by using a *non-linear inductor* to introduce *anharmonicity*. A Josephson junction is a very versatile non-linear inductor. A Josephson junction is just a metal/insulator/metal stack and it only functions as intended when its metal regions become superconducting. Moreover, as demonstrated in Example 15.3, an ambient at milli-kelvin is required to distinguish the states in an LC-tank. At this temperature, metals on a quantum chip usually become superconducting. Therefore, in this Chapter, we will introduce the superconductor and Josephson junction. For details on superconductivity, readers may refer to [1]. Readers may also refer to Section A1 in the Supplementary Information of [2] to gain a deeper understanding of Josephson junction theory.

16.1.1 Learning Outcomes

Be able to distinguish superconductor from perfect conductor; be able to explain the meanings of Josephson equations; understand the fabrication, structures, and parameters of a Josephson junction.

16.1.2 Teaching Videos

- Search for Ch16 in this playlist
  - https://tinyurl.com/3yhze3jn
- Other videos
  - https://youtu.be/h3Vy2Z1mxVU
  - https://youtu.be/rBlrBCZFaBo
  - https://youtu.be/MUXGTnilvII

16.2 Superconductors

16.2.1 Definition of Superconductors

A **superconductor** is *not* just a **perfect conductor**. A perfect conductor is a conductor with zero resistivity. In principle, many metals will have their resistivity, ρ, approach zero when their temperature approaches absolute zero ($0K$). But they are not necessarily superconductors. When a metal becomes a superconductor at a temperature below the **critical temperature**, T_c, an abrupt decrease in resistivity to zero will happen (Fig. 16.1).

Besides having zero resistivity, the second criterion for it to be superconducting is that it needs to expel the magnetic field that already exists inside the metal. This is called the **Meissner effect**. To make this clear, let us consider two cases. The first case is that the metal already has zero resistivity, and there is no external magnetic field penetrating the metal ($\vec{B} = 0$) at the beginning. Now, a magnetic field is turned on. For both a perfect conductor and a superconductor, they do not allow the magnetic field to penetrate them. This is because, with zero resistivity, a change of magnetic field inside the metal is not allowed due to Maxwell's equations (require $\frac{\partial \vec{B}}{\partial t} = 0$). In this case, we cannot distinguish a superconductor from a perfect conductor.

In the second case, we assume the metals are at high temperatures with finite resistivity and a non-zero magnetic field has penetrated them at the beginning (Fig. 16.2). As the temperature reduces, if it is a perfect conductor, after it reaches zero resistivity, the magnetic field will stay there as it does not violate the requirement of $\frac{\partial \vec{B}}{\partial t} = 0$ mentioned earlier. On the other hand, if it is a superconductor, it will expel the magnetic field *that is already inside* the metal when it is cooled down below T_c and becomes superconducting. Figure 16.2 illustrates the difference between a perfect conductor and a superconductor when the temperature is low enough to give them zero resistivity. This can be used to distinguish a superconductor from a perfect conductor.

16.2 Superconductors

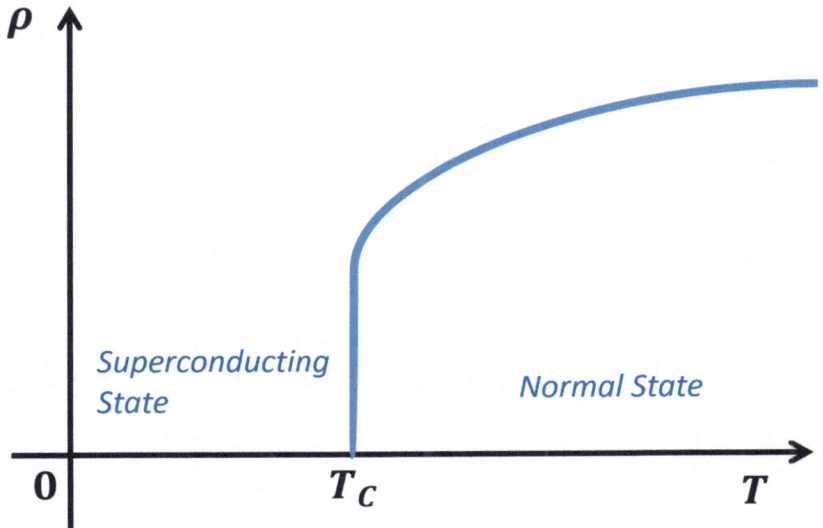

Fig. 16.1 Changes of metal resistivity, ρ, as a function of temperature. When its temperature decreases below its critical temperature, T_c, the resistivity decreases to zero abruptly

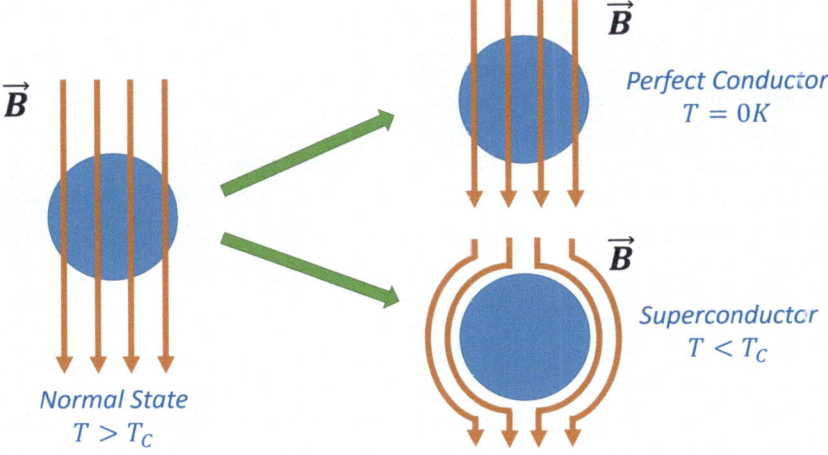

Fig. 16.2 Change of magnetic field of a metal at its normal state (left) when it is cooled down. For a perfect conductor, the magnetic field will not change even if it has zero resistivity at a low enough temperature (top right). For a superconductor, it will expel the magnetic field when it becomes superconducting at $T < T_c$ (bottom right). This is the Meissner effect

16.2.2 Superconductor Physics

Let us now have a brief summary of superconductor physics. Many of the discussions are not rigorous. The intention is to bring up the essential aspects that might help you understand the discussions in superconducting qubit papers in the future. Readers may refer to [1] for details.

A superconductor is formed because the electrons are paired to form **Cooper pairs** or **Bardeen-Cooper-Schrieffer pairs** (**BCS pairs**) which is explained by the Bardeen-Cooper-Schrieffer theory (**BCS theory**). The electrons in a Cooper pair have opposite momentum and opposite spin (Fig. 16.3). How can two electrons both with negative charge pair with each other? This does not seem to be energetically favorable due to the repulsion of the Coulomb force. In the BCS theory, it is shown that lattice vibration (**phonon**) enables the pairing. It can be hand-wavingly understood in this way. While the electrons are free in a metal, they are also attracted to the positively charged lattice (array of ionized atoms). In Fig. 16.3, due to the attraction, the atoms move closer to electron B. As a result, the region near electron B has a larger positive charge density. Electron A is then attracted to this relatively more positive region and forms a Cooper pair with electron B. Therefore, the interaction between the lattice (and thus the lattice vibration, phonon) and the electron is the cause of the superconducting phenomenon in some metals. This is why some high-conductivity metals, such as copper, cannot be superconductors. Such metals have high conductivity because they have small phonon-electron interactions. As a result, they cannot form Cooper pairs at low temperatures.

The significance of Cooper pair formation is that it turns the carriers from **fermions** to **bosons**. Electrons are fermions because they have half-spin. Fermions obey **Fermi-Dirac distribution** which says that no electron should have the *same* state as the other. This results in **Pauli exclusion principle** which we learned in chemistry. The probability of having an electron occupying a certain state at energy E (or the average number of electrons in a state) is thus

$$f_{FD}(E) = \frac{1}{e^{\frac{E-\mu}{kT}} + 1}, \tag{16.1}$$

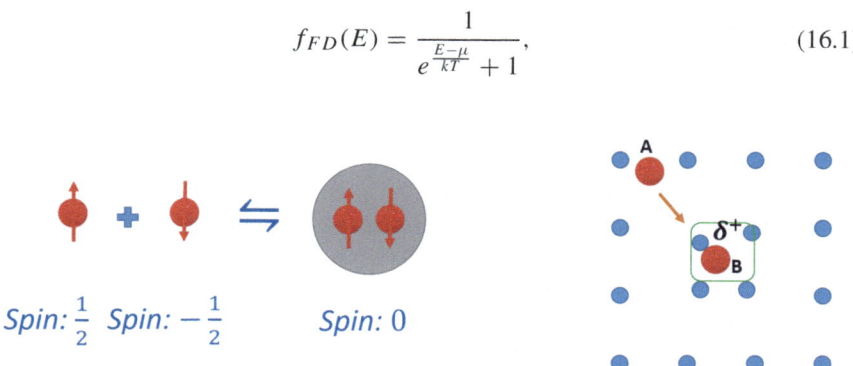

Fig. 16.3 Left: Forming and breaking of Cooper pairs. Right: Illustration of Cooper pair formation mediated by phonon-electron interaction. Red particles are electrons and blue particles are ionized atoms in a crystal

16.2 Superconductors

where k is the Boltzmann constant and μ is the **chemical potential** or **Fermi level**. Here, we assume *no degeneracy* for simplicity. This means that for each energy level E, there is only one state. At very low temperatures, if $E > \mu$, $f_{FD} \sim 0$. If $E < \mu$, $f_{FD} \sim 1$. This means that the electrons will occupy all energy levels up to the chemical potential at a very low temperature.

Cooper pairs have an integral spin because it is a combination of two electrons with opposite spins (Fig. 16.3). Therefore, a Cooper pair is a boson. Bosons can occupy the same state as others. Photons are bosons. We are allowed to create photons at the same state (same momentum, same wavelength, same polarization) and, thus, create a very powerful LASER! We cannot do that for electrons as they are fermions and, therefore, there is no electron laser. Like other bosons, Cooper pairs thus obey **Bose-Einstein distribution** and the expected number of Cooper pairs in an energy state is

$$f_{BE}(E) = \frac{1}{e^{\frac{E-\mu}{kT}} - 1}, \quad (16.2)$$

where k is the Boltzmann constant, μ is the chemical potential and no degeneracy is assumed. We assume, $\mu = E_0$, where E_0 is the ground state energy. Therefore, at low temperatures, *most* Cooper pairs will occupy the same ground state because for $E > \mu$, $f_{BE} \sim 0$ when T is very small. This means that all Cooper pairs have the same ground state wavefunction and together they have the following wavefunction as a **macroscopic quantum object** (with a size larger than μm) like a matching army:

$$\Psi = \sqrt{n_s} e^{i\phi}, \quad (16.3)$$

where ϕ is the phase of the quantum object and $n_s \approx 10^{23}$ cm^{-3} because the density of electron in a metal is in the order of 10^{23} cm^{-3}.

Apparently, if the Cooper pairs are broken, it will lose its superconductivity. One Cooper pair will create two electrons (known as **quasiparticles**). The energy required to break a Cooper pair is $2\Delta(T)$. $\Delta(T)$ is the energy of the gap formed in the conduction band of a metal when it becomes superconducting. Based on the BCS theory [1]

$$\Delta(T = 0) = 1.764 k T_c. \quad (16.4)$$

The reason that the energy required to break a Copper pair is two times the gap energy is because there are two electrons in a Cooper pair.

When there is a strong enough magnetic field, the superconducting state will cease to exist. The magnetic field above which the superconducting state becomes a normal state is called the **critical magnetic field**, B_c. There is a trade-off between T_c and B_c. Applying a larger B_c reduces T_c, and thus it needs to be cooled to a lower temperature to be superconducting and vice versa.

16.3 Josephson Junction

16.3.1 Structure and Fabrication

Josephson junctions are superconductor/insulator/superconductor ($S - i - S$) structures. Yes, it is the same as a capacitor except that the insulator is so thin that it allows tunneling of Cooper pairs. Figure 16.4 shows the schematic of a Josephson junction and the cross section of a Josephson junction fabricated on a highly resistive substrate (e.g., silicon wafer). Usually, aluminum (Al) is used as the electrode, which becomes superconducting below $1.2K$, and nonstoichiometric aluminum oxide (AlO_x) is used as the insulator. This forms an $Al/AlO_x/Al$ stack. There are also other types of stack proposed such as using niobium (Nb) as the electrodes sandwiching a thin layer of Al and AlO_x [3], which forms a $Nb/Al/AlO_x/Nb$ stack. However, $Al/AlO_x/Al$ stacks usually have a better quality.

To form an $Al/AlO_x/Al$ stack, the bottom layer of Al will be deposited first followed by controlled oxidation of Al to form a thin layer of AlO_x, which is then covered by another layer of Al, also through deposition. The metal is usually deposited through evaporation techniques [4–6]. To obtain a high-quality Josephson junction, it is important to perform the deposition and oxidation without breaking the vacuum. This requires a single mask for the top and bottom layers, which is impossible through traditional lithography techniques. As can be seen in Fig. 16.4, the top and bottom electrodes have different patterns and require two masks if the traditional lithography process is used.

There are two common methods to achieve this. The first one is called the **Dolan bridge** method [4, 6]. The second one is the **Manhattan-style** method [5, 6].

Figure 16.5 illustrates a Dolan process using the cross sections of the structure. The mask is patterned to have two openings. The mask is suspended and anchored to the substrate outside of the picture and in the direction perpendicular to the cross section. The suspension is obtained from an isotropic etch with an undercut. Then, angled deposition of Al is performed from one side (solid red lines). This

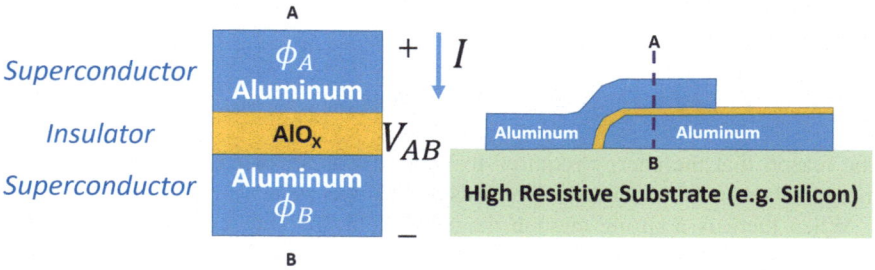

Fig. 16.4 Left: Schematic of a Josephson junction. Right: Cross section of a Josephson junction fabricated on a resistive substrate. The $A - B$ cutline corresponds to the structure shown on the left

16.3 Josephson Junction

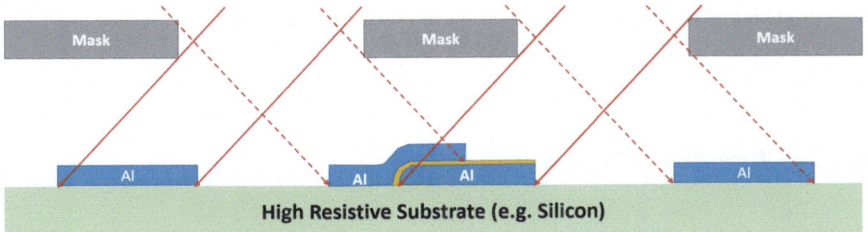

Fig. 16.5 Dolan bridge process used to fabricate Josephson junction with one mask. Angled evaporation is used to perform directional deposition (solid red lines) of the first layer of metal followed by oxidation. After that, another directional deposition (dashed red lines) is performed to form the second layer, and thus complete the junction. The masks are suspended with anchors not shown in the picture

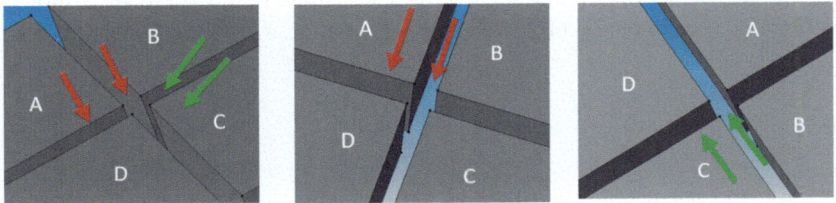

Fig. 16.6 Manhattan-style process for Josephson junction fabrication. Masks with a high aspect ratio are used. The first directional deposition is performed in the $A - D$ direction (red arrows in the left and middle pictures). The evaporation beams outside of the trench in the $A - D$ direction will be blocked and cannot reach the bottom. Then, oxidation is performed. Afterward, a second directional deposition is performed in the $C - D$ direction (green arrows in the left and right pictures). Only the middle crossover will receive both depositions and the junction is formed there

transfers the mask pattern to the Al deposited on the substrate. Then, oxidation is performed without breaking the vacuum. After that, an angled deposition from another direction (dashed red lines) is performed to form the second Al layer. Due to angled depositions, a Josephson junction is formed in the middle.

Figure 16.6 shows the Manhattan-style process. The masks are patterned into horizontal and vertical openings (like the Manhattan streets with tall buildings). High aspect ratio masks (photoresist) are used. As a result, the trench looks deep and narrow. In the first directional deposition, it is along one of the "streets" ($A - D$ direction). Note that it is still an angled implant not perpendicular to the wafer. Only the beams hitting the trench along the $A - D$ direction can reach the substrate (street) and form a layer of Al. Beams at other locations are blocked by the "tall buildings." Then, oxidation is performed. After that, a second directional deposition is performed in the direction of the orthogonal trench ($C - D$ direction). Again, it is still an angled implant that is not perpendicular to the wafer. Only the beams hitting the trench along the $C - D$ direction can reach the substrate (street) and form the second layer of *Al*. As a result, the crossing of the trench (or "junction of the streets") becomes a Josephson junction.

16.3.2 Josephson Equations

The physics of Josephson junction is pretty profound. For the purpose of this book, we will not go into the details. But we will learn and emphasize the essential and fundamental concepts. Firstly, we will start with Eq. (16.3). As mentioned, the Cooper pairs in a superconducting metal are bosons, and they share the same ground state. Therefore, they can be described as a macroscopic wavefunction in a superconducting region. Before region A and region B in Fig. 16.4 are brought near to each other, the Cooper pairs in each of them can be described as the following wavefunctions based on Eq. (16.3):

$$\Psi_A = \sqrt{n_A} e^{i\phi_A}, \tag{16.5}$$

$$\Psi_B = \sqrt{n_B} e^{i\phi_B}, \tag{16.6}$$

where n_A and n_B are the densities of Cooper pairs in region A and region B, respectively. ϕ_A and ϕ_B are the phases of the wavefunctions in region A and region B, respectively. As we bring them closer to each other, the wavefunctions in both regions will couple to each other, and the Cooper pairs will tunnel to the other region through the insulator. Therefore, the insulator acts as a **weak link** through which the wavefunctions have a small overlap. Then, we define a new quantity called the **Josephson phase**:

$$\varphi = \phi_A - \phi_B. \tag{16.7}$$

If a positive voltage is applied across the two regions (V_{AB} in Fig. 16.4), the energy of the negative-charge Cooper pairs in region B will be higher than that in region A by $2eV_{AB}$ with $e = 1.6 \times 10^{-19} C > 0$. Putting this into the Schrödinger equation and introducing a coupling factor for the weak link, one will obtain the following Josephson equations (see Chapter 4 in [7] for the derivation). The **first Josephson equation** is

$$I(t) = I_c \sin \varphi(t), \tag{16.8}$$

where I_c is the **critical current**, a constant specific to the junction for a given temperature and magnetic field. Like the concepts of critical temperature and critical field, the critical current is the maximum current that a Josephson junction can still operate as an $S - i - S$ structure. More generally, the **critical current density**, J_c, is used. This can be used not just in Josephson junction but also in bulk

16.3 Josephson Junction

superconductors. Critical current density is the current density above which the superconductivity will be destroyed.

The **second Josephson equation** is

$$\frac{\partial \varphi(t)}{\partial t} = \frac{2e}{\hbar} V_{AB},$$

$$= \frac{2\pi}{\Phi_0} V_{AB},$$

$$= \frac{V_{AB}}{\Phi_0'}, \qquad (16.9)$$

where we have introduced **magnetic flux quantum**, Φ_0, and **reduced magnetic flux quantum**, Φ_0' (see Sect. 16.4 and Eq. (16.34) for its meaning). They are defined as

$$\Phi_0 = \frac{h}{2e}, \qquad (16.10)$$

$$\Phi_0' = \frac{\Phi_0}{2\pi} = \frac{\hbar}{2e}. \qquad (16.11)$$

The first Josephson equation is also called the **current-phase relation** because it gives the relationship between the current of the junction and the Josephson phase. It shows that the current depends on the phase. If the phase is a constant, the current will be a constant. But the phase depends on the voltage. Assume V_{AB} is a constant (V_0), if we integrate Eq. (16.9) to obtain the change of phase from time 0 to time t and add it to the initial phase, $\varphi(0)$, we obtain:

$$\varphi(t) = \varphi(0) + \int_0^t \frac{2\pi}{\Phi_0} V_{AB} d\tau,$$

$$= \varphi(0) + \int_0^t \frac{2\pi}{\Phi_0} V_0 d\tau,$$

$$= \varphi(0) + \frac{2\pi}{\Phi_0} V_0 t. \qquad (16.12)$$

This shows that if the applied voltage is zero ($V_{AB} = V_0 = 0$), $\varphi(t) = \varphi(0)$ and I is a constant based on the first equation. If the applied voltage is a nonzero constant, the phase will change linearly with time, and thus the current will oscillate sinusoidally because $I = I_c \sin(\varphi(0) + \frac{2\pi}{\Phi_0} V_0 t)$. Figure 16.7 shows the current for zero and nonzero constant V_{AB}.

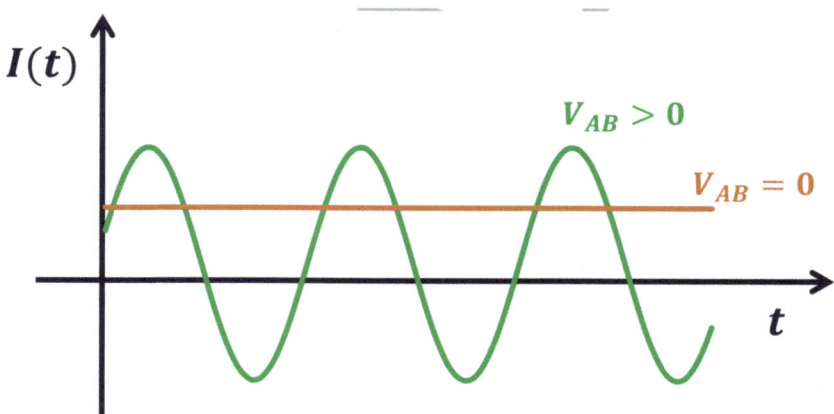

Fig. 16.7 Current in Josephson junction as a function of time when zero and nonzero constant voltage is applied

16.3.2.1 Definition of Standard Volt

The second Josephson equation applies to Josephson junctions made of any superconductors and insulators. When V_{AB} is a nonzero constant, we have:

$$I(t) = I_c \sin(\varphi(0) + \frac{2\pi}{\Phi_0} V_0 t),$$
$$= I_c \sin(\omega_{JJ} t + \varphi(0)), \quad (16.13)$$

where we have set $\omega_{JJ} = \frac{2\pi}{\Phi_0} V_0$. The current oscillates at an angular frequency, ω_{JJ}, which depends on Φ_0. And Φ_0 is only dependent on the fundamental constants e and h (Eq. (16.11)). Therefore, the Josephson junction is used to define the unit *volt* since the corresponding frequency can be reproduced universally.

Example 16.1 Find the current oscillation frequency when $1V$ is applied across a Josephson junction.

Using Eq. (16.13)

$$\omega_{JJ} = \frac{2\pi}{\Phi_0} V_0,$$
$$2\pi f_{JJ} = \frac{2\pi}{\Phi_0} V_0,$$
$$f_{JJ} = \frac{2e}{h} V_0,$$
$$f_{JJ} = \frac{2 \times 1.6 \times 10^{-19}}{6.626 \times 10^{-34}} \times 1 \text{ Hz},$$

$$f_{JJ} = 4.83 \times 10^{14} \text{ Hz}. \qquad (16.14)$$

It is also common to quote that the frequency is 4.83×10^{14} Hz/V. A rule-of-thumb is that it is about 5 GHz per $10\,\mu$V. ∎

16.3.3 Josephson Inductance, Kinetic Inductance, and Josephson Energy

Let us recall the relationship between voltage and current in an inductor from Eq. (15.4) and is quoted here for convenience:

$$V_L = L\frac{dI_L}{dt},$$

$$V_{AB} = L\frac{dI}{dt}. \qquad (16.15)$$

Now, let us substitute the first Josephson junction equation to express it in terms of the Josephson phase:

$$V_{AB} = L\frac{d(I_c \sin \varphi(t))}{dt},$$

$$= LI_c \cos\varphi(t)\frac{d\varphi(t)}{dt},$$

$$= LI_c \cos\varphi(t)\frac{2\pi}{\Phi_0}V_{AB}, \qquad (16.16)$$

where we used the chain rule to get line two ($\frac{d\sin\varphi}{dt} = \frac{d\sin\varphi}{d\varphi}\frac{d\varphi}{dt}$). In line three, we used the second Josephson equation. Rearranging the terms, we get:

$$L = \frac{V_{AB}}{I_c \cos\varphi(t)\frac{2\pi}{\Phi_0}V_{AB}},$$

$$= L_J \frac{1}{\cos\varphi(t)}, \qquad (16.17)$$

where we have defined **Josephson inductance** as

$$L_J = \frac{1}{I_c \frac{2\pi}{\Phi_0}},$$

$$= \frac{\Phi_0}{2\pi I_c}. \qquad (16.18)$$

Therefore, Josephson junction has a **nonlinear inductance** which depends on φ.

It is worthwhile to spend some time to understand what we have obtained. Firstly, Josephson junction has an inductance (Eq. (16.17)). Its value depends on its critical current, I_c, magnetic flux quantum, Φ_0, and the Josephson phase, φ. If the phase is close to $\pi/2$, one will get a very large inductance.

Secondly, we have not seen the role of the magnetic field when we derive the inductance. Indeed, there is no magnetic flux associated with it. Then why do we have an inductance? This is because of the **kinetic inductance**. Inductance has the effect of storing energy and resisting changes under AC voltage (i.e., with inertia). At room temperature, inductance usually originates from the magnetic field which stores the energy as magnetic energy. There is another form of energy storage or inertia. It is the moving Cooper pairs that see no friction in superconductors. Instead of storing energy as a magnetic field in one direction, the energy is stored as the kinetic energy of the moving Cooper pairs in the superconducting current. This is the source of inertia and thus the inductance. The inductance in Eq. (16.17) is due to kinetic inductance. In principle, normal metals also have kinetic inductance, but they are small compared to the regular inductance due to flux because the inertia dies out quickly due to resistance. Moreover, in regular inductors, a large geometry is needed to obtain a large inductance. Josephson junction usually can have a large kinetic inductance despite its small geometry.

Like regular inductors which can store energy, a Josephson junction also stores energy through its kinetic inductance. We may derive the energy it stores by computing the integral of the power when its phase changes from $\varphi = \varphi_1$ to $\varphi = \varphi_2$. We set $V = V_{AB}$:

$$\begin{aligned} E &= \int_{\varphi_1}^{\varphi_2} IV dt, \\ &= \int_{\varphi_1}^{\varphi_2} I \frac{\Phi_0}{2\pi} d\varphi, \\ &= \int_{\varphi_1}^{\varphi_2} I_c \sin\varphi \frac{\Phi_0}{2\pi} d\varphi, \\ &= \frac{I_c \Phi_0}{2\pi}(\cos\varphi_1 - \cos\varphi_2), \end{aligned} \qquad (16.19)$$

where, in line 2, we have used the second Josephson equation to obtain $V dt = \frac{\Phi_0}{2\pi} d\varphi$. In line 3, we substituted the first Josephson equation.

Assume $\varphi_1 = 0$ and set $\varphi_2 = \varphi$, we get:

$$\begin{aligned} E &= \frac{I_c \Phi_0}{2\pi}(1 - \cos\varphi), \\ &= E_J(1 - \cos\varphi), \end{aligned} \qquad (16.20)$$

16.3 Josephson Junction

where we have defined **Josephson energy**, E_J, as

$$E_J = \frac{I_c \Phi_0}{2\pi}. \tag{16.21}$$

Finally, it is useful to relate Josephson energy to Josephson inductance:

$$L_J = \frac{\Phi_0^2}{(2\pi)^2 E_J} = \frac{\Phi_0'^2}{E_J}. \tag{16.22}$$

16.3.4 Phase, Flux, and Magnetic Flux Quantum

Reduced magnetic flux quantum, $\Phi_0' = \frac{\Phi_0}{2\pi}$, has appeared many times in the equations we have discussed. Magnetic flux quantum, Φ_0, is related to the quantized magnetic flux in a loop where the magnetic flux can only increase by multiples of magnetic flux quantum. The second Josephson equation can be rewritten as

$$\frac{\partial \varphi(t)}{\partial t} = \frac{V_{AB}}{\Phi_0'},$$

$$\frac{\partial \Phi_0' \varphi(t)}{\partial t} = V_{AB},$$

$$V_{AB} = \frac{\partial \Phi_0' \varphi(t)}{\partial t}. \tag{16.23}$$

Now we introduce a new quantity, Φ, defined as,

$$\Phi = \Phi_0' \varphi(t). \tag{16.24}$$

Then Eq. (16.23) becomes:

$$V_{AB} = \frac{\partial \Phi}{\partial t}. \tag{16.25}$$

This looks like Lenz's rule where the induced back electromotive force is proportional to the change of flux and it is the same as Eq. (15.22). Therefore, we may treat Φ as a flux quantity associated with a Josephson junction. Of course, this is not a real flux.

If we expand the energy equation in Eq. (16.20), by substituting $\cos x = 1 - \frac{x^2}{2!} + \frac{x^4}{4!} - \cdots$

$$E = E_J(1 - \cos\varphi),$$
$$= E_J\left(1 - \left(1 - \frac{\varphi^2}{2!} + \frac{\varphi^4}{4!} - \cdots\right)\right),$$
$$= E_J\left(\frac{\varphi^2}{2!} - \frac{\varphi^4}{4!} + \cdots\right),$$
$$\approx E_J \frac{\varphi^2}{2!},$$
$$= \frac{\Phi_0'^2}{L_J}\frac{\varphi^2}{2!},$$
$$= \frac{(\Phi_0'\varphi)^2}{2L_J},$$
$$= \frac{1}{2L_J}\Phi^2, \tag{16.26}$$

where we have used the relationship between E_J and L_J in line 5 (Eq. (16.22)) and the definition of Φ in the last line. In this approximation, we only take up to the second term of the cosine function. This is equivalent to linearizing the inductance to only contain L_J. Now the energy resembles the energy of a linearized inductor in Eq. (15.15). This again shows that Φ plays the role of the flux of a regular inductor.

16.3.5 Josephson Junction Design Considerations

In a superconducting quantum computing circuit, the Josephson junction acts as a **nonlinear inductor**. E_J and L_J are important design criteria. Both of them depend on the critical current, I_c. Therefore, in the initial phase of design, one needs to pick the correct critical current. After fabrication, it is also important to make sure the critical current meets the specification. However, the measurement of critical current is expensive as the chip needs to be cooled to cryogenic temperature. Fortunately, the critical current of a Josephson junction is related to its normal state resistance (R_n) through the **Ambegaokar-Baratoff formula** [7, 8]. The normal state resistance is the resistance of the Josephson junction at room temperature due to electron tunneling. It is given that

$$I_c = \frac{\pi\Delta}{2R_n e},$$

16.3 Josephson Junction

$$I_c R_n = \frac{\pi \Delta}{2e}, \quad (16.27)$$

where Δ is the superconducting gap of the electrode (Eq. (16.4)). This is a simplified equation by assuming that the temperature of interest is much lower than T_c. One may also refer to [8] for the formula applicable to a junction with different electrode materials on each side (so it has different Δ_1 and Δ_2).

Equation (16.27) also shows that $I_c R_n$ is a constant for a given electrode material. Therefore, a designer may increase the area to reduce R_n which will proportionally increase I_c for a given fabrication process. Or one may reduce R_n by reducing the tunneling barrier thickness to increase I_c which needs approval from the process developer.

After picking I_c, we also want to have E_C to be smaller than kT_c [5]. For example, the T_c of Al and Nb are $1.2K$ and $9.3K$, respectively.

Example 16.2 To achieve a critical current density, J_c, of 30 A/cm^2 on a 150 nm × 150 nm $Al/AlO_x/Al$ Josephson junction, what should be the corresponding resistance, R_n, at room temperature?

Firstly, let us find the critical current, I_c:

$$I_c = 30 \text{ A/cm}^2 \times 150 \text{ nm} \times 150 \text{ nm},$$
$$= 6.75 \text{ nA}. \quad (16.28)$$

Based on Eq. (16.4), the superconducting gap of Al is

$$\Delta(T = 0) = 1.764 kT_c,$$
$$= 1.764 \times 1.38 \times 10^{-23} \text{ J/K} \times 1.2 \text{ K},$$
$$= 2.92 \times 10^{-23} \text{ J}. \quad (16.29)$$

Therefore, the corresponding normal state resistance is

$$R_n = \frac{\pi \Delta}{2 I_c e}$$
$$= \frac{3.14 \times 2.92 \times 10^{-23} J}{2 \times 6.75 \times 10^{-9} A \times 1.6 \times 10^{-19} C}$$
$$= 42.4 \text{ k}\Omega \quad (16.30)$$

∎

Example 16.3 Find the relationship between the Josephson junction inductance, L_J, and its normal state resistance, R_n.

From Eq. (16.18), we have:

$$L_J = \frac{\Phi_0}{2\pi I_c},$$

$$I_c = \frac{\Phi_0}{2\pi L_J}. \quad (16.31)$$

Substituting into Eq. (16.27), we obtain:

$$R_n = \frac{\pi \Delta}{2 I_c e},$$

$$= \frac{\pi^2 L_J \Delta}{\Phi_0 e}. \quad (16.32)$$

Therefore, using Δ obtained in Example 16.2 for Al junction and the definition of magnetic flux quantum in Eq. (16.10)

$$\frac{L_J}{R_n} = \frac{\Phi_0 e}{\pi^2 \Delta},$$

$$= \frac{\frac{h}{2e} e}{\pi^2 \Delta},$$

$$= \frac{h}{2\pi^2 \Delta},$$

$$= \frac{6.626 \times 10^{-34}\,\text{J} \cdot \text{Hz}^{-1}}{2 \times 3.1415^2 \times 2.92 \times 10^{-23}\,\text{J}},$$

$$= 1.15 \times 10^{-12}\,\text{Hz}^{-1},$$

$$= 1.15 \times 10^{-12} \frac{H}{\Omega},$$

$$= 1.15 \frac{nH}{k\Omega}. \quad (16.33)$$

∎

16.4 Flux-Tunable Josephson Junction Loop

If a Josephson junction is used to replace the inductor in an LC tank, its energy determines the potential energy due to the conjugate coordinate, φ or Φ. The Josephson junction energy, E_J, thus has a direct impact on the resonant frequency

16.4 Flux-Tunable Josephson Junction Loop

of a qubit. Of course, E_J depends on I_c (Eq. (16.21)). It is thus desirable to fabricate a Josephson junction with tunable I_c and, thus, **tunable** E_J. This cannot be done easily using a single Josephson junction. On the other hand, one can build a superconducting loop with two Josephson junctions. The circuit then has an effective critical current depending on the external magnetic flux, Φ_{ext}. This is also called a **superconducting quantum interference device, SQUID**. Note that we have been emphasizing that there is no real magnetic flux associated with a Josephson junction despite using the concept of circuit branch flux as the conjugate coordinate. But here Φ_{ext} is the real magnetic flux externally applied to the circuit which can be adjusted during a quantum chip operation.

If such a qubit is used to form a transmon qubit (which will be discussed in the following chapters), it is called the **split transmon qubit**. Once we find the effective E_J, the treatment is the same as in a transmon formed by a single Josephson junction from Chaps. 18 to 21.

We will not go into the details of the derivations of the equations of the flux-tunable Josephson junction loop. They are based on [7] and [9]. We will highlight the outcomes and the methodologies. *Readers might be able to appreciate the equations below better after going through the superconducting qubit discussions in the following chapters.*

Firstly, it is instructive to understand that in a loop formed by a superconductor, if the superconductor is thick enough that there is a continuous loop on which there is no current, the total magnetic flux, Φ, passing through the loop is quantized (see Chapter 3.10 in [7]). Figure 16.8 illustrates the concept. The total flux is

$$\Phi = n\Phi_0,$$
$$= \frac{nh}{2e}. \tag{16.34}$$

Therefore, it is not surprising that if we go through the basic equations, a loop formed by two Josephson junctions should also be quantized in *a certain way* depending on the total magnetic flux, Φ, in its loop. Figure 16.8 shows such a loop based on Section 5.10 in [7]. It has two junctions. Each of them has E_{J1}, I_{c1} and E_{J2}, I_{c2}, respectively. The quantization process results in the following equation:

$$\varphi_1 - \varphi_2 = 2\pi \frac{\Phi}{\Phi_0}, \tag{16.35}$$

where a $2n\pi$ term is ignored and absorbed into Φ by following the treatment in [7]. This equation tells us that the relationship between the phase of the two junctions is linked by the total magnetic flux through the loop. Therefore, it only has one degree of freedom and should act like a single Josephson junction. Following the approach in [9], we define the following variables:

$$\varphi = \frac{\varphi_1 + \varphi_2}{2}, \tag{16.36}$$

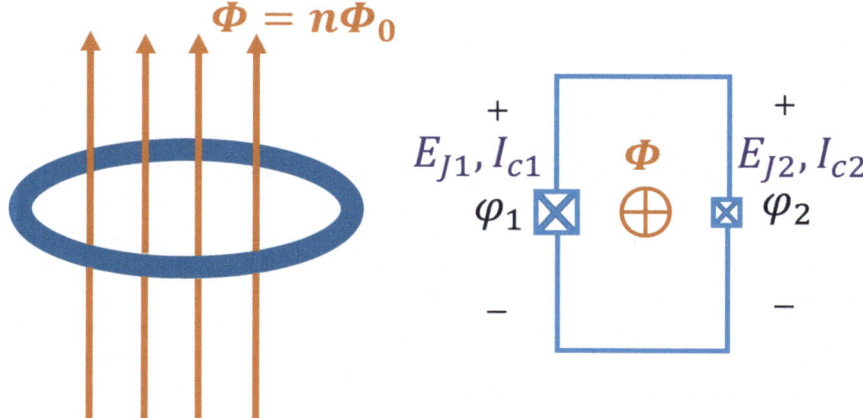

Fig. 16.8 Left: Quantization of magnetic flux through a superconducting loop. The wire is thick enough that there is no current in the center of the metal. Right: Flux-tunable Josephson junction loop with two asymmetric Josephson junctions

$$E_{J\Sigma} = E_{J1} + E_{J2}, \tag{16.37}$$

$$d = \frac{E_{J2} - E_{J1}}{E_{J2} + E_{J1}}. \tag{16.38}$$

Also, we assume Φ is a constant (DC flux) and perform a transformation. Then, after operator promotion, the total energy due to the Josephson junctions can be expressed as

$$\begin{aligned}\hat{H}_J &= -E_{J1}\cos\hat{\varphi}_1 - E_{J2}\cos\hat{\varphi}_2, \\ &= -E_{J\Sigma}\cos\left(\frac{\pi\Phi}{\Phi_0}\right)\sqrt{1 + d^2\tan^2\left(\frac{\pi\Phi}{\Phi_0}\right)}\cos\hat{\varphi}.\end{aligned} \tag{16.39}$$

We can then fine-tune the **effective Josephson junction energy**, E_{JT}, by adjusting Φ through the following equation:

$$E_{JT} = E_{J\Sigma}\cos\left(\frac{\pi\Phi}{\Phi_0}\right)\sqrt{1 + d^2\tan^2\left(\frac{\pi\Phi}{\Phi_0}\right)}. \tag{16.40}$$

Finally, we should note that Φ is the total flux which is a sum of the external flux Φ_{ext} and the flux caused by the current flowing in the loop. In our case, we can assume $\Phi = \Phi_{ext}$. Also, if the junctions are identical (symmetric), $d = 0$, we have:

$$E_{JT} = E_{J\Sigma}\cos\left(\frac{\pi\Phi}{\Phi_0}\right). \tag{16.41}$$

16.5 Summary

In this chapter, we have learned the basics of superconductor and Josephson junction. A superconductor expels a magnetic field in addition to having zero resistance. It can only keep its superconductivity when it operates below the critical current density, critical magnetic field, and critical temperature. Josephson junction is an $S - i - S$ stack. It has a constant current even if there is zero voltage. The phase of the junction determines the current and is dependent on the applied voltage. The Josephson phase is a very important quantity and is involved in both Josephson equations. As a nonlinear inductor, Josephson junction has inductance due to kinetic inductance. Josephson inductance and Josephson energy depend on its critical current. We showed that the product of phase and reduced magnetic flux quantum plays a role similar to magnetic flux in a regular inductor. This will be useful in the following chapters when we quantize the circuits. We also discuss the two common methods for fabricating the Josephson junction, namely, the Dolan-bridge method and the Manhattan-style method.

Problems

16.1 Carrier Distribution
Use Excel or Python to plot the Fermi-Dirac distribution and Bose-Einstein distribution at $10\,\text{mK}$, $1K$, and $300K$. Discuss your observations.

16.2 Cooper Pair Gap Energy
Find the critical temperatures of Al and Nb. Use Eq. (16.4) to find the energy required to break a Cooper pair in Al and Nb. Remember the factor of 2.

16.3 Critical Current Density
Repeat Example 16.2 but apply to an $Nb/insulator/Nb$ structure.

16.4 Flux-Tunable Qubit
Prove Eq. (16.39). The original form of Eq. (16.39) should be

$$\hat{H}_J = -E_{J1}\cos\hat{\varphi}_1 - E_{J2}\cos\hat{\varphi}_2$$
$$= -E_{J\Sigma}\cos\left(\frac{\pi\Phi}{\Phi_0}\right)\sqrt{1 + d^2\tan^2\left(\frac{\pi\Phi}{\Phi_0}\right)}\cos(\hat{\varphi} - \varphi_0), \quad (16.42)$$

where it has an extra φ_0 in the last *cosine*. When the magnetic flux is constant, we can perform a shift of variables to get Eq. (16.39) [9]. So, we actually need to prove Eq. (16.42). To do so, it is also given that $\tan\varphi_0 = d\tan\frac{\pi\Phi}{\Phi_0}$.

Firstly, use Eqs. (16.36) and (16.35) to express the first line of Eq. (16.42) in terms of $\hat{\varphi}$ and use trigonometric identities to expand it. Then substitute the definitions of d and φ_0 and use trigonometric identities in the second line of Eq. (16.42) to prove that it is the same as the first line.

References

1. Michael Tinkham. *Introduction To Superconductivity*. McGraw-Hill College, 1995.
2. Z.K. Minev, Z. Leghtas, S.O. Mundhada, L. Christakis, I.M. Pop, and M.H. Devoret. Energy-participation quantization of josephson circuits. *npj Quantum Inf.*, 7(131), 2021.
3. Leif Grönberg, Mikko Kiviranta, Visa Vesterinen, Janne Lehtinen, Slawomir Simbierowicz, Juho Luomahaara, Mika Prunnila, and Juha Hassel. Side-wall spacer passivated sub-um josephson junction fabrication process. *Superconductor Science and Technology*, 30(12):125016, nov 2017.
4. G. J. Dolan. Offset masks for lift-off photoprocessing. *Applied Physics Letters*, 31(5):337–339, 09 1977.
5. A. Potts, G.J. Parker, J.J. Baumberg, and P.A.J. de Groot. Cmos compatible fabrication methods for submicron josephson junction qubits. *IEE Proceedings - Science, Measurement and Technology*, 148:225–228(3), September 2001.
6. Nandini Muthusubramanian, Matvey Finkel, Pim Duivestein, Christos Zachariadis, Sean L. M. van der Meer, Hendrik M. Veen, Marc W. Beekman, Thijs Stavenga, Alessandro Bruno, and Leonardo DiCarlo. Wafer-scale uniformity of Dolan-bridge and bridgeless Manhattan-style Josephson junctions for superconducting quantum processors. *Quantum Science and Technology*, 9(2):025006, April 2024.
7. Theodore Van Duzer. *Principles of Superconductive Devices and Circuits*. Pearson, 2008.
8. Vinay Ambegaokar and Alexis Baratoff. Tunneling between superconductors. *Phys. Rev. Lett.*, 10:486–489, Jun 1963.
9. Jens Koch, Terri M. Yu, Jay Gambetta, A. A. Houck, D. I. Schuster, J. Majer, Alexandre Blais, M. H. Devoret, S. M. Girvin, and R. J. Schoelkopf. Charge-insensitive qubit design derived from the cooper pair box. *Phys. Rev. A*, 76:042319, Oct 2007.

Chapter 17
Cooper Pair Box Qubit: Hamiltonian

17.1 Introduction

In Chap. 13, we introduced classical Lagrangian and Hamiltonian mechanics. Lagrangian is used to construct Hamiltonian, which connects classical mechanics to quantum mechanics through the promotion of conjugate variables to operators. We demonstrated this through the quantization of a simple harmonic oscillator (SHO) in Chap. 14. In Chap. 15, we applied the same methodology to quantize an LC tank but showed that it could not be used as a qubit due to the lack of anharmonicity. We, therefore, introduced the Josephson junction, which can serve as a non-linear inductor to introduce anharmonicity, and its equations in Chap. 16. In this chapter, we will construct a circuit using a Josephson junction and a capacitor, known as a Cooper pair box, which can be used as a *superconducting qubit*. We will first study its property when it is fully isolated. Then, we will study a Cooper pair box capacitively connected to an external source. In this process, we will review the systematic approach to quantizing a circuit. We will go through the derivation step-by-step and understand its physics and implications.

17.1.1 Learning Outcomes

Be familiar with the procedure for obtaining the Hamiltonian of a given circuit with a Josephson junction; be able to explain the meaning of the parameters of a Cooper pair box circuit.

17.1.2 Teaching Videos

- Search for Ch17 in this playlist
    - https://tinyurl.com/3yhze3jn
- Other videos
    - https://youtu.be/jMS0mJkvjrY
    - https://youtu.be/ncBc5gqPgcE

17.2 Isolated Cooper Pair Box

In Chap. 15, we quantized an LC tank and showed that it is not a suitable qubit due to its lack of anharmonicity since its inductor is linear (with a constant inductance). Here, we replace the inductor with a Josephson junction to form a **Cooper pair box**. Figure 17.1 shows the circuit. It is the same as the circuit in Fig. 15.1 except that the inductor is replaced by a Josephson junction. Figure 17.1 also shows the circuit symbol of a Josephson junction. Note that a Josephson junction has intrinsic capacitance, C_J, because it has a metal-insulator-metal structure. It is common to draw this explicitly by showing C_J in parallel with the "bare"/"irreducible" junction, which represents the Josephson junction actions governed by the Josephson equations (Eqs. (16.8) and (16.9)).

In Sect. 15.3, we discussed that the **generalized coordinate, generalized velocity**, and **generalized momentum** of an electrical circuit with capacitors and inductors are Φ, V, and Q, respectively. In Eq. (16.24), we introduced a flux quantity, Φ, associated with Josephson phase, φ, and is repeated here for convenience:

$$\Phi = \Phi_0' \varphi(t),$$
$$\varphi(t) = \frac{\Phi}{\Phi_0'}, \qquad (17.1)$$

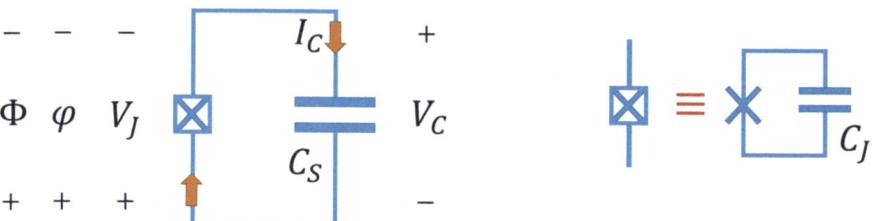

Fig. 17.1 Left: An isolated Cooper pair box obtained by replacing the inductor in an LC tank with Josephson junction. Right: Circuit symbols of Josephson junction

17.2 Isolated Cooper Pair Box

where Φ_0' is the **reduced magnetic flux quantum**. We discussed that Φ plays the role of the flux in a regular geometric inductor. Therefore, we will treat it as the general coordinate in Josephson junction.

We now construct the Hamiltonian directly by replacing the inductor energy in Eq. (15.15) with the energy in the Josephson junction in Eq. (16.20). Here, we do this for simplicity. In the next section, we will go through a formal approach for a Cooper pair box coupled to a voltage source. Therefore,

$$\begin{aligned}\mathcal{H} &= \frac{1}{2}\frac{Q^2}{C} + E_J(1 - \cos\varphi), \\ &= \frac{1}{2}\frac{Q^2}{C} + E_J\left(1 - \cos\frac{\Phi}{\Phi_0'}\right), \\ &= \frac{1}{2}\frac{Q^2}{C} + E_J - E_J\cos\frac{\Phi}{\Phi_0'},\end{aligned} \quad (17.2)$$

where $C = C_J + C_S$.

Since Hamiltonian is the total energy of the system and any constant offset has no physical significance, we can ignore the E_J term. This can be understood by recognizing that any constant term will disappear after taking derivatives in Hamilton's equations, which are used to derive the equations of motion of a system. Therefore,

$$\mathcal{H} = \frac{1}{2}\frac{Q^2}{C} - E_J\cos\frac{\Phi}{\Phi_0'}. \quad (17.3)$$

We can then promote $\mathcal{H}$, Q and Φ to operators like what we did in Chap. 15:

$$\hat{\mathcal{H}} = \frac{1}{2}\frac{\hat{Q}^2}{C} - E_J\cos\frac{\hat{\Phi}}{\Phi_0'}. \quad (17.4)$$

17.2.1 Number Operator and Phase Operator

However, we will stop here because Q and Φ are not convenient conjugate variables to use in our problem. Rather, we will introduce a new **number operator**, $\hat{n}$, and phase operator, $\hat{\varphi}$. This number operator is *not* the number operator we used in the LC tank ($\hat{N}$). $\hat{N}$ is the number of energy quanta or photons with the resonant frequency of the LC tank (Eq. (15.38)). In other words, the eigenstate of $\hat{N}$ is $|N\rangle$ with $N \geq 0$. This is also called the **Fock state** as it represents a state with a well-defined number of particles. But $\hat{n}$ corresponds to the number of Cooper pairs tunneling across the junction and stored at the capacitor (at the superconducting island between the capacitor and the Josephson junction), and it can be negative (tunneling in the opposite direction). $\hat{\varphi}$ corresponds to the Josephson phase. They are defined as

$$\hat{n} = -\frac{\hat{Q}}{2e}, \tag{17.5}$$

$$\hat{\varphi} = \frac{\hat{\Phi}}{\Phi'_0}, \tag{17.6}$$

where we used Eq. (17.1) to define the phase. The Hamiltonian becomes:

$$\begin{aligned}\hat{\mathcal{H}} &= \frac{1}{2}\frac{\hat{Q}^2}{C} - E_J \cos\frac{\hat{\Phi}}{\Phi'_0}, \\ &= \frac{1}{2}\frac{(2e\hat{n})^2}{C} - E_J \cos\hat{\varphi}, \\ &= 4\left(\frac{1}{2}\frac{e^2}{C}\right)\hat{n}^2 - E_J \cos\hat{\varphi}, \\ &= 4E_c\hat{n}^2 - E_J \cos\hat{\varphi}, \end{aligned} \tag{17.7}$$

where we defined the **charging energy** of *one* electron to the capacitor as E_c

$$E_c = \frac{1}{2}\frac{e^2}{C}. \tag{17.8}$$

This should *not* be confused with the total capacitive energy in Eq. (15.11). **Note that, in the literature, sometimes E_c is defined as the charging energy of a Cooper pair (i.e., two electrons),** $E_c = \frac{1}{2}\frac{(2e)^2}{C}$. Then, Eq. (17.7) would not have the factor 4.

$\hat{\varphi}$ and $\hat{n}$ are conjugated pairs, and they are the conjugate coordinate and the conjugate momentum, respectively (we take this as given). They obey the following commutation relation:

$$[\hat{\varphi}, \hat{n}] = i. \tag{17.9}$$

Note that there is *no* $\hbar$ on the right-hand side (unlike Eq. (15.28)).

Example 17.1 Given that $\left[e^{i\hat{\varphi}}, \hat{n}\right] = -e^{i\hat{\varphi}}$, show that $e^{i\hat{\varphi}}|n\rangle = |n+1\rangle$, where $|n\rangle$ is the eigenvector of $\hat{n}$.

Since

$$\left[e^{i\hat{\varphi}}, \hat{n}\right] = -e^{i\hat{\varphi}}, \tag{17.10}$$

17.2 Isolated Cooper Pair Box

applying it to $|n\rangle$,

$$\left[e^{i\hat{\varphi}}, \hat{n}\right]|n\rangle = -e^{i\hat{\varphi}}|n\rangle,$$

$$e^{i\hat{\varphi}}\hat{n}|n\rangle - \hat{n}e^{i\hat{\varphi}}|n\rangle = -e^{i\hat{\varphi}}|n\rangle,$$

$$e^{i\hat{\varphi}}n|n\rangle - \hat{n}e^{i\hat{\varphi}}|n\rangle = -e^{i\hat{\varphi}}|n\rangle,$$

$$\hat{n}(e^{i\hat{\varphi}}|n\rangle) = (n+1)(e^{i\hat{\varphi}}|n\rangle). \tag{17.11}$$

The equation shows that when $\hat{n}$ is applied to vector $e^{i\hat{\varphi}}|n\rangle$, it does not change the vector but only multiplies it by $n+1$. Therefore, $e^{i\hat{\varphi}}|n\rangle$ must be an eigenvector of $\hat{n}$ with eigenvalue $n+1$, which is $|n+1\rangle$. That is

$$e^{i\hat{\varphi}}|n\rangle = |n+1\rangle. \tag{17.12}$$

Similarly, we can show that

$$e^{-i\hat{\varphi}}|n\rangle = |n-1\rangle. \tag{17.13}$$

∎

17.2.2 Hamiltonian in Number Basis

Now, we will represent the Hamiltonian in number basis (i.e., the basis formed by the eigenvectors of the number operator, $|n\rangle$). Firstly,

$$\hat{\mathcal{H}} = 4E_c\hat{n}^2 - E_J \cos\hat{\varphi},$$
$$= 4E_c\hat{n}^2 - E_J\frac{1}{2}(e^{i\hat{\varphi}} + e^{-i\hat{\varphi}}). \tag{17.14}$$

We need to find the m-th-row n-th-column element of $\hat{\mathcal{H}}$. This is just $\langle m|\hat{\mathcal{H}}|n\rangle$. That is,

$$\langle m|\hat{\mathcal{H}}|n\rangle = 4E_c\langle m|\hat{n}^2|n\rangle - E_J\frac{1}{2}(\langle m|e^{i\hat{\varphi}}|n\rangle + \langle m|e^{-i\hat{\varphi}}|n\rangle). \tag{17.15}$$

Note that $|n\rangle$ is orthonormal and most of the terms (e.g. $m = n+2$) are zero except for the following conditions. When $m = n$, we have:

$$4E_c\langle m|\hat{n}^2|n\rangle = 4E_c\langle n|\hat{n}^2|n\rangle,$$
$$= 4E_c\langle n|n^2|n\rangle,$$
$$= 4E_c n^2\langle n|n\rangle,$$

$$= 4E_c n^2, \tag{17.16}$$

and when $m = n + 1$, we have:

$$\langle m| e^{i\hat{\varphi}} |n\rangle = \langle n+1| e^{i\hat{\varphi}} |n\rangle,$$
$$= \langle n+1|n+1\rangle,$$
$$= 1, \tag{17.17}$$

where we have used Eq. (17.12). When $m = n - 1$, we have:

$$\langle m| e^{-i\hat{\varphi}} |n\rangle = \langle n-1| e^{-i\hat{\varphi}} |n\rangle,$$
$$= \langle n-1|n-1\rangle,$$
$$= 1, \tag{17.18}$$

where we have used Eq. (17.13). Therefore,

$$\hat{\mathcal{H}} = \sum_{m=-\infty}^{\infty} \sum_{n=-\infty}^{\infty} \langle m| \hat{\mathcal{H}} |n\rangle |m\rangle \langle n|,$$
$$= \sum_{n=-\infty}^{\infty} \left(4E_c n^2 |n\rangle \langle n| - E_J \frac{1}{2}(|n+1\rangle \langle n| + |n-1\rangle \langle n|)\right). \tag{17.19}$$

The summation over m is eliminated because only $m = n$, $m = n+1$, and $m = n-1$ are nonzero. Also note that, unlike N in LC tank which must be nonnegative, n here refers to the number of extra Cooper pairs stored in the superconducting island and can be negative (which means positive charges are left over on the superconductor).

Example 17.2 Write $\hat{\mathcal{H}}$ in $|n\rangle$ basis. Assume n can only take either 0 or 1.

This means that there are either 0 or 1 extra Cooper pairs. Therefore, it is a 2×2 matrix with basis $|n = 0\rangle$ and $|n = 1\rangle$:

$$\hat{\mathcal{H}} = 4E_c 0^2 |0\rangle \langle 0| + 4E_c 1^2 |1\rangle \langle 1| - E_J \frac{1}{2} |1\rangle \langle 0| - E_J \frac{1}{2} |0\rangle \langle 1|,$$
$$= 0 |0\rangle \langle 0| + 4E_c |1\rangle \langle 1| + \frac{-E_J}{2} |1\rangle \langle 0| + \frac{-E_J}{2} |0\rangle \langle 1|,$$
$$= \begin{pmatrix} 0 & \frac{-E_J}{2} \\ \frac{-E_J}{2} & 4E_c \end{pmatrix}. \tag{17.20}$$

∎

17.3 Cooper Pair Box Coupled to a Voltage Source

A Cooper pair box needs to be connected to the outside world so that the qubit can be manipulated. Now, we will study a Cooper pair box which is capacitively coupled to a gate voltage. We will also follow the formal quantization method we have learned so far. Parts of the flow are based on [1].

Figure 17.2 shows the flowchart to quantize a system with electrical circuit variables as an example. Now, we will follow the flowchart step-by-step.

The circuit to be quantized is shown in Fig. 17.3. This is the same as the circuit in Fig. 17.1 but now is coupled to a *gate* voltage through C_g (note that some variable names and polarity definitions have been changed).

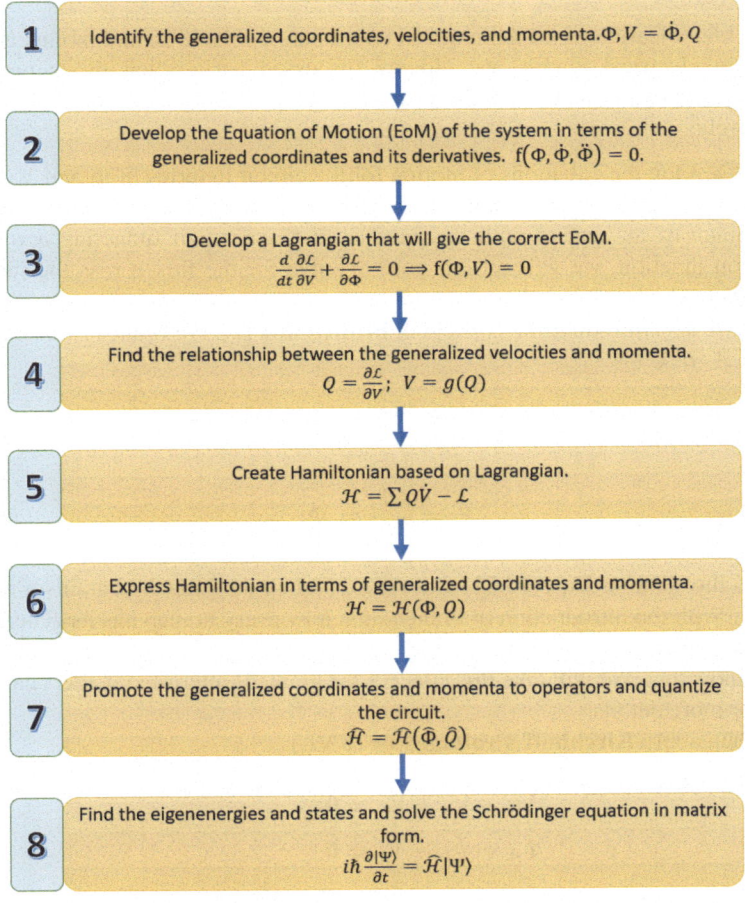

Fig. 17.2 Quantization flow using the electrical circuit as an example

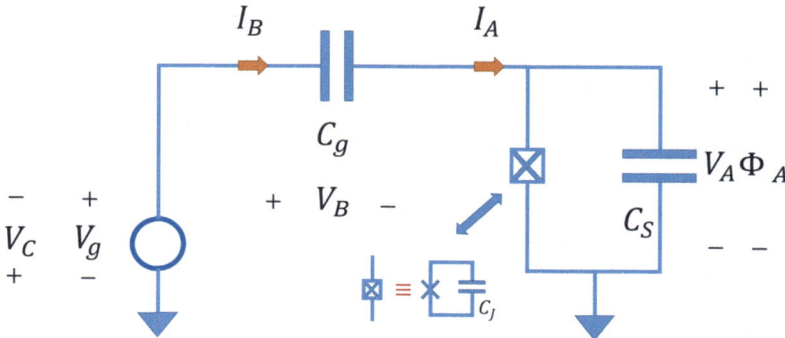

Fig. 17.3 Circuit of Cooper pair box with an external gate voltage

Step 1 Identification of the generalized coordinates, velocities, and momenta.

For an electrical circuit, we already discussed earlier that the generalized coordinate, velocity, and momentum are flux (Φ), voltage (V), and charge (Q), respectively.

Step 2 Develop the equations of motion for the circuit in terms of Φ and V.

This is just to solve the circuit classically. Since our final goal is to obtain the Hamiltonian of the circuit with the number of Cooper pair tunneling across the Josephson junction, we will set our goal to solve for the flux across the junction (Eq. (17.1)) which is labeled as $\Phi = \Phi_A$.

Here we also introduce the concept of **branch flux** [2]. For branch i in the circuit (for $i = A, B, C$)

$$\Phi_i(t) = \int_{-\infty}^{t} V_i(\tau) d\tau, \tag{17.21}$$

$$V_i = \frac{d\Phi_i}{dt} = \dot{\Phi}_i, \tag{17.22}$$

which is the same as Eqs. (15.21) and (15.22) when the branch contains a regular inductor. With the introduction of branch flux, now every branch has its generalized coordinate, even if it might not have a real magnetic flux (such as the one with a regular inductor). We will take this concept for granted. Interested readers can refer to [2] for more details.

We then apply **Kirchhoff's voltage law** (KVL):

$$V_A + V_B + V_C = 0,$$
$$\dot{\Phi}_A + \dot{\Phi}_B - V_g = 0,$$
$$\dot{\Phi}_B = V_g - \dot{\Phi}_A,$$
$$\ddot{\Phi}_B = \dot{V}_g - \ddot{\Phi}_A, \tag{17.23}$$

17.3 Cooper Pair Box Coupled to a Voltage Source

where we have used Eq. (17.22) and the fact that $V_C = -V_g$ as labeled in the figure. In the last line, we also take one more time derivative to prepare for future use.

Our goal is to eliminate $\dot{\Phi}_B$ so that we have only Φ_A and its first and second derivatives in the equation. Therefore, we also use **Kirchhoff's current law** (KCL):

$$I_B = I_A. \tag{17.24}$$

Since $I_B = \frac{dQ_B}{dt}$ (the displacement current passing through the capacitor) and $Q_B = C_g V_B$, we have:

$$\begin{aligned} I_B &= \dot{Q}_B, \\ &= C_g \dot{V}_B, \\ &= C_g \ddot{\Phi}_B. \end{aligned} \tag{17.25}$$

For I_A, it is the sum of the current passing through the "bare" Josephson junction (I_J) and the displacement currents passing through C_J and C_S. Using the same approach we used for I_B, we have:

$$\begin{aligned} I_A &= I_J + C_J \ddot{\Phi}_A + C_S \ddot{\Phi}_A, \\ &= I_J + C_\Sigma \ddot{\Phi}_A, \\ &= I_c \sin \frac{\Phi_A}{\Phi_0'} + C_\Sigma \ddot{\Phi}_A, \end{aligned} \tag{17.26}$$

where we have defined $C_\Sigma = C_J + C_S$ and used the first Josephson equation (Eq. (16.8)). Now we substitute Eqs. (17.26) and (17.25) into Eq. (17.24):

$$\begin{aligned} I_B &= I_A, \\ C_g \ddot{\Phi}_B &= I_c \sin \frac{\Phi_A}{\Phi_0'} + C_\Sigma \ddot{\Phi}_A. \end{aligned} \tag{17.27}$$

To eliminate $\ddot{\Phi}_B$, we substitute Eq. (17.23) (the result from KVL) into Eq. (17.27) and obtain:

$$C_g (\dot{V}_g - \ddot{\Phi}_A) = I_c \sin \frac{\Phi_A}{\Phi_0'} + C_\Sigma \ddot{\Phi}_A. \tag{17.28}$$

This is the equation of motion for this circuit, which contains only Φ_A and its derivatives.

Step 3 Construct the Lagrangian that gives the correct equations of motion.

We need to guess the Lagrangian which is kinetic energy minus potential energy (Eq. (13.2)). Based on our experience with the LC tank and also what we assumed in Eq. (17.2), we know that the capacitor stores the kinetic energy and the Josephson junction stores the potential energy. The energy across C_Σ is $\frac{1}{2}C_\Sigma V_A^2$. The energy across C_g is $\frac{1}{2}C_g(V_A + V_C)^2$ (note that $-V_C - V_A$ is the potential difference across C_g due to the signs in the figure). And the energy stored in the Josephson junction is $-E_J \cos \frac{\Phi_A}{\Phi_0'}$ (Eq. (17.4)). Therefore, we *guess*:

$$\mathcal{L} = T - V,$$
$$= \frac{1}{2}C_\Sigma V_A^2 + \frac{1}{2}C_g(V_A + V_C)^2 + E_J \cos \frac{\Phi_A}{\Phi_0'},$$
$$= \frac{1}{2}C_\Sigma \dot{\Phi}_A^2 + \frac{1}{2}C_g(\dot{\Phi}_A + \dot{\Phi}_C)^2 + E_J \cos \frac{\Phi_A}{\Phi_0'}, \quad (17.29)$$

where we have used Eq. (17.22). Now apply Lagrange's equations to check if our guess is correct:

$$\frac{d}{dt}\frac{\partial \mathcal{L}}{\partial \dot{q}_i} - \frac{\partial \mathcal{L}}{\partial q_i} = 0,$$

$$\frac{d}{dt}\frac{\partial \left(\frac{1}{2}C_\Sigma \dot{\Phi}_A^2 + \frac{1}{2}C_g(\dot{\Phi}_A + \dot{\Phi}_C)^2 + E_J \cos \frac{\Phi_A}{\Phi_0'} \right)}{\partial \dot{\Phi}_A}$$
$$- \frac{\partial \left(\frac{1}{2}C_\Sigma \dot{\Phi}_A^2 + \frac{1}{2}C_g(\dot{\Phi}_A + \dot{\Phi}_C)^2 + E_J \cos \frac{\Phi_A}{\Phi_0'} \right)}{\partial \Phi_A} = 0. \quad (17.30)$$

Here we only have one generalized coordinate, Φ_A and, thus, we only have one equation. Note that $\dot{\Phi}_C = -V_g$ is given as an external voltage source. In the first term, the Josephson junction energy part does not depend on the generalized velocity, $\dot{\Phi}_A$. In the second term, the capacitive energy part does not depend on the generalized coordinate, Φ_A. Therefore,

$$\frac{d}{dt}\frac{\partial \left(\frac{1}{2}C_\Sigma \dot{\Phi}_A^2 + \frac{1}{2}C_g(\dot{\Phi}_A + \dot{\Phi}_C)^2 \right)}{\partial \dot{\Phi}_A} - \frac{\partial \left(E_J \cos \frac{\Phi_A}{\Phi_0'} \right)}{\partial \Phi_A} = 0,$$

$$\frac{d}{dt}\left(C_\Sigma \dot{\Phi}_A + C_g(\dot{\Phi}_A + \dot{\Phi}_C) \right) + \frac{E_J}{\Phi_0'} \sin \frac{\Phi_A}{\Phi_0'} = 0,$$

$$C_\Sigma \ddot{\Phi}_A + C_g(\ddot{\Phi}_A + \ddot{\Phi}_C) + \frac{E_J}{\Phi_0'} \sin \frac{\Phi_A}{\Phi_0'} = 0,$$

$$C_\Sigma \ddot{\Phi}_A + C_g(\ddot{\Phi}_A - \dot{V}_g) + \frac{E_J}{\Phi_0'} \sin \frac{\Phi_A}{\Phi_0'} = 0,$$

$$I_c \sin \frac{\Phi_A}{\Phi_0'} + C_\Sigma \ddot{\Phi}_A = C_g(\dot{V}_g - \ddot{\Phi}_A), \quad (17.31)$$

17.3 Cooper Pair Box Coupled to a Voltage Source

where we used the definition of E_J in the last line (Eq. (16.21)). This is the same as Eq. (17.28). Therefore, we have constructed a valid Lagrangian in Eq. (17.29).

Step 4 Now, we want to find the generalized momentum from the Lagrangian and express the generalized velocity in terms of the momentum.

Based on Eq. (13.20)

$$p_i = \frac{\partial \mathcal{L}}{\partial \dot{q}_i},$$

$$Q_A = \frac{\partial \left(\frac{1}{2} C_\Sigma \dot{\Phi}_A^2 + \frac{1}{2} C_g (\dot{\Phi}_A + \dot{\Phi}_C)^2 + E_J \cos \frac{\Phi_A}{\Phi_0'} \right)}{\partial \dot{\Phi}_A},$$

$$= C_\Sigma \dot{\Phi}_A + C_g (\dot{\Phi}_A + \dot{\Phi}_C). \tag{17.32}$$

Our goal is to replace generalized velocity with generalized momentum later. Therefore, we rearrange it as

$$\dot{\Phi}_A = \frac{Q_A - C_g \dot{\Phi}_C}{C_\Sigma + C_g}. \tag{17.33}$$

Steps 5 and 6 Now we can create the Hamiltonian and express it in terms of the generalized coordinate and momentum using Eq. (13.23).

Again, we only have one generalized coordinate Φ_A:

$$\mathcal{H} = p_0 \dot{q}_0 - \mathcal{L},$$

$$= Q_A \dot{\Phi}_A - \left(\frac{1}{2} C_\Sigma \dot{\Phi}_A^2 + \frac{1}{2} C_g (\dot{\Phi}_A + \dot{\Phi}_C)^2 + E_J \cos \frac{\Phi_A}{\Phi_0'} \right),$$

$$= Q_A \dot{\Phi}_A - \dot{\Phi}_A^2 \left(\frac{C_\Sigma + C_g}{2} \right) - \frac{2 C_g \dot{\Phi}_A \dot{\Phi}_C}{2} - \frac{C_g}{2} \dot{\Phi}_C^2 - E_J \cos \frac{\Phi_A}{\Phi_0'},$$

$$= -\dot{\Phi}_A^2 \left(\frac{C_\Sigma + C_g}{2} \right) + \dot{\Phi}_A (Q_A - C_g \dot{\Phi}_C) - \frac{C_g}{2} \dot{\Phi}_C^2 - E_J \cos \frac{\Phi_A}{\Phi_0'},$$

$$= -\left(\frac{Q_A - C_g \dot{\Phi}_C}{C_\Sigma + C_g} \right)^2 \left(\frac{C_\Sigma + C_g}{2} \right) + \left(\frac{Q_A - C_g \dot{\Phi}_C}{C_\Sigma + C_g} \right) (Q_A - C_g \dot{\Phi}_C)$$

$$\quad - \frac{C_g}{2} \dot{\Phi}_C^2 - E_J \cos \frac{\Phi_A}{\Phi_0'},$$

$$= -\frac{(Q_A - C_g \dot{\Phi}_C)^2}{2 (C_\Sigma + C_g)} + \frac{(Q_A - C_g \dot{\Phi}_C)^2}{C_\Sigma + C_g} - \frac{C_g}{2} \dot{\Phi}_C^2 - E_J \cos \frac{\Phi_A}{\Phi_0'},$$

$$= \frac{(Q_A - C_g \dot{\Phi}_C)^2}{2 (C_\Sigma + C_g)} - \frac{C_g}{2} \dot{\Phi}_C^2 - E_J \cos \frac{\Phi_A}{\Phi_0'}, \tag{17.34}$$

where we have substituted Eq. (17.33) in line 5. $-\frac{C_g}{2}\dot{\Phi}_C^2 = -\frac{C_g}{2}V_g^2$ provides an energy shift to the whole system (although can vary as a function of time). Regardless of the basis used, it adds to the Hamiltonian matrix by $-\frac{C_g}{2}V_g^2 I$ which is a diagonal matrix with constant diagonal elements. It, thus, adds an additional total phase shift to the unitary matrix constructed from $e^{-i\frac{\mathcal{H}t}{\hbar}}$. Therefore, we can ignore $-\frac{C_g}{2}\dot{\Phi}_C^2$. Also we replace $\dot{\Phi}_C$ by $-V_g$ in the first term to get the final Hamiltonian as

$$\mathcal{H} = \frac{(Q_A + C_g V_g)^2}{2(C_\Sigma + C_g)} - E_J \cos\frac{\Phi_A}{\Phi_0'}. \tag{17.35}$$

This is very similar to Eq. (17.3). Particularly, if C_g is zero, which means the Cooper pair box is not coupled to the outside world, Eq. (17.35) becomes Eq. (17.3).

Step 7 Now we will promote generalized coordinate, Φ_A, and generalized momentum, Q_A, to operators. That is,

$$\hat{\mathcal{H}} = \frac{(\hat{Q}_A + C_g V_g)^2}{2(C_\Sigma + C_g)} - E_J \cos\frac{\hat{\Phi}_A}{\Phi_0'}. \tag{17.36}$$

Step 8 Find the eigenvalues and solve the Schrödinger equation.

Similar to what we did in Sect. 17.2.1, we want to use the number and phase operators as the generalized variables to continue to solve this problem. Besides having $\hat{n} = -\frac{\hat{Q}_A}{2e}$ as in Eq. (17.5), we also define:

$$n_g = \frac{C_g V_g}{2e}. \tag{17.37}$$

For the given setup in Fig. 17.3, if $V_g > 0$, it will induce negative charges (i.e., positive number of Cooper pairs) in the Cooper pair box. This becomes the **offset charge**, which is due to the coupling from the gate. Therefore, there is no negative sign in Eq. (17.37). As a result

$$\hat{\mathcal{H}} = \frac{(2e)^2(-\hat{n} + n_g)^2}{2(C_\Sigma + C_g)} - E_J \cos\frac{\hat{\Phi}_A}{\Phi_0'},$$

$$\approx \frac{(2e)^2(-\hat{n} + n_g)^2}{2C_\Sigma} - E_J \cos\frac{\hat{\Phi}_A}{\Phi_0'},$$

$$= 4E_c(\hat{n} - n_g)^2 - E_J \cos\hat{\varphi}, \tag{17.38}$$

where we have assumed $C_\Sigma \gg C_g$ from line 1 to line 2. This is the same as Eq. (17.7) except that it has an offset charge n_g. This is a very important aspect of a Cooper pair box. The offset charge can also be induced through capacitive coupling from the environment (this is easy to understand if you treat V_g as an unknown noise source). Therefore, the charge qubit formed by the Cooper pair box is very sensitive to **charge noise**.

We will defer the eigenvalue finding and Schrödinger equation solving to the next chapter.

17.4 Summary

In this chapter, we study the Cooper pair box which is formed by a Josephson junction and a capacitor. We find the Hamiltonian and introduce the number and phase operators. The number operator has its eigenvalues, and eigenvectors correspond to the number of extra Cooper pairs in the superconducting island (or equivalently tunneling through the junction). The zero and one states are used as the 2-D qubit basis states. *We note that the number state here, $|n\rangle$, is not the number state of an SHO, $|N\rangle$*. We then go through a more formal procedure to quantize a Cooper pair box circuit, which is capacitively coupled to an external voltage source. We obtain a result that is dependent on the offset charge due to capacitively coupled-external voltages that can be noise. In the next chapter, we will diagonalize the Hamiltonian to understand its eigenenergy states.

Problems

17.1 Commutation Relationship of Phase and Number Operators 1
Prove Eq. (17.10). Hints: First expand $e^{i\hat{\varphi}}$ using Taylor expansion so it is a sum of the powers of $\hat{\varphi}$. Then apply the commutation relation in Eq. (17.9) to show that $\left[\hat{\varphi}^k, \hat{n}\right] = ik\hat{\varphi}^{k-1}$ (which can be proved using induction).

17.2 Commutation Relationship of Phase and Number Operators 2
Prove Eq. (17.13).

17.3 Cooper Pair Box Hamiltonian
Following the approach in Example 17.2, write the Hamiltonian matrix with $n = -2$ to $n = 2$, which is a 5×5 matrix.

References

1. Alexandre Blais, Arne L. Grimsmo, S. M. Girvin, and Andreas Wallraff. Circuit quantum electrodynamics. *Rev. Mod. Phys.*, 93:025005, May 2021.
2. M. H. Devoret. Quantum Fluctuations in Electrical Circuits. In S. Reynaud, E. Giacobino, and J. Zinn-Justin, editors, *Fluctuations Quantiques/Quantum Fluctuations*, page 351, January 1997.

Chapter 18
Cooper Pair Box: Analytical Solution

18.1 Introduction

In Chap. 17, we derived the Hamiltonian of a Cooper pair box coupled to an external voltage and found that it depended on the offset charge capacitively induced by the external voltage. In this Chapter, we will diagonalize the Hamiltonian analytically to understand the nature of its eigenstates. Since obtaining an analytical solution is difficult, we will only consider a two-level system which in principle *does not give the correct solution*. However, it will give us a lot of insight and it is instructive. We will discuss how to reduce its charge sensitivity by operating at the sweet spot. To further reduce the charge sensitivity, we discuss the creation of a transmon qubit by increasing the capacitance of the shunt capacitor.

18.1.1 Learning Outcomes

Understand why Cooper pair box is sensitive to charge noise; be able to describe the characteristics of a transmon qubit; be skillful in matrix diagonalization; understand the limitation of two-basis-state approximation.

18.1.2 Teaching Videos

- Search for Ch18 in this playlist
  - https://tinyurl.com/3yhze3jn
- Other videos
  - https://youtu.be/hFiHRZ-s8H0

18.2 General Hamiltonian

In Chap. 17, we derived the Hamiltonian of a Cooper pair box coupled to an external voltage source (circuit given in Fig. 17.3). The Hamiltonian is repeated here for convenience (Eq. (17.38)):

$$\hat{\mathcal{H}} = 4E_c(\hat{n} - n_g)^2 - E_J \cos\hat{\varphi}, \tag{18.1}$$

where E_c is the charging energy of an electron and n_g is the number of **offset charge** (number of extra Cooper pairs due to the external voltage source). We will follow the approach in Sect. 17.2.2 to find its eigenenergies and eigenstates. The readers are encouraged to try themselves first as it is very similar to Sect. 17.2.2. This step corresponds to *Step 8* in Fig. 17.2.

Firstly,

$$\begin{aligned}\hat{\mathcal{H}} &= 4E_c(\hat{n} - n_g)^2 - E_J \cos\hat{\varphi}, \\ &= 4E_c(\hat{n} - n_g)^2 - E_J \frac{1}{2}(e^{i\hat{\varphi}} + e^{-i\hat{\varphi}}).\end{aligned} \tag{18.2}$$

To find the m-th-row n-th-column element of $\hat{\mathcal{H}}$, which is $\langle m|\hat{\mathcal{H}}|n\rangle$, we have:

$$\langle m|\hat{\mathcal{H}}|n\rangle = 4E_c \langle m|(\hat{n}-n_g)^2|n\rangle - E_J \frac{1}{2}(\langle m|e^{i\hat{\varphi}}|n\rangle + \langle m|e^{-i\hat{\varphi}}|n\rangle). \tag{18.3}$$

Note that $|n\rangle$ is orthonormal and most of the terms (e.g. $m = n+2$) are zero except for the following conditions. When $m = n$, we have:

$$\begin{aligned}4E_c \langle m|(\hat{n}-n_g)^2|n\rangle &= 4E_c \langle n|(\hat{n}^2 - 2\hat{n}n_g + n_g^2)|n\rangle, \\ &= 4E_c \langle n|(n^2 - 2nn_g + n_g^2)|n\rangle, \\ &= 4E_c \langle n|(n-n_g)^2|n\rangle, \\ &= 4E_c(n-n_g)^2 \langle n|n\rangle, \\ &= 4E_c(n-n_g)^2,\end{aligned} \tag{18.4}$$

and when $m = n+1$, we have:

$$\begin{aligned}\langle m|e^{i\hat{\varphi}}|n\rangle &= \langle n+1|e^{i\hat{\varphi}}|n\rangle, \\ &= \langle n+1|n+1\rangle, \\ &= 1,\end{aligned} \tag{18.5}$$

18.3 Hamiltonian with Two-Basis-State Approximation

where we have used Eq. (17.12). When $m = n - 1$, we have:

$$\langle m | e^{-i\hat{\varphi}} | n \rangle = \langle n - 1 | e^{-i\hat{\varphi}} | n \rangle ,$$
$$= \langle n - 1 | n - 1 \rangle ,$$
$$= 1, \quad (18.6)$$

where we have used Eq. (17.13). Therefore,

$$\hat{\mathcal{H}} = \sum_{m=-\infty}^{\infty} \sum_{n=-\infty}^{\infty} \langle m | \hat{\mathcal{H}} | n \rangle | m \rangle \langle n | ,$$

$$= \sum_{n=-\infty}^{\infty} \left(4E_c (n - n_g)^2 | n \rangle \langle n | - E_J \frac{1}{2} (| n + 1 \rangle \langle n | + | n - 1 \rangle \langle n |) \right). \quad (18.7)$$

The summation over m is eliminated because only $m = n$, $m = n+1$, and $m = n-1$ are nonzero.

18.3 Hamiltonian with Two-Basis-State Approximation

In this section, we follow the symbols in [4]. We assume that n can only take either 0 or 1. This means that there can only be 0 or 1 extra Cooper pairs. Then, it is a 2×2 matrix with basis $|n = 0\rangle$ and $|n = 1\rangle$. *Note that in principle, we should diagonalize the matrix from $|n = -\infty\rangle$ to $|n = \infty\rangle$. Limiting it to only two basis states will introduce* **pretty significant errors in some parts of the solution.** *But this is useful for instructional purposes.* We have:

$$\hat{\mathcal{H}} = 4E_c (0 - n_g)^2 |0\rangle \langle 0| + 4E_c (1 - n_g)^2 |1\rangle \langle 1| - E_J \frac{1}{2} |1\rangle \langle 0| - E_J \frac{1}{2} |0\rangle \langle 1|,$$

$$= 4E_c n_g^2 |0\rangle \langle 0| + 4E_c (1 - n_g)^2 |1\rangle \langle 1| + \frac{-E_J}{2} |1\rangle \langle 0| + \frac{-E_J}{2} |0\rangle \langle 1|,$$

$$= E_{cc} n_g^2 |0\rangle \langle 0| + E_{cc} (1 - n_g)^2 |1\rangle \langle 1| + \frac{-E_J}{2} |1\rangle \langle 0| + \frac{-E_J}{2} |0\rangle \langle 1|,$$

$$= \begin{pmatrix} E_{cc} n_g^2 & \frac{-E_J}{2} \\ \frac{-E_J}{2} & E_{cc} (1 - n_g)^2 \end{pmatrix}, \quad (18.8)$$

where we have defined $E_{cc} = 4E_c$, which is the charging energy of a Cooper pair (instead of that of a single electron).

We now shift the energy by a constant so that the diagonal elements are symmetric about zero. We first find their average, which is $E_{av} = \frac{1}{2}(E_{cc} n_g^2 +$

$E_{cc}(1-n_g)^2) = E_{cc}n_g^2 - E_{cc}n_g + \frac{E_{cc}}{2}$. Then, we shift the energy reference by E_{av}:

$$\hat{\mathcal{H}}' = \hat{\mathcal{H}} - E_{av}\mathbf{I}$$

$$= \begin{pmatrix} E_{cc}n_g^2 & \frac{-E_J}{2} \\ \frac{-E_J}{2} & E_{cc}(1-n_g)^2 \end{pmatrix} - \begin{pmatrix} E_{av} & 0 \\ 0 & E_{av} \end{pmatrix},$$

$$= \begin{pmatrix} -E_{cc}(\frac{1}{2} - n_g) & \frac{-E_J}{2} \\ \frac{-E_J}{2} & E_{cc}(\frac{1}{2} - n_g) \end{pmatrix}. \tag{18.9}$$

We then define:

$$E = E_{cc}(1 - 2n_g). \tag{18.10}$$

Then we have:

$$\hat{\mathcal{H}}' = \frac{1}{2}\begin{pmatrix} -E & -E_J \\ -E_J & E \end{pmatrix}. \tag{18.11}$$

18.4 Eigenenergies in Copper Pair Box

Let us now find the eigenenergies for the Hamiltonian in Eq. (18.11). *We need to remind ourselves again that some results will be wrong because of the two-basis-state approximation.* We need to diagonalize it. Let the eigenenergy be λ, we can find it by setting the following determinant equation (Eq. (9.7) in [1]):

$$\begin{vmatrix} -\frac{E}{2} - \lambda & -\frac{E_J}{2} \\ -\frac{E_J}{2} & \frac{E}{2} - \lambda \end{vmatrix} = 0. \tag{18.12}$$

Solving the equation, we obtain:

$$\lambda = \pm\frac{\sqrt{E_J^2 + E^2}}{2}, \tag{18.13}$$

$$= \pm\frac{\sqrt{E_J^2 + E_{cc}^2(1-2n_g)^2}}{2}. \tag{18.14}$$

We will denote the ground state energy as E_0 (negative case) and the excited state as E_1 (positive case). We see that the eigenenergies are a function of n_g. But note that this is a result of the diagonalization of the 2×2 matrix in Eq. (18.11). If we diagonalize the full Hamiltonian in Eq. (18.7) by including other values of n, we

18.4 Eigenenergies in Copper Pair Box

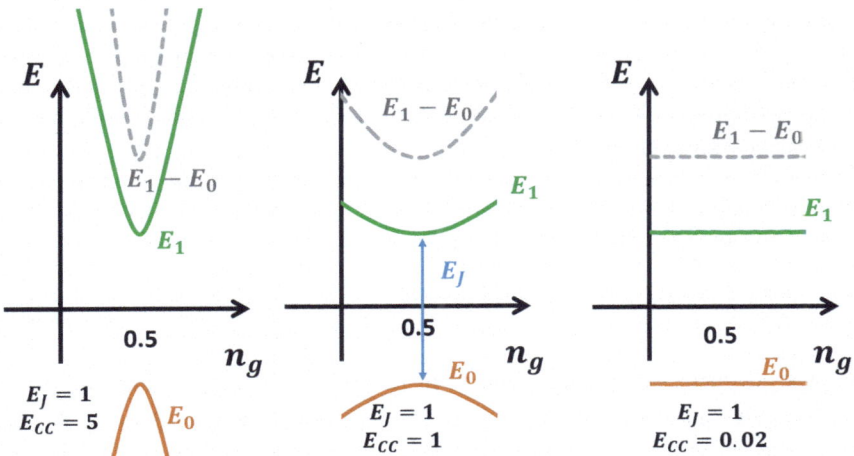

Fig. 18.1 Plots of the two eigenenergies of $|n=0\rangle$ and $|n=1\rangle$ of a Cooper pair box as a function of n_g (limited to $0 \leq n_g \leq 1$). Note that a two-basis-state approximation is used to obtain the result. Therefore, *some parts of the solution are wrong*. Different $\frac{E_{cc}}{E_J}$ ratios are shown with $E_J = 1$

will get a periodic function of n_g with a period of 1 [2, 4]. Therefore, *we restrict the discussion for* $0 \leq n_g \leq 1$ in the following discussion.

Figure 18.1 plots the ground state energy, E_0, the excited state energy, E_1, and their difference, $E_1 - E_0$ as a function of n_g for three cases. It should also be noted that when $n_g = \frac{1}{2}$, $E_1 - E_0 = E_J$. This can be seen in Eq. (18.14) also. When n_g moves away from $\frac{1}{2}$, $E_1 - E_0$ increases.

In the first case, the Josephson energy is smaller than the charging energy with $\frac{E_{cc}}{E_J} = 5:1$. We can see that the energy level separation depends strongly on n_g. Note that the charging energy is that of a Cooper pair, E_{cc}. In the literature, it is often that the charging energy of an electron, E_c, is used. Indeed, when $E_J \ll E_{cc}$, we have:

$$\lambda \approx \pm \frac{\sqrt{E_{cc}^2(1-2n_g)^2}}{2},$$

$$= \pm \frac{E_{cc}(1-2n_g)}{2}. \quad (18.15)$$

But note that this approximation is not valid when $n_g \approx \frac{1}{2}$. Since n_g is induced by the external voltage, it may be similarly induced by external voltage noise through capacitive coupling. As a result, $E_1 - E_0$ changes due to noise, and the two-level system is not well-defined and reduces the T_2 decoherence time (to be introduced in Chap. 25). To minimize the noise effect, we want to operate at $n_g = \frac{1}{2}$ where the slope is zero (i.e., $\frac{d(E_1 - E_0)}{dn_g} = 0$). This is called the **sweet spot**. Note that n_g

is the number of excess Cooper pairs. Therefore, at the sweet spot, it is equivalent to having one excess electron charge. Of course, at a superconducting state, if the temperature is low enough, all electrons should be paired. This is just an equivalency concept as simple (or absurd) as allowing a state to be in the superposition of having 0 and 1 excess Cooper pair.

However, if the noise is large, the problem is still there. One solution is to reduce $\frac{E_{cc}}{E_J}$. For example, when $\frac{E_{cc}}{E_J} = 1:1$, the slope is more gentle, and thus, $E_1 - E_0$ is less sensitive to noise. In the extreme case, when $E_J \gg E_{cc}$, we have:

$$\lambda \approx \pm \frac{E_J}{2}. \tag{18.16}$$

Now the eigenenergies are independent of n_g and can be seen in Fig. 18.1 when $\frac{E_{cc}}{E_J} = 1:50$. Since $E_{cc} = 4E_c = 2\frac{|e|^2}{C}$ (Eq. (17.8)), the larger the C is, the smaller the E_{cc} will be. Therefore, a large shunt C_s can be used to almost eliminate charge sensitivity. This is called the **transmon** qubit. It stands for "transmission-line shunted plasma oscillation qubit" [3].

18.5 Eigenstates of a Cooper Pair Box

Let us find the eigenstates of a Cooper pair box with the two-basis-state approximation. Using λ found in Eq. (18.14), we can solve for the eigenstates. But we will only look at three special cases.

In the first case, assume $E_J \ll E_c$ and $n_g = 0$. In this case, Eq. (18.11) becomes:

$$\hat{\mathcal{H}}' = \frac{1}{2} \begin{pmatrix} -E & -E_J \\ -E_J & E \end{pmatrix},$$

$$\approx \frac{1}{2} \begin{pmatrix} -E & 0 \\ 0 & E \end{pmatrix},$$

$$= \frac{1}{2} \begin{pmatrix} -E_{cc} & 0 \\ 0 & E_{cc} \end{pmatrix}. \tag{18.17}$$

This is a diagonalized matrix, and the eigenvalues are $-\frac{E_{cc}}{2}$ and $\frac{E_{cc}}{2}$ which is exactly what we expect from Eq. (18.14) when $E_J \ll E_c$ and $n_g = 0$. Therefore, its eigenstates are the eigenvectors, $|n\rangle$, of Cooper pair number operator, $\hat{n}$, because the number basis is what was used to construct the matrix (Eq. (18.7)). This means that the ground state is $|0\rangle$ (this is correct) and the first excited state is $|1\rangle$ (this is wrong due to the two-basis-state approximation; see Sect. 19.3.1).

18.5 Eigenstates of a Cooper Pair Box

For the second case, assume $E_J \ll E_c$ and $n_g = \frac{1}{2}$. Therefore, it is operating at the sweet spot. In this case, Eq. (18.14) becomes:

$$\lambda = \pm \frac{\sqrt{E_J^2 + E_{cc}^2(1-2n_g)^2}}{2},$$

$$= \pm \frac{\sqrt{E_J^2}}{2},$$

$$= \pm \frac{E_J}{2}, \tag{18.18}$$

and Eq. (18.11) becomes:

$$\hat{\mathcal{H}}' = \frac{1}{2}\begin{pmatrix} 0 & -E_J \\ -E_J & 0 \end{pmatrix}, \tag{18.19}$$

because $E = 0$. We can solve for the eigenstates. The solutions are $\frac{|0\rangle + |1\rangle}{\sqrt{2}}$ and $\frac{|0\rangle - |1\rangle}{\sqrt{2}}$ (this is correct). Therefore, *the eigenstates are the equal superposition of Cooper pair number states!* We can check this easily.

Example 18.1 Show that the ground state of a Cooper pair box with $E_J \ll E_c$ and $n_g = \frac{1}{2}$ is $\frac{|0\rangle + |1\rangle}{\sqrt{2}}$.

We just need to check,

$$\hat{\mathcal{H}}'(\frac{|0\rangle + |1\rangle}{\sqrt{2}}) = -\frac{E_J}{2}\left(\frac{|0\rangle + |1\rangle}{\sqrt{2}}\right). \tag{18.20}$$

Representing it in matrix form, we have:

$$\hat{\mathcal{H}}'(\frac{|0\rangle + |1\rangle}{\sqrt{2}}) = \frac{1}{2}\begin{pmatrix} 0 & -E_J \\ -E_J & 0 \end{pmatrix}\frac{1}{\sqrt{2}}\begin{pmatrix} 1 \\ 1 \end{pmatrix},$$

$$= \frac{1}{2\sqrt{2}}\begin{pmatrix} -E_J \\ -E_J \end{pmatrix},$$

$$= -\frac{E_J}{2}\frac{1}{\sqrt{2}}\begin{pmatrix} 1 \\ 1 \end{pmatrix},$$

$$= -\frac{E_J}{2}\left(\frac{|0\rangle + |1\rangle}{\sqrt{2}}\right). \tag{18.21}$$

∎

In the third case, let us consider the eigenstates of a transmon qubit where $E_J \gg E_{cc}$. In this case, Eq. (18.11) becomes:

$$\hat{\mathcal{H}}' = \frac{1}{2}\begin{pmatrix} -E & -E_J \\ -E_J & E \end{pmatrix},$$

$$\approx \frac{1}{2}\begin{pmatrix} 0 & -E_J \\ -E_J & 0 \end{pmatrix}. \tag{18.22}$$

This is the same Hamiltonian as the general Cooper pair box at the sweet spot in Eq. (18.19). And the eigenvalues, which are given in Eq. (18.16), are the same as Eq. (18.18). Therefore, it has the same eigenstates as case two. The solutions are $\frac{|0\rangle+|1\rangle}{\sqrt{2}}$ and $\frac{|0\rangle-|1\rangle}{\sqrt{2}}$ (this is wrong due to the two-basis-state approximation; see Sect. 19.3.1). And again, *the eigenstates are the equal superposition of Cooper pair number states!* Note that transmon is insensitive to n_g and thus this is true at every n_g value.

18.6 Summary

In this chapter, we derive the general Hamiltonian of a Cooper pair box capacitively coupled to an external voltage source. We try to solve it analytically by limiting it to two levels. We see that when the shunt capacitor is small, the Josephson energy is much smaller than the charging energy ($E_J \ll E_{cc}$), and it is very sensitive to charge noise which appears as the variation of n_g. One can operate it at the sweet spot ($n_g = \frac{1}{2}$) to eliminate the first-order variation as the sweet spot is where the derivative is zero. To further reduce the second-order variation, transmon qubit is introduced. It has $E_J \gg E_{cc}$ by adding a very large shunt capacitor. It is almost insensitive to charge noise when $\frac{E_J}{E_{cc}} > 50$. Studying the two-level system is instructive. However, it is incomplete and it is inaccurate in some aspects. It also cannot give us insight into the anharmonicity of the qubit. In the next chapter, we will study multi-level systems using numerical methods. We will see that our findings in this chapter are mostly correct and insightful.

Problems

18.1 Sweet Spot
Explain why Eq. (18.15) is not valid when $n_g \approx \frac{1}{2}$. If so, is the minimum $E_1 - E_0 = E_J$?

18.2 Eigenstate at Sweet Spot 1
Show that the excited state of a Cooper pair box with $E_J \ll E_c$ and $n_g = \frac{1}{2}$ is $\frac{|0\rangle - |1\rangle}{\sqrt{2}}$.

18.3 Eigenstate at Sweet Spot 2
Find the eigenstates at the sweet spot by solving the equation with the Hamiltonian in Eq. (18.19) and eigenenergies in Eq. (18.18).

References

1. Hiu-Yung Wong. *Introduction to Quantum Computing*. Springer, 2024.
2. Alexandre Blais, Arne L. Grimsmo, S. M. Girvin, and Andreas Wallraff. Circuit quantum electrodynamics. *Rev. Mod. Phys.*, 93:025005, May 2021.
3. Jens Koch, Terri M. Yu, Jay Gambetta, A. A. Houck, D. I. Schuster, J. Majer, Alexandre Blais, M. H. Devoret, S. M. Girvin, and R. J. Schoelkopf. Charge-insensitive qubit design derived from the cooper pair box. *Phys. Rev. A*, 76:042319, Oct 2007.
4. V Bouchiat, D Vion, P Joyez, D Esteve, and M H Devoret. Quantum coherence with a single cooper pair. *Physica Scripta*, 1998(T76):165, jan 1998.

Chapter 19
Cooper Pair Box: Numerical Solution

19.1 Introduction

In Chap. 18, we limit the Cooper pair box to a two-level system to solve for its eigenenergies and eigenstates analytically. We also derive its eigenstates under certain conditions. Particularly, we see that it has the least charge sensitivity when it is operated at the sweet spot and we can further reduce its charge sensitivity by increasing the Josephson-to-charging energy ratio. This can be achieved by having a large shunt capacitor and it is called the transmon qubit. However, we are not sure if the conclusions from the two-level system analytical solutions are correct because a two-basis-state approximation is used. Moreover, without studying higher levels, we do not understand its anharmonicity and cannot appreciate its role as an *artificial atom* due to its *uneven energy spacings* like in a natural atom. In this chapter, we will diagonalize the Hamiltonian numerically so that we can include more levels. We will compare the results to the conclusions we drew in the previous chapter. Finally, we will perform expansion for Transmon to extract its parameters in analytical form which can be used to design superconducting circuits.

19.1.1 Learning Outcomes

Able to write a Python program to diagonalize a Hamiltonian and plot the energy as a function of background charge; understand the meaning of anharmonicity; understand the trade-off between anharmonicity and charge insensitivity in transmon qubit; be able to relate anharmonicity and qubit frequency to Josephson energy and charging energy.

19.1.2 Teaching Videos

- Search for Ch19 in this playlist
  - https://tinyurl.com/3yhze3jn
- Other videos
  - https://youtu.be/ncBc5gqPgcE

19.2 Anharmonicity

As mentioned, the introduction of a Josephson junction is to replace the linear inductor in an LC tank as a nonlinear inductor to increase the **anharmonicity**. This can be seen in Eq. (17.7) and Fig. 19.1 in which the potential energy is no longer parabolic.

Figure 19.1 also shows the first three eigenenergy levels of the Cooper pair box, and they do not have uniform spacings. We define **absolute anharmonicity** as

$$\alpha = E_{21} - E_{10}, \tag{19.1}$$

where E_{21} is the energy spacing between the third level and the second level and E_{10} is the energy spacing between the second level and the first level (Fig. 19.1). If

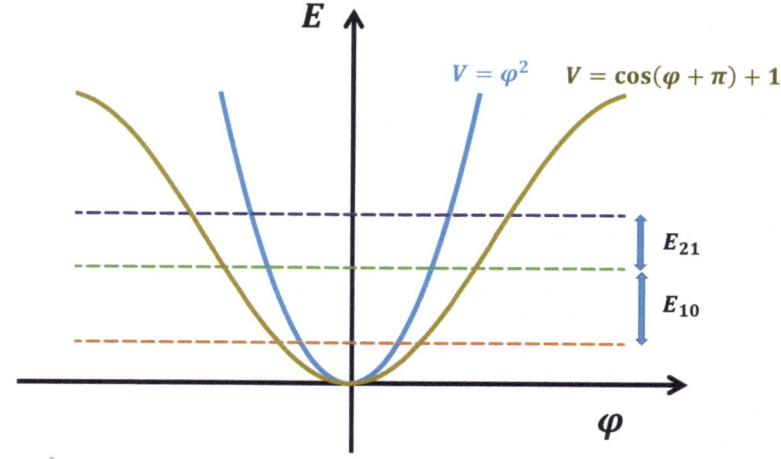

Fig. 19.1 Plots of the potential energy term of the Hamiltonian of an LC tank (Eq. (15.15) by setting $\Phi = \varphi$ for the sake of simplicity) and a Cooper pair box (Eq. (17.7) with shifting). The first three eigenenergy levels of the Cooper pair box are also shown

they are the same (uniform spacing), then $\alpha = 0$ and this is the case for an LC tank (Fig. 15.3).

We may also define **relative anharmonicity** as

$$\alpha_r = \frac{E_{21} - E_{10}}{E_{10}},$$
$$= \frac{\alpha}{E_{10}}. \tag{19.2}$$

Example 19.1 A qubit has E_0, E_1, and E_2 being 0 meV, 0.02 meV, 0.041 meV, respectively. Find its absolute anharmonicity in the unit of Hz.

We have, $E_{10} = E_1 - E_0 = 0.02\,\text{meV} - 0\,\text{meV} = 0.02\,\text{meV}$ and $E_{21} = E_2 - E_1 = 0.041\,\text{meV} - 0.02\,\text{meV} = 0.021\,\text{meV}$.

It is common to express the energy in the unit of Hz because $E = \hbar\omega$. Therefore,

$$\alpha(in\ \text{Hz}) = \frac{E_{21} - E_{10}}{2\pi\hbar},$$
$$= \frac{0.021\,\text{meV} - 0.02\,\text{meV}}{2 \times 3.14 \times 6.58 \times 10^{-16}\,\text{eV}\cdot\text{s}},$$
$$= 242\,\text{MHz} \tag{19.3}$$

Note that this has a positive α which is usually *not* the case for a superconducting qubit (Fig. 19.1). ■

19.3 Python Code for Finding Eigenvalues and Eigenvectors

In the previous chapter, we have limited the Hamiltonian to a two-level system, which is not exact, although we have been able to gain insight into the idea of **sweet spot** and **transmon qubit**. One needs to include more basis states and study levels beyond the first excited state in order to understand its **anharmonicity**. We can do this numerically. In principle, we need to diagonalize a matrix in Cooper pair number basis, $|n\rangle$, from $-\infty$ to $+\infty$. The Hamiltonian is given in Eq. (18.7) and repeated here for convenience:

$$\hat{\mathcal{H}} = \sum_{n=-\infty}^{\infty} \left(4E_c(n - n_g)^2 |n\rangle\langle n| - E_J \frac{1}{2}(|n+1\rangle\langle n| + |n-1\rangle\langle n|) \right). \tag{19.4}$$

However, for practical purposes, we can just include about 100 levels to get fairly accurate solutions. Moreover, we want to shift the reference energy as in Eq. (18.9) so that we can compare it to the two-basis-state approximated analytical solution in

Chap. 18. Therefore, we will diagonalize the following Hamiltonian:

$$\hat{\mathcal{H}}' = \sum_{n=-\infty}^{\infty} \left[\left(4E_c(n-n_g)^2 \right. \right.$$
$$\left. \left. -E_{av}\right) |n\rangle \langle n| - E_J \frac{1}{2}(|n+1\rangle \langle n| + |n-1\rangle \langle n|) \right]. \quad (19.5)$$

The Python code is shown as the following and can be downloaded from https://github.com/hywong2/Quantum_Computing_Architecture. Firstly, we import the necessary libraries:

```
import numpy as np
import matplotlib.pyplot as plt
```

Then, we define a function to construct the Hamiltonian based on the given Josephson energy, E_J, charging energy, E_{CC}, offset charge, n_g, lowest charge basis, $N1$, and the highest charge basis, $N2$. It will return the eigenvalues and eigenvectors. Note that *the eigenvectors are normalized to unity* (i.e., with a length of one).

```
def Diagonalization(EJ, ECC, ng, N1, N2):
    """
    Definitions of arguments
    EJ=1
    #Charging energy (normalized)
    ECC = 1
    # Offset Charge
    ng = 0.3
    #Lowest Charge Basis
    N1 = 0
    #Highest Charge Basis
    N2 = 1
    """
    # The matrix will be indexed from 0 to N
    N = N2-N1+1
    # Initialize the Hamiltonian
    H_dash = np.zeros((N)*(N)).reshape((N,N))
    # Calculate Eav for shifting
    Eav = ECC*ng*ng-ECC*ng+ECC/2

    for i in range(N):
        for j in range(N):
          if(i==j):
            H_dash[i,j]=ECC*((i+N1)-ng)*((i+N1)-ng)-Eav
          if((i==j+1)):
            H_dash[i,j]=-EJ/2
          if((i==j-1)):
            H_dash[i,j]=-EJ/2
    # The returned i-th eigenvector is in eigenvectors[:,i]
    eigenvalues, eigenvectors = np.linalg.eig(H_dash)
    # Sort from low eigenvalues to high values
    idx = eigenvalues.argsort()
    eigenvalues = eigenvalues[idx]
```

19.3 Python Code for Finding Eigenvalues and Eigenvectors

```
    eigenvectors = eigenvectors[:,idx]
    # return with transpose
    return eigenvalues, np.transpose(eigenvectors)
```

Then, we sweep through n_g and plot the curves. One can choose to plot the energy levels, E_0, E_1, and E_2, the energy spacings, E_{10} and E_{21}, and energy levels referenced to the ground state, E_{00}, E_{10}, and E_{20}. For example, $E_{20} = E_2 - E_0$. In the following example, we solve a 201×201 matrix with $|n=-100\rangle$ to $|n=100\rangle$:

```
# def Diagonalization(EJ, ECC, ng, N1, N2)
# Definitions of arguments
EJ=1
#Charging energy (normalized)
ECC = 5
# Offset Charge
ng = 0
#Lowest Charge Basis
N1 = -50
#Highest Charge Basis
N2 = 50

# ng range (extend to -value value)
ng_range = 1
# discretization points in the positive size (so double overall)
discretization = 100

# Initialize eigenenergy arrays
E0=[]
E1=[]
E2=[]
for i in range(0, 2*discretization+1, 1):
    eigenvalues, eigenvectors = Diagonalization\
        (EJ, ECC, -ng_range+2*ng_range*i/2/discretization, N1, N2)
    E0.append(eigenvalues[0])
    E1.append(eigenvalues[1])
    E2.append(eigenvalues[2])

"""
#Plot each level
plt.plot(np.linspace(-ng_range,ng_range,2*discretization+1), E0)
plt.plot(np.linspace(-ng_range,ng_range,2*discretization+1), E1)
plt.plot(np.linspace(-ng_range,ng_range,2*discretization+1), E2)
plt.legend(["E0","E1","E2"],fontsize=13)
"""


#Plot spacing
plt.plot(np.linspace(-ng_range,ng_range,2*discretization+1), \
    np.array(E1)-np.array(E0))
plt.plot(np.linspace(-ng_range,ng_range,2*discretization+1), \
    np.array(E2)-np.array(E1))
plt.legend(["E10","E21"],fontsize=13)
```

```
"""
#Plot each level offset to E0
plt.plot(np.linspace(-ng_range,ng_range,2*discretization+1), \
         np.array(E0)-np.array(E0))
plt.plot(np.linspace(-ng_range,ng_range,2*discretization+1), \
         np.array(E1)-np.array(E0))
plt.plot(np.linspace(-ng_range,ng_range,2*discretization+1), \
         np.array(E2)-np.array(E0))
plt.legend(["E00","E10","E20"],fontsize=13)
"""

plt.grid()
plt.xlabel('$n_g$', fontsize=20)
plt.ylabel('Energy (a.u.)', fontsize=20)
#plt.ylim(0,5)
plt.tick_params(labelsize=16)
```

19.3.1 Validity of Two-Basis-State Approximation

In Fig. 19.2, using the code, we recompute the **charge dispersion** graphs corresponding to those in Fig. 18.1. Firstly, in the **charge qubit regime** ($E_J < E_{CC}$), the two-basis-state approximation in Fig. 18.1 is fairly accurate. Even when $E_J = E_{CC}$, at the sweet spot ($n_g = \frac{1}{2}$), the qubit energy, E_{10}, is about E_J. However, when $n_g = 0$, there is a large discrepancy in the qubit energy (we will see that for the eigenstate also).

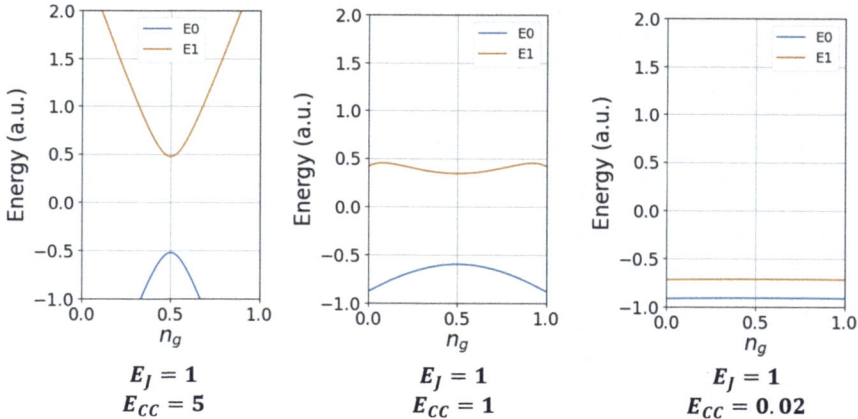

Fig. 19.2 Plots of the two eigenenergies of $|n = 0\rangle$ and $|n = 1\rangle$ of a Cooper pair box as a function of n_g (limited to $0 \leq n_g \leq 1$). Different $\frac{E_{cc}}{E_J}$ ratios are shown with $E_J = 1$. This is set up in the same way as in Fig. 18.1 except that it is computed with 201 charge basis states

However, in the **transmon regime** ($E_J \gg E_{CC}$), the qubit energy is different from E_J, although it still shows that it is insensitive to n_g.

How about the eigenvectors? By printing the components of the eigenvectors, one can find its decomposition in the charge basis. In the charge qubit regime ($E_J < E_{CC}$) and $n_g = 0$, it is found that the ground state is still $|n=0\rangle$, but the first excited state is $\frac{|n=-1\rangle - |n=1\rangle}{2}$ instead of $|n=1\rangle$! This is understandable because the two-level model does not include $|n=-1\rangle$ in the basis.

When it is at the sweet spot ($n_g = \frac{1}{2}$), the ground state is $\frac{|0\rangle + |1\rangle}{\sqrt{2}}$ and the first excited state is $\frac{|0\rangle - |1\rangle}{\sqrt{2}}$. This is the same as what we obtained using the two-basis-state approximation method in Sect. 18.5.

In the transmon region, we expect the eigenstates to be different from the two-basis-state approximation (i.e., the two-basis-state approximation is inaccurate) also since the qubit frequency is shown to be wrong already. We will discuss this in the next section.

In summary, *two-basis-state approximation is fairly accurate at the sweet spot in the charge qubit regime.*

19.4 Transmon Qubit

Now using the code, we want to study various properties of a transmon qubit.

19.4.1 Anharmonicity

To study the anharmonicity, we now plots E_{00}, E_{10}, and E_{20} from $n_g = -2$ to $n_g = 2$ for the three conditions (Fig. 19.3). Energy spacings are the difference between adjacent lines. It can be seen that, for most n_g values, energy spacings are very different, particularly at the sweet spots. The anharmonicity, $\alpha = E_{21} - E_{10}$ (Eq. (19.1)), is the largest in the charge qubit regime. In the transmon regime, while charge sensitivity is reduced, the anharmonicity is also reduced. This is a trade-off. Luckily, with the introduction of a large shunt capacitor to make a transmon qubit, its charge sensitivity is suppressed *exponentially* with $e^{-\sqrt{E_J/E_C}}$ and the anharmonicity only reduces as $\sqrt{E_C/E_J}$ [1].

19.4.2 Hamiltonian of Transmon Qubit

We know that a Cooper pair box has a large anharmonicity due to the introduction of a Josephson junction. In transmon, it has a smaller anharmonicity (assume we make $E_J \gg E_C$). That means it behaves like a linear inductor with some small nonlinear

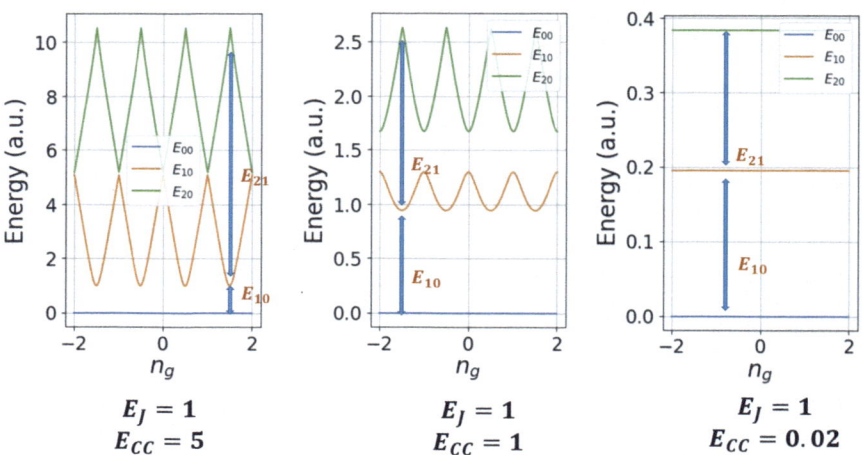

Fig. 19.3 Plots of eigenenergies referenced to the ground state (E_{00}, E_{10}, and E_{20}) as a function of n_g. Different $\frac{E_{cc}}{E_J}$ ratios are shown with $E_J = 1$. This is set up in the same way as in Fig. 19.2

terms. This also means that $\hat{\varphi}$ does not deviate from 0 too much (otherwise, we should have a large anharmonicity due to the large nonlinearity of the inductance). Therefore, we can perform Taylor expansion on the *cosine* term in the Hamiltonian in Eq. (18.1) like what was done in Eq. (16.26) by only keeping the fourth-power term. Also, for simplicity, we set $n_g = 0$. Then,

$$\hat{\mathcal{H}} = 4E_c\hat{n}^2 - E_J \cos\hat{\varphi},$$
$$\approx 4E_c\hat{n}^2 - E_J \left(1 - \frac{\hat{\varphi}^2}{2!} + \frac{\hat{\varphi}^4}{4!}\right),$$
$$= 4E_c\hat{n}^2 + E_J\frac{\hat{\varphi}^2}{2!} - E_J\frac{\hat{\varphi}^4}{4!},$$
$$= 4E_c\hat{n}^2 + E_J\frac{\hat{\varphi}^2}{2} - E_J\frac{\hat{\varphi}^4}{24}, \qquad (19.6)$$

where from line 2 to line 3, the constant term E_J is dropped like what we did in Eq. (17.3). As expected, when $\hat{\varphi}$ is small, the *cosine* term can be approximated by a parabola very well. Now we introduce the annihilation operator, $\hat{b}$, and creation operator, $\hat{b}^\dagger$, based on the conjugate variables $\hat{\varphi}$ and $\hat{n}$. The methodology should be the same as what we did in Chap. 14 for a mechanical SHO or what we did in Chap. 15 for an *LC* tank. *In this process, we ignore the fourth-order term which is treated as a perturbation.*

However, in the commutation relation of $\hat{\varphi}$ and $\hat{n}$ in Eq. (17.9), it does not have $\hbar$ as in Eq. (15.28). Therefore, in order to map to the variables seamlessly, we will

19.4 Transmon Qubit

rewrite Eq. (19.6) as

$$\hat{\mathcal{H}} = 4\frac{E_c}{\hbar}(\sqrt{\hbar}\hat{n})^2 + \frac{E_J}{\hbar}\frac{(\sqrt{\hbar}\hat{\varphi})^2}{2} - E_J\frac{\hat{\varphi}^4}{24}. \quad (19.7)$$

We will ignore the last term for now as it is treated as a perturbation to the system. With that, the system is just a simple harmonic oscillator (an LC-tank). And we have the following commutation relation,

$$\left[\sqrt{\hbar}\hat{\varphi}, \sqrt{\hbar}\hat{n}\right] = i\hbar. \quad (19.8)$$

Then, we can perform a similar mapping as what we did in Eqs. (15.17)–(15.20). We have the following correspondence:

$$\sqrt{\hbar}\hat{n} \iff p, \quad (19.9)$$

$$\sqrt{\hbar}\hat{\varphi} \iff q, \quad (19.10)$$

$$4\frac{E_c}{\hbar} \iff \frac{1}{2m},$$

$$\frac{\hbar}{8E_c} \iff m, \quad (19.11)$$

$$\frac{E_J}{2\hbar} \iff \frac{k}{2},$$

$$\frac{E_J}{\hbar} \iff k. \quad (19.12)$$

Therefore, the "resonant frequency" for this problem is

$$\omega_0 = \sqrt{\frac{\frac{E_J}{\hbar}}{\frac{\hbar}{8E_c}}} = \frac{\sqrt{8E_c E_J}}{\hbar} \iff \sqrt{\frac{k}{m}}. \quad (19.13)$$

This is consistent with what we have in the right of Fig. 19.3 where $E_{10} \approx \sqrt{8 \times 1 \times \frac{0.02}{4}} = 0.2$ (note that $E_{cc} = 4E_c$).

We can define $\hat{b}$ and $\hat{b}^\dagger$ like before and also the number operator $\hat{N} = \hat{b}^\dagger \hat{b}$. By adding back the perturbation term, we eventually convert the Hamiltonian to be

$$\hat{\mathcal{H}} = \hbar\omega_0(\hat{N} + \frac{1}{2}) - E_J\frac{\hat{\varphi}^4}{24},$$

$$= \sqrt{8E_c E_J}\left(\hat{b}^\dagger\hat{b} + \frac{1}{2}\right) - E_J\frac{\hat{\varphi}^4}{24}. \quad (19.14)$$

We can also define $\sqrt{\hbar}\hat{\varphi}$ (the generalized coordinate) and $\sqrt{\hbar}\hat{n}$ (the generalized momentum) as a function of $\hat{b}$ and $\hat{b}^\dagger$ like in Eqs. (14.19) and (14.20). We have:

$$\sqrt{\hbar}\hat{\varphi} = \sqrt{\frac{\hbar}{2m\omega_0}}(\hat{b}+\hat{b}^\dagger),$$

$$= \sqrt{\frac{\hbar}{2\frac{\hbar}{8E_c}\frac{\sqrt{8E_cE_J}}{\hbar}}}(\hat{b}+\hat{b}^\dagger),$$

$$= \sqrt{\hbar}\left(\frac{2E_c}{E_J}\right)^{\frac{1}{4}}(\hat{b}+\hat{b}^\dagger), \qquad (19.15)$$

where we kept m from the mechanical SHO for comparison in line 1. Therefore,

$$\hat{\mathcal{H}} = \sqrt{8E_cE_J}\left(\hat{b}^\dagger\hat{b}+\frac{1}{2}\right) - E_J\frac{\hat{\varphi}^4}{24},$$

$$= \sqrt{8E_cE_J}\left(\hat{b}^\dagger\hat{b}+\frac{1}{2}\right) - E_J\frac{\left(\left(\frac{2E_c}{E_J}\right)^{\frac{1}{4}}(\hat{b}+\hat{b}^\dagger)\right)^4}{24},$$

$$= \sqrt{8E_cE_J}\left(\hat{b}^\dagger\hat{b}+\frac{1}{2}\right) - \frac{E_c}{12}(\hat{b}+\hat{b}^\dagger)^4. \qquad (19.16)$$

This is a very important equation. Firstly, the first term is just that of a **simple harmonic oscillator**. Without the fourth-power term, it tells us that the *qubit energy spacing* or **qubit energy** is just what we had in Eq. (19.13):

$$E_{10} = \hbar\omega_0,$$
$$= \sqrt{8E_cE_J}. \quad (approximation) \qquad (19.17)$$

More precisely, the qubit energy is $\sqrt{8E_cE_J} - E_c$ because the anharmonicity reduces E_1 by E_c (see Problem 19.4). Therefore,

$$E_{10} = \hbar\omega_0,$$
$$= \sqrt{8E_cE_J} - E_c. \qquad (19.18)$$

The perturbation term will introduce anharmonicity. We can solve it using perturbation theory [2]. It can be shown that (see Problem 19.4) the anharmonicity is

$$\alpha = E_{21} - E_{10},$$
$$= -E_c. \qquad (19.19)$$

19.4 Transmon Qubit

Note that this is the negative of the charging energy of an electron, E_c, not of the charging energy of a Cooper pair, E_{cc}.

Example 19.2 Express $\sqrt{\hbar}\hat{n}$ as a function of $\hat{b}$ and $\hat{b}^\dagger$.

We will use Eq. (14.20) as $\sqrt{\hbar}\hat{n}$ is the conjugate momentum:

$$\sqrt{\hbar}\hat{n} = i\sqrt{\frac{m\omega_0\hbar}{2}}(-\hat{b}+\hat{b}^\dagger),$$

$$= i\sqrt{\frac{\frac{\hbar}{8E_c}\frac{\sqrt{8E_cE_J}}{\hbar}\hbar}{2}}(-\hat{b}+\hat{b}^\dagger),$$

$$= i\frac{\sqrt{\hbar}}{2}\left(\frac{E_J}{2E_c}\right)^{\frac{1}{4}}(-\hat{b}+\hat{b}^\dagger),$$

$$= i\sqrt{\hbar}n_{zpf}(-\hat{b}+\hat{b}^\dagger). \tag{19.20}$$

Again, we keep the mechanical SHO symbols in the first line and then substitute them with Eq. (19.11). We also defined the **zero-point fluctuation of the charge number** as $n_{zpf} = \frac{1}{2}\left(\frac{E_J}{2E_c}\right)^{\frac{1}{4}}$. ∎

19.4.3 Eigenvectors of Transmon Qubit

Using the code, we may also plot the eigenvectors of a transmon in the linear combination of the charge basis, $|n\rangle$. The first three eigenvectors are shown in Fig. 19.4. Note that $\hat{n}$ is the conjugate momentum. Therefore, we are projecting

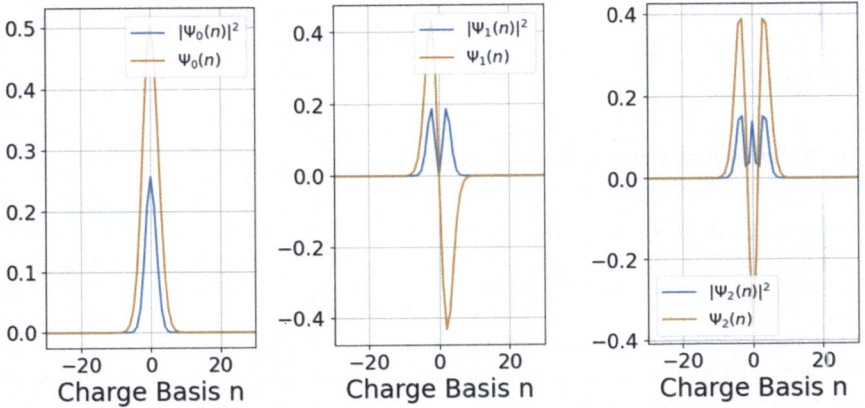

Fig. 19.4 The wavefunctions of the first three states of a transmon with $\frac{E_J}{E_{CC}} = 50$ projected to the charge basis ($|n\rangle$)

the eigenvector to the conjugate momentum space. In Fig. 14.3, we projected the eigenvector of a mechanical simple harmonic oscillator to the *coordinate* space. We see that their wavefunctions have very similar shapes as those in Fig. 19.4 (up to a negative sign). This is expected because it is known that the wavefunction in the coordinate space has the same shape as the wavefunction in the momentum space for an SHO. This further confirms that the eigenvectors of a transmon (when the perturbation due to the nonlinearity is small) are those of an SHO.

In summary, the eigenstate of a transmon is $|N\rangle$ like a SHO and is *nonnegative*. It should not be confused with the charge state (Cooper pair number), $|n\rangle$, which runs from $-\infty$ to ∞.

19.5 Summary

In this chapter, we study the definition of anharmonicity and show that a Cooper pair box has a large anharmonicity when it is in the charge qubit regime with $E_J \ll E_{cc}$. As we increase $\sqrt{\frac{E_J}{E_c}}$, the anharmonicity reduces linearly, but the charge sensitivity reduces exponentially. We created a Python script to study higher excited states in a Cooper pair box. By comparing to the results from the Python script which takes 201 basis states into account, we show that in the charge qubit regime, the two-basis-state approximation in Chap. 18 is fairly accurate in terms of qubit energy ($E_{10} = E_J$) and eigenvector calculation at the sweet spot. However, the two-basis-state approximation fails in the transmon qubit regime. For example, the qubit energy is now $\sqrt{8E_cE_J} - E_c$. We also emphasize that while in the charge qubit regime, the first two eigenstates are the linear combinations of $|n = 0\rangle$ and $|n = 1\rangle$, the eigenstates of a transmon are Fock states, $|N\rangle$, like a regular SHO which has many components of $|n\rangle$. It is also shown that the anharmonicity of transmon qubit is $-E_c$.

Problems

19.1 Relative Anharmonicity
Find the relative anharmonicity in Example 19.1.

19.2 Transmon Qubit Hamiltonian
Following the approach in mechanical SHO, derive the Hamiltonian of transmon qubit by neglecting the fourth-power term in Eq. (19.6).

19.3 Cooper Pair Box Numerical Solution
Modify the given script to study how the anharmonicity of a transmon qubit changes with the number of the basis states used.

19.4 Qubit Frequency and Anharmonicity of Transmon Qubit.
Show that the anharmonicity of a transmon qubit is $-E_c$.

Hints: Expand $(\hat{b} + \hat{b}^\dagger)^4$ in Eq.(19.16). There are 16 terms (make sure to follow the commutation relation). Find $\langle N = 0| (\hat{b} + \hat{b}^\dagger)^4 |N = 0\rangle$, $\langle N = 1| (\hat{b} + \hat{b}^\dagger)^4 |N = 1\rangle$, and $\langle N = 2| (\hat{b} + \hat{b}^\dagger)^4 |N = 2\rangle$ by applying it to each term. Then the anharmonicity is (why?)

$$\alpha = -\frac{E_c}{12}\Big[(\langle N = 2| (\hat{b} + \hat{b}^\dagger)^4 |N = 2\rangle - \langle N = 1| (\hat{b} + \hat{b}^\dagger)^4 |N = 1\rangle)$$
$$-(\langle N = 1| (\hat{b} + \hat{b}^\dagger)^4 |N = 1\rangle - \langle N = 0| (\hat{b} + \hat{b}^\dagger)^4 |N = 0\rangle)\Big]$$

Also note that, if the numbers of $\hat{b}$ and $\hat{b}^\dagger$ are not the same, it will be evaluated to be zero. Only six terms are balanced. Use Eqs. (14.36) and (14.37). Indeed, in [2],

$$\langle m| (\hat{b} + \hat{b}^\dagger)^4 |m\rangle = 6m^2 + 6m + 3$$

Using what you have obtained, also show that the qubit energy is $\sqrt{8E_cE_J} - E_c$.

19.5 Conjugate Coordinate and Momentum for Transmon
Verify Eqs. (19.15) and (19.20) by substituting them into Eq. (19.8).

References

1. Alexandre Blais, Arne L. Grimsmo, S. M. Girvin, and Andreas Wallraff. Circuit quantum electrodynamics. *Rev. Mod. Phys.*, 93:025005, May 2021.
2. L.S. Bishop. *Circuit Quantum Electrodynamics*. BiblioBazaar, 2011.

Chapter 20
Charge Qubit Dynamics: Precession and 1-Qubit Gate

20.1 Introduction

In Chaps. 18 and 19, we analyzed the Hamiltonian and its eigenstates and eigenvectors of a Cooper pair box using two-basis-state approximation and numerical calculation. Regardless of its operating regime (small $\frac{E_J}{E_c}$ as a charge qubit or large $\frac{E_J}{E_c}$ as a transmon qubit), it has well-defined levels and high enough anharmonicity. Therefore, it fulfills one of DiVincenzo's criteria to have a "well-characterized qubit" (Sect. 1.3). In this chapter, we will study how to map the qubit states to the Bloch sphere and apply voltage pulses to manipulate the qubit to obtain a general 1-qubit gate. We will work on the charge qubit because it is simple and instructive, despite the fact that transmon qubit, which is much less sensitive to charge noise, is the dominant qubit being used as a superconducting qubit. Transmon qubit will be discussed in the following chapters.

20.1.1 Learning Outcomes

Appreciate how to map charge qubit physics to the Bloch sphere; be able to use the concepts learned in spin qubit to assist the understanding of charge qubit dynamics.

20.1.2 Teaching Videos

- Search for Ch20 in this playlist
    - https://tinyurl.com/3yhze3jn

- Other videos
    - https://youtu.be/hFiHRZ-s8H0
    - https://youtu.be/7E6ISrP7nCc
    - https://youtu.be/vXReRIsbQ90

20.2 Mapping Charge Qubit to Spin Qubit

In Chap. 19, through numerical calculation of 201 charge basis ($|n = -100\rangle$ to $|n = 100\rangle$), we showed that the two-basis-state approximation in Chap. 18 is valid for a **charge qubit** with $E_J \ll E_{cc}$ and *at the sweet spot*. Therefore, we will use Eqs. (18.10) and (18.11) to study the dynamics of a charge qubit in this chapter. They are repeated here for convenience. *Moreover, we are not interested in quantization now. We will demote the Hamiltonian operator back to a variable.* We continue to follow some of the notations and approaches in [1]:

$$\mathcal{H}' = \frac{1}{2}\begin{pmatrix} -E & -E_J \\ -E_J & E \end{pmatrix}, \tag{20.1}$$

with

$$E = E_{cc}(1 - 2n_g). \tag{20.2}$$

20.2.1 Mapping to Bloch Sphere

Equation (20.1) is a 2×2 matrix. We definitely wonder if it can share the same mathematical procedures we have learned from spin qubit. To do so, we want to express it as a linear combination of **Pauli matrices** (Sect. 5.4). Since Hamiltonian is Hermitian (as energy is an observable), this is possible as shown in Eq. (6.3). Therefore, we rewrite it as

$$\mathcal{H}' = \frac{1}{2}\begin{pmatrix} -E & -E_J \\ -E_J & E \end{pmatrix},$$
$$= -\frac{E_J}{2}\begin{pmatrix} 0 & 1 \\ 1 & 0 \end{pmatrix} - \frac{E}{2}\begin{pmatrix} 1 & 0 \\ 0 & -1 \end{pmatrix},$$
$$= -\frac{E_J}{2}\sigma_x - \frac{E}{2}\sigma_z,$$

20.2 Mapping Charge Qubit to Spin Qubit

$$= -\frac{1}{2}(\sigma_x \sigma_y \sigma_z)\begin{pmatrix} E_J \\ 0 \\ E \end{pmatrix},$$

$$= -\frac{1}{2}\vec{\sigma} \cdot \vec{B}_E, \tag{20.3}$$

where we have defined an *effective fictitious* external magnetic field, $\vec{B}_E$, with

$$\vec{B}_E = \begin{pmatrix} E_J \\ 0 \\ E \end{pmatrix}, \tag{20.4}$$

and its magnitude is

$$|\vec{B}_E| = \sqrt{E_J^2 + E^2}. \tag{20.5}$$

It should also be emphasized that $\vec{B}_E$ *has the unit of energy* as can be seen below. Therefore, the Hamiltonian of a charge qubit is equivalent to a spin qubit under an external magnetic with scaling. This can be compared to Eqs. (8.1), (9.2), and (9.3), and we can map $\gamma\frac{\hbar}{2}$ in those equations to $\frac{1}{2}$ in Eq. (20.3). Or we can *substitute γ with $\frac{1}{\hbar}$ in spin qubit equations to obtain charge qubit equations*. For clarity, they are copied and demonstrated below:

$$\mathcal{H}' = -\frac{1}{2}\vec{\sigma} \cdot \vec{B}_E, \quad Charge\ Qubit,$$

$$H = -\vec{B} \cdot \gamma S,$$

$$= -\gamma\frac{\hbar}{2}\vec{\sigma} \cdot \vec{B}, \quad Spin\ Qubit. \tag{20.6}$$

In Chap. 8, we demonstrated that a spin qubit will **precess** about the external magnetic field (along $\hat{z}$ direction) at **Larmor frequency** of $2B_0|\gamma\vec{S}|/\hbar = B_0|\gamma|$ (Eq. (8.11)). This was obtained with $\gamma\hbar$ in the spin qubit Hamiltonian. In the charge qubit Hamiltonian, since $\gamma\hbar$ is missing, if we had gone through the same derivation as for the spin qubit, we would have obtained a frequency scaled down by $|\gamma|\hbar$. Therefore, we also expect that the charge qubit will perform **Larmor precession** about $\vec{B}_E$ at a frequency of

$$\omega_L = \frac{|\vec{B}_E||\gamma|}{|\gamma|\hbar},$$

$$= \frac{|\vec{B}_E|}{\hbar}. \tag{20.7}$$

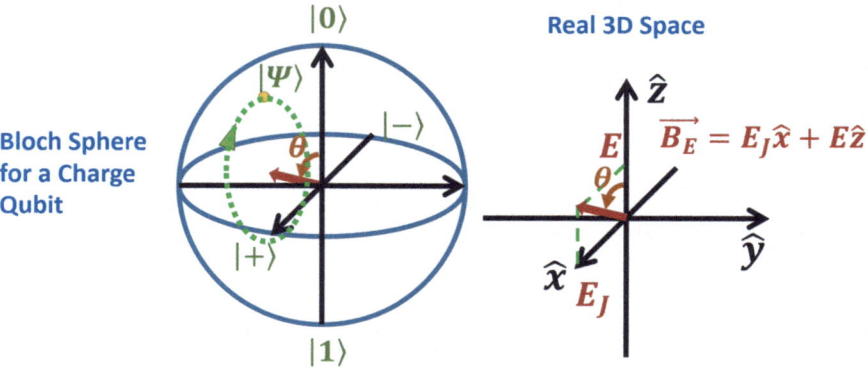

Fig. 20.1 The Bloch sphere representation of a charge qubit and the real 3D space coordinate system in which the direction of the effective fictitious magnetic field is shown. *Note that the so-called magnetic field has the unit of energy*

Note that although we only show the spin qubit precession equation when the field is in the $\hat{z}$ direction, it is true for a field in an arbitrary direction as the space is isotropic.

In Chap. 8, an electron qubit was used. In Eq. (20.3), it has an effective positive gyromagnetic ratio. Therefore, a charge qubit on the Bloch sphere acts as a hole spin qubit, and the precession direction will be opposite to that of an electron qubit for the *same magnetic field direction*.

Figure 20.1 shows the charge qubit represented on the Bloch sphere. Firstly, the effective fictitious magnetic field is only on the $\hat{x} - \hat{z}$ plane as it does not have a $\hat{y}$ component. Secondly, the higher the E_J is, the more the fictitious magnetic field points along the $\hat{x}$ direction. This is the case for a **transmon qubit**, and it explains why $|+\rangle$ and $|-\rangle$ are the eigenstates when we used two-basis-state approximation in Chap. 18 for a transmon (although two-basis-state approximation is inaccurate for transmon). When E_J is small, we expect it to point more vertically. For a charge qubit but not at the sweet spot, in the extreme case, it only points vertically, and, thus, its eigenstates are $|0\rangle$ and $|1\rangle$ which was the result from two-basis-state approximation. However, when it is at the sweet spot, $E = 0$, and it points along the $\hat{x}$ direction and has $|+\rangle$ and $|-\rangle$ as the eigenstates as shown in two-basis-state approximation and numerical calculation. Moreover, since it is equivalent to a spin qubit with a positive charge, it precesses in the opposite direction of an electron spin qubit for the *same magnetic field direction*.

Example 20.1 Express the charge qubit Hamiltonian in terms of θ (see Fig. 20.1) and $|B_E|$:

$$\mathcal{H}' = \frac{1}{2}\begin{pmatrix} -E & -E_J \\ -E_J & E \end{pmatrix},$$

20.2 Mapping Charge Qubit to Spin Qubit

$$= -\frac{\sqrt{E^2+E_J^2}}{2}\begin{pmatrix} \frac{E}{\sqrt{E^2+E_J^2}} & \frac{E_J}{\sqrt{E^2+E_J^2}} \\ \frac{E_J}{\sqrt{E^2+E_J^2}} & -\frac{E}{\sqrt{E^2+E_J^2}} \end{pmatrix},$$

$$= -\frac{|\vec{B}_E|}{2}\begin{pmatrix} \frac{E}{|\vec{B}_E|} & \frac{E_J}{|\vec{B}_E|} \\ \frac{E_J}{|\vec{B}_E|} & -\frac{E}{|\vec{B}_E|} \end{pmatrix},$$

$$= -\frac{|\vec{B}_E|}{2}\begin{pmatrix} \cos\theta & \sin\theta \\ \sin\theta & -\cos\theta \end{pmatrix}, \tag{20.8}$$

where we have used Eq. (20.5). ∎

20.2.2 Charge Qubit at the Sweet Spot

At the sweet spot, $n_g = \frac{1}{2}$. Therefore, $E = E_{cc}(1 - 2n_g) = 0$. The fictitious magnetic field becomes $\vec{B}_E = E_J \hat{x}$ which is along $\hat{x}$ with $\theta = \frac{\pi}{2}$ (Fig. 20.2). The Hamiltonian becomes:

$$\mathcal{H}' = \frac{1}{2}\begin{pmatrix} -E & -E_J \\ -E_J & E \end{pmatrix},$$

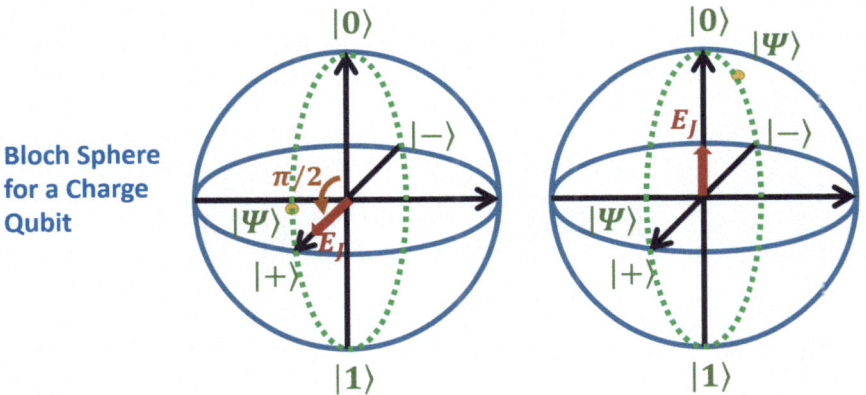

Fig. 20.2 Left: Bloch sphere representation of the charge qubit at the sweet spot before transformation. A vector state $|\Psi\rangle$ is shown as an example. Right: Bloch sphere representation of the charge qubit at the sweet spot after transformation with the rotated $|\Psi\rangle$ (rotated by $\frac{-\pi}{2}$ about the y-axis). The fictitious magnetic field in 3-D real space is shown by overlapping with the Bloch sphere

$$= \frac{-E_J}{2}\begin{pmatrix} 0 & 1 \\ 1 & 0 \end{pmatrix},$$

$$= \frac{-\hbar\omega_L}{2}\begin{pmatrix} 0 & 1 \\ 1 & 0 \end{pmatrix}. \tag{20.9}$$

This is because, based on Eq. (20.7), $\omega_L = \frac{|\vec{B_E}|}{\hbar} = \frac{E_J}{\hbar}$. Again this is expected because we found the energy spacing to be E_J for a charge qubit at the sweet spot in Eq. (18.18).

It is instructive to perform a transformation exercise so that we can work on the system we are familiar with. That is having the effective $\vec{B}_E$ pointing in $\hat{z}$ direction. To do so, we can transform it so that $|+\rangle$ and $|-\rangle$ are the computational basis vectors (after which they will be called $|0\rangle$ and $|1\rangle$ as we always denote $|0\rangle$ and $|1\rangle$ as the computational basis vectors). After the transformation, $\vec{B}_E$ will be pointing in the $\hat{z}$ direction.

Example 20.2 Transform the coordinate so that $\vec{B}_E$ is pointing in the $\hat{z}$-direction for a charge qubit at the sweep spot. What does its Hamiltonian look like?

The question is equivalent to transforming the system from the old basis of $|0\rangle / |1\rangle$ to the new basis $|+\rangle / |-\rangle$. Firstly, we can use Eq. (3.24) in Sect. 3.3.5 to construct the transformation matrix, U. This is the same as Eq. (3.25):

$$U = \begin{pmatrix} \langle +|0\rangle & \langle +|1\rangle \\ \langle -|0\rangle & \langle -|1\rangle \end{pmatrix},$$

$$= \begin{pmatrix} \frac{1}{\sqrt{2}} & \frac{1}{\sqrt{2}} \\ \frac{1}{\sqrt{2}} & -\frac{1}{\sqrt{2}} \end{pmatrix} = \frac{1}{\sqrt{2}}\begin{pmatrix} 1 & 1 \\ 1 & -1 \end{pmatrix}. \tag{20.10}$$

As a result,

$$\mathcal{H}'' = U\mathcal{H}'U^\dagger$$

$$= -\frac{\hbar\omega_L}{2}\frac{1}{\sqrt{2}}\begin{pmatrix} 1 & 1 \\ 1 & -1 \end{pmatrix}\begin{pmatrix} 0 & 1 \\ 1 & 0 \end{pmatrix}\frac{1}{\sqrt{2}}\begin{pmatrix} 1 & 1 \\ 1 & -1 \end{pmatrix},$$

$$= -\frac{\hbar\omega_L}{2}\begin{pmatrix} 1 & 0 \\ 0 & -1 \end{pmatrix},$$

$$= -\frac{\hbar\omega_L}{2}\sigma_z. \tag{20.11}$$

This is the same as Eq. (8.4) which is the Hamiltonian of an electron spin qubit with the magnetic B-field pointing down in Fig. 8.1. More importantly, they both have negative coefficients which means that they have the same motion on the Bloch sphere. This is consistent with what we said in the earlier part of this section that we can treat the charge qubit as a hole (positive charge) spin qubit. Note that after

20.3 One-Qubit Gate

transformation the fictitious magnetic field is pointing *up*. Therefore, it will precess in the same direction as the electron with the magnetic field pointing *down*. The right part of Fig. 20.2 shows the transformed Bloch sphere. ∎

20.3 One-Qubit Gate

Now, let us study how to manipulate a charge qubit at the sweet spot by applying an appropriate external voltage. Since a charge qubit behaves like a positive charge spin qubit with an external magnetic field of $\vec{B}_E = E_J \hat{x}$, we guess that by applying a small AC voltage, we might be able to achieve **Rabi oscillation** as what we did in Chap. 9 in which a small AC magnetic field is applied in the orthogonal direction to the external DC magnetic field (Fig. 9.1). Figure 20.3 shows the Cooper pair box circuit again, and we assume the external voltage is given by $V_g = V_0 - V_1 \cos(\omega_1 t + \phi)$. V_0 is a constant voltage. It is used to set the background charge, n_g, to $\frac{1}{2}$ when there is no AC component so that it is at the sweet spot. V_1 is the amplitude of a sinusoidal voltage which is expected to be small. The angular frequency is ω_1. ϕ is the initial phase of the voltage.

Now, let us look at the Hamiltonian and then determine how it will rotate a state on the Bloch sphere. From Eq. (17.37), the offset charge is

$$n_g = \frac{C_g V_g}{2e},$$
$$= \frac{C_g(V_0 - V_1 \cos(\omega_1 t + \phi))}{2e},$$
$$= \frac{C_g V_0}{2e} - \frac{C_g V_1 \cos(\omega_1 t + \phi)}{2e},$$

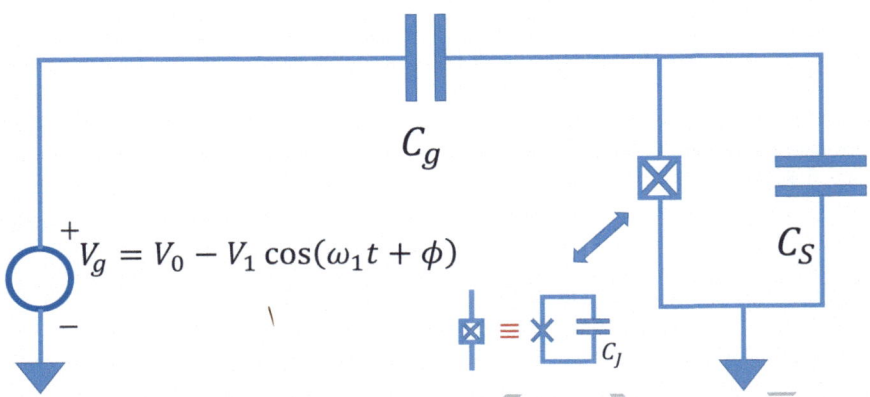

Fig. 20.3 Cooper pair box circuit with an AC external voltage. The circuit is the same as that in Fig. 17.3 except that the AC external voltage formula is explicitly shown

$$= \frac{C_g V_0}{2e} - \eta\cos(\omega_1 t + \phi),$$
$$= n_{g0} - \tilde{n}_g(t), \tag{20.12}$$

where we defined $n_{g0} = \frac{C_g V_0}{2e}$, $\eta = \frac{C_g V_1}{2e}$, and $\tilde{n}_g(t) = \eta\cos(\omega_1 t + \phi)$.

From Eq. (20.2)

$$E = E_{cc}(1 - 2n_g),$$
$$= E_{cc}(1 - 2n_{g0} + 2\tilde{n}_g(t)),$$
$$= 2E_{cc}\tilde{n}_g(t), \tag{20.13}$$

where $n_{g0} = \frac{1}{2}$ is used as we are studying a charge qubit at the sweep spot. Then, the Hamiltonian becomes:

$$\mathcal{H}' = \frac{1}{2}\begin{pmatrix} -E & -E_J \\ -E_J & E \end{pmatrix},$$
$$= \frac{1}{2}\begin{pmatrix} -2E_{cc}\tilde{n}_g(t) & -E_J \\ -E_J & 2E_{cc}\tilde{n}_g(t) \end{pmatrix},$$
$$= \frac{1}{2}\begin{pmatrix} -2E_{cc}\tilde{n}_g(t) & 0 \\ 0 & 2E_{cc}\tilde{n}_g(t) \end{pmatrix} + \frac{1}{2}\begin{pmatrix} 0 & -E_J \\ -E_J & 0 \end{pmatrix},$$
$$= \frac{-2E_{cc}\tilde{n}_g(t)}{2}\begin{pmatrix} 1 & 0 \\ 0 & -1 \end{pmatrix} + \frac{-E_J}{2}\begin{pmatrix} 0 & 1 \\ 1 & 0 \end{pmatrix},$$
$$= -E_{cc}\tilde{n}_g(t)\sigma_z - \frac{E_J}{2}\sigma_x,$$
$$= -E_{cc}\eta\cos(\omega_1 t + \phi)\sigma_z - \frac{E_J}{2}\sigma_x,$$
$$= -E_{cc}\eta\cos(\omega_1 t + \phi)\sigma_z - \frac{\hbar\omega_L}{2}\sigma_x. \tag{20.14}$$

If we compare this to the spin qubit case, this is very similar to Eq. (9.6) (copied here for convenience):

$$H = \frac{e\hbar}{2m}B_1\cos(\omega_1 t)\sigma_x - \frac{\hbar\omega_L}{2}\sigma_z, \tag{20.15}$$

except that they have different coefficients and σ_x and σ_z are swapped. Therefore, for the given AC voltage, its effect on the Hamiltonian is the same as the effect of a small oscillating magnetic field on a spin qubit (Fig. 9.1). As a result, we can say that, besides the effective fictitious DC magnetic field due to V_0 ($\vec{B}_E = E_J\hat{x}$), there is also a fictitious oscillating magnetic field, $\vec{B}_{E1}$, oscillating in the orthogonal

20.3 One-Qubit Gate

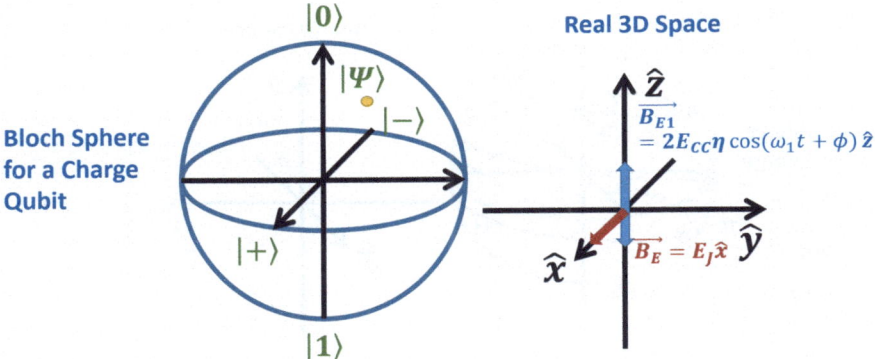

Fig. 20.4 The Bloch sphere representation of a charge qubit at the sweet spot and the real 3D space coordinate system in which the directions of the effective fictitious magnetic fields are shown. Due to the AC voltage shown in Fig. 20.3, the fictitious magnetic field has a DC and also an oscillating part. This is similar to Fig. 9.1

direction ($\hat{z}$). Figure 20.4 illustrates the equivalency. The amplitude of the fictitious oscillating magnetic field can be obtained by comparing the coefficients of the *cosine* term and realizing that the charge qubit does not need $\gamma\hbar$ as discussed before. Therefore, the amplitude of $\vec{B}_{E1}$ is $2E_{cc}\eta$ (with the unit of energy).

Like the DC case, we prefer to work in a system that we are familiar with. That is to have the effective fictitious DC magnetic field pointing in the $\hat{z}$-direction. To do this, we will perform the same basis transformation as in Example 20.2. Applying the transformation to the Hamiltonian, we have:

$$\mathcal{H}'' = U\mathcal{H}'U^\dagger$$
$$= U\left(-E_{cc}\eta\cos(\omega_1 t + \phi)\sigma_z - \frac{\hbar\omega_L}{2}\sigma_x\right)U^\dagger$$
$$= -E_{cc}\eta\cos(\omega_1 t + \phi)U\sigma_z U^\dagger - \frac{\hbar\omega_L}{2}U\sigma_x U^\dagger. \tag{20.16}$$

Since

$$U\sigma_z U^\dagger = \frac{1}{\sqrt{2}}\begin{pmatrix}1 & 1\\ 1 & -1\end{pmatrix}\begin{pmatrix}1 & 0\\ 0 & -1\end{pmatrix}\frac{1}{\sqrt{2}}\begin{pmatrix}1 & 1\\ 1 & -1\end{pmatrix},$$
$$= \sigma_x, \tag{20.17}$$

and $U\sigma_x U^\dagger = \sigma_z$ by using the same approach, we have:

$$\mathcal{H}'' = -E_{cc}\eta\cos(\omega_1 t + \phi)\sigma_x - \frac{\hbar\omega_L}{2}\sigma_z. \tag{20.18}$$

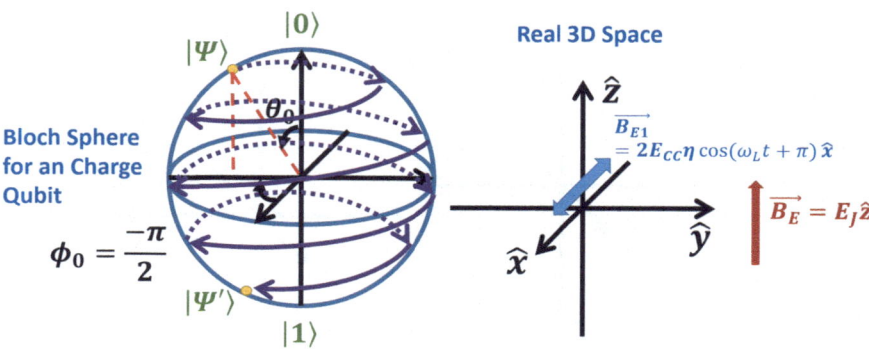

Fig. 20.5 Rabi oscillation for a charge qubit at the sweet spot after coordinate transformation (Eq. (20.16)). The horizontal field is oscillating at Larmor frequency ($\omega_1 = \omega_L$). The left shows how the state moves on the Bloch sphere due to Larmor precession and Rabi oscillation. The right shows the fictitious magnetic fields (with energy unit) due to the external AC voltage source

This is the same as the electron spin qubit Hamiltonian in Eq. (20.15) except, again, the electron qubit's *cosine* term has a positive coefficient while it is negative for the charge qubit. This is again expected as the charge qubit is like a hole spin qubit. With the appropriate initial phase $\phi = \pi$ of the oscillating field, we will observe the same Rabi oscillation path as in Fig. 9.2 because this will make the *cosine* term positive. This is shown in Fig. 20.5. By comparing to Eq. (9.32), we can find the **Rabi frequency** without going through the derivation. Since the electron spin qubit has $\omega_R = \frac{e}{2m} B_1$ when the prefactor of the *cosine* term is $\frac{e\hbar}{2m} B_1$ (ratio of $\hbar$). Now, the prefactor of the charge qubit is $E_{cc}\eta$; then, we have:

$$\omega_R = \frac{E_{cc}\eta}{\hbar}. \tag{20.19}$$

Therefore, we may also rewrite the Hamiltonian as

$$\mathcal{H}'' = -\hbar\omega_R \cos(\omega_1 t + \phi)\sigma_x - \frac{\hbar\omega_L}{2}\sigma_z. \tag{20.20}$$

20.3.1 Rotating Frame and Rotating Wave Approximation (RWA)

So far, we have demonstrated the possibility of implementing Larmor precession and Rabi oscillation using a sinusoidal voltage signal to a Cooper pair box. But this is not convenient because the mathematics of qubit state movement on the Bloch sphere is complex as demonstrated in Chap. 9. This is because the Hamiltonian is

20.3 One-Qubit Gate

time-dependent. Therefore, we want to work in a rotating frame as in Chap. 10. Firstly, we decompose the Hamiltonian into

$$\mathcal{H}'' = -\hbar\omega_R \cos(\omega_1 t + \phi)\sigma_x - \frac{\hbar\omega_L}{2}\sigma_z,$$

$$= -\frac{\hbar\omega_R}{2}\left[\cos(\omega_1 t + \phi)\sigma_x + \sin(\omega_1 t + \phi)\sigma_y\right.$$

$$\left. + \cos(\omega_1 t + \phi)\sigma_x - \sin(\omega_1 t + \phi)\sigma_y\right] - \frac{\hbar\omega_L}{2}\sigma_z,$$

$$= -\frac{\hbar\omega_R}{2}(A_+ + A_-) - \frac{\hbar\omega_L}{2}\sigma_z. \quad (20.21)$$

This equation reminds us of the rotating magnetic field in Chap. 10. The Hamiltonian of the setup in Fig. 10.1 is (Eq. (10.3))

$$H = \frac{\hbar\Omega_R}{2}\cos(\omega_1 t + \phi_B)\sigma_x - \frac{\hbar\Omega_R}{2}\sin(\omega_1 t + \phi_B)\sigma_y - \frac{\hbar\omega_L}{2}\sigma_z,$$

$$= \frac{\hbar\Omega_R}{2}A_- - \frac{\hbar\omega_L}{2}\sigma_z. \quad (20.22)$$

In the charge qubit case (Eq. (20.21)), it has an extra term of $A_+ = \cos(\omega_1 t + \phi)\sigma_x + \sin(\omega_1 t + \phi)\sigma_y$. On the other hand, we know that we obtained the common term $A_- = \cos(\omega_1 t + \phi)\sigma_x - \sin(\omega_1 t + \phi)\sigma_y$ due to a rotating magnetic field rotating clockwise (looking from top) in the spin qubit case. If we had applied a rotating magnetic field rotating anticlockwise in the spin qubit case, it would have given us A_+ instead. Therefore, in the charge qubit case, its Hamiltonian is equivalent to applying two fictitious rotating magnetic fields rotating in the opposite direction. This is expected because this is the decomposition of its linearly oscillating fictitious magnetic field. Figure 20.6 shows that the fictitious magnetic field $\vec{B}_{E1}$ is decomposed into two rotating magnetic fields rotating in opposite directions but with half of the original amplitude, $\vec{B}_{E1+}$ and $\vec{B}_{E1-}$. $\vec{B}_{E1+}$ and $\vec{B}_{E1-}$ contribute to A_+ and A_- in the charge qubit Hamiltonian, respectively.

Now we will work in the rotating frame by following the rotation of $\vec{B}_{E1-}$. Note that we are rotating at ω_1, not ω_L with reference to the **laboratory frame**. In this case, $\vec{B}_{E1+}$ is rotating at an angular velocity of $2\omega_1$ from us, and the quantum state only feels its average effect which is essentially zero as it is rotating (oscillating) rapidly. Therefore, we can ignore the effect of $\vec{B}_{E1+}$ in the Hamiltonian! This is called the **rotating wave approximation** which was discussed in Sect. 9.6. Now working on the rotating frame and using RWA, we are then having the same problem as that in Chap. 10 *after working in the rotating frame in* Sect. 10.4. However, the signs in front of σ_x and σ_y in the charge qubit case are negative of the corresponding sign in the electron spin qubit case. And therefore, we can use its result directly

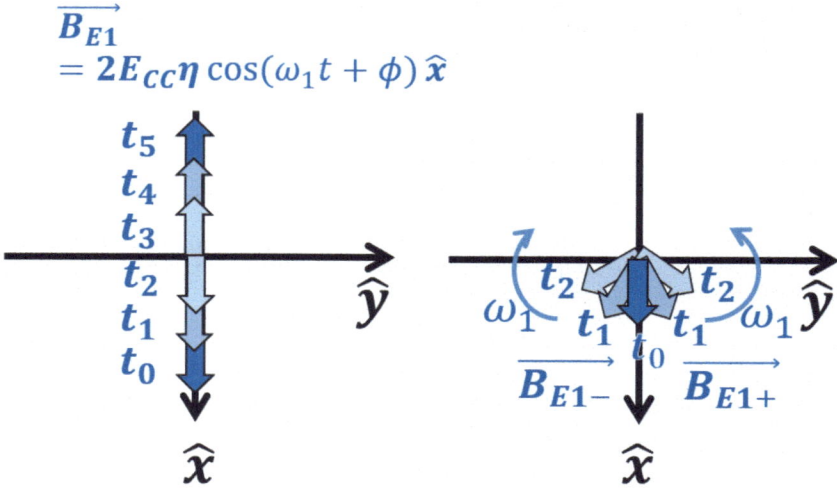

Fig. 20.6 The linearly oscillating fictitious magnetic field of a charge qubit as the one shown in the right of Fig. 20.5 can be decomposed into two rotating magnetic fields rotating in opposite directions but with half of the original amplitude. This idea is similar to the spin qubit case in Fig. 9.4

(Eq. (10.27)) after adjusting the signs. The Hamiltonian becomes:

$$\mathcal{H}''_{RF} = \frac{\hbar}{2}\vec{\Omega}_R \cdot \vec{\sigma}, \tag{20.23}$$

where

$$\vec{\Omega}_R = \begin{pmatrix} -\Omega_R \cos\phi \\ \Omega_R \sin\phi \\ -\Delta \end{pmatrix}, \tag{20.24}$$

where **detuning** is $\Delta = \omega_L - \omega_1$, and

$$\Omega_R = \frac{E_{cc}\eta}{\hbar} = \omega_R. \tag{20.25}$$

For a given **angular frequency vector**, $\vec{\Omega}_R$, the qubit will rotate about it with the generalized Rabi frequency, $\Omega'_R = |\vec{\Omega}_R| = \sqrt{\Omega_R^2 + \Delta^2}$ as in Fig. 10.3. But note that whether it is a positive or negative charge, once the angular frequency vector is given/found, the qubit will rotate about it in the same direction using the right-hand rule.

20.3.2 Recap: How Voltage Source Parameters Affect the Hamiltonian

We have come a long way. Let us summarize how to design an electrical circuit to realize an artificial atom as a charge qubit and how to realize a one-qubit gate for it.

Firstly, the charge qubit has a qubit frequency of $\omega_L = \frac{E_J}{\hbar}$ (Eqs. (20.7) and (20.9)). Therefore, based on Eq. (16.21), we want to set the Josephson junction critical current to be

$$I_c = \frac{E_J}{\Phi_0'},$$
$$= \frac{\hbar \omega_L}{\Phi_0'}, \quad (20.26)$$

and pick the right I_c for the desired Larmor frequency.

We also apply a voltage pulse of this form:

$$V_g = V_0 - V_1 \cos(\omega_1 t + \phi), \quad (20.27)$$

to obtain this Hamiltonian in the rotating frame

$$\begin{aligned}
\mathcal{H}''_{RF} &= \frac{\hbar}{2} \vec{\Omega}_R \cdot \vec{\sigma}, \\
&= \frac{\hbar}{2} \left(-\Omega_R \cos\phi \, \sigma_x + \Omega_R \sin\phi \, \sigma_y - \Delta \sigma_z \right), \\
&= -\frac{\hbar \Omega_R}{2} \cos\phi \, \sigma_x + \frac{\hbar \Omega_R}{2} \sin\phi \, \sigma_y - \frac{\hbar \Delta}{2} \sigma_z, \\
&= -\frac{E_{cc}\eta}{2} \cos\phi \, \sigma_x + \frac{E_{cc}\eta}{2} \sin\phi \, \sigma_y - \frac{\hbar(\omega_L - \omega_1)}{2} \sigma_z, \\
&= -\frac{eC_g V_1}{2C} \cos\phi \, \sigma_x + \frac{eC_g V_1}{2C} \sin\phi \, \sigma_y - \frac{E_J - \hbar \omega_1}{2} \sigma_z, \quad (20.28)
\end{aligned}$$

where the definitions of η in Eq. (20.12) and the charging energy of a Cooper pair have been used.

The Hamiltonian is the summation of weighted σ_x, σ_y, and σ_z, and each of them controls the qubit state rotation on the Bloch sphere about $\hat{x}$, $\hat{y}$, and $\hat{z}$, respectively (see Eqs. (5.23)–(5.25)). Therefore, C_g, V_1, $C = C_\Sigma + C_g$, and ϕ are used to control the rotations about $\hat{x}$ and $\hat{y}$. ω_1 is used to control the rotation about $\hat{z}$. Of course, V_0 needs to be set to a value so that the circuit is at its sweet spot. That is

$$n_{g0} = \frac{1}{2} = \frac{C_g V_0}{2e}. \quad (20.29)$$

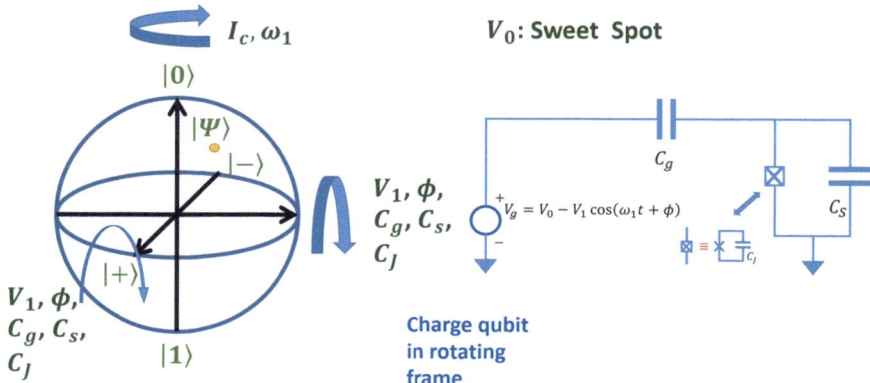

Fig. 20.7 Illustration of how each circuit parameter is used to control qubit rotation in a charge qubit at the sweep spot

Figure 20.7 shows the effect of the circuit parameters on qubit rotations.

20.4 Summary

In this chapter, we show that a charge qubit at the sweep spot behaves like a hole spin qubit under an external magnetic field. The fictitious magnetic field is determined by its charging energy and the Josephson energy. We map the Hamiltonian to the one in spin qubit and are able to "reuse" the equations without repeating the derivation. When an external sinusoidal voltage is applied and coupled to the qubit, the AC voltage acts like a linearly oscillating fictitious magnetic field resulting in Rabi oscillation. More importantly, if we work in the rotating frame with the angular frequency of that AC voltage, using rotating wave approximation, the problem becomes a spin qubit problem under a rotating magnetic field. We are able to describe the rotation of the qubit by the rotation about an angular velocity vector. This angular velocity vector is determined by the circuit and pulse parameters.

Problems

20.1 Coordinate Transformation for Charge Qubit
In Example 20.2, it is tempted to use a rotation of the state about $\hat{y}$ by $\frac{-\pi}{2}$ to achieve the goal (instead of transforming the basis in the way it was done). If we are willing to use that system, then it is fine. We may use Eq. (5.24) to find the transformation

matrix, U:

$$U = R_y\left(\frac{-\pi}{2}\right) = \begin{pmatrix} \cos\frac{-\pi}{4} & -\sin\frac{-\pi}{4} \\ \sin\frac{-\pi}{4} & \cos\frac{-\pi}{4} \end{pmatrix},$$

$$= \frac{1}{\sqrt{2}}\begin{pmatrix} 1 & 1 \\ -1 & 1 \end{pmatrix}.$$

Therefore,

$$U^\dagger = \frac{1}{\sqrt{2}}\begin{pmatrix} 1 & -1 \\ 1 & 1 \end{pmatrix}.$$

As a result,

$$\mathcal{H}'' = U\mathcal{H}'U^\dagger$$
$$= -\frac{\hbar\omega_L}{2}\frac{1}{\sqrt{2}}\begin{pmatrix} 1 & 1 \\ -1 & 1 \end{pmatrix}\begin{pmatrix} 0 & 1 \\ 1 & 0 \end{pmatrix}\frac{1}{\sqrt{2}}\begin{pmatrix} 1 & -1 \\ 1 & 1 \end{pmatrix},$$
$$= -\frac{\hbar\omega_L}{2}\begin{pmatrix} 1 & 0 \\ 0 & -1 \end{pmatrix},$$
$$= -\frac{\hbar\omega_L}{2}\sigma_z.$$

Try to draw the figures and appreciate their differences from Example 20.2. Apply this to the AC case, how will the equation change, and why?

20.2 Rotating Frame
Repeat the derivation in Chap. 10 to obtain Eq. (20.23) from Eq. (20.21).

20.3 One-Qubit Gates
How should the circuit parameters be set to implement a NOT gate and a Hadamard gate?

Reference

[1]. V Bouchiat, D Vion, P Joyez, D Esteve, and M H Devoret. Quantum coherence with a single cooper pair. *Physica Scripta*, 1998(T76):165, jan 1998.

Chapter 21
Transmon Qubit: One-Qubit and Two-Qubit Gates

21.1 Introduction

In the last chapter, we studied the dynamics of the charge qubit at the sweet spot. We used the Hamiltonian from the two-basis-state approximation, which is fairly accurate for a charge qubit at the sweet spot, and demonstrated that by changing the amplitude and phase of the AC driving voltage, we can perform rotation about $\hat{x}$ and $\hat{y}$ on the Bloch sphere. By using detuning (i.e. different driving frequency than the qubit frequency), we can also perform rotation about $\hat{z}$ on the Bloch sphere. These are the results of working on the rotating frame. But charge qubit is not a promising superconducting qubit due to its sensitivity to charge noise. A transmon qubit, which has the same setup as a charge qubit but with its Josephson energy E_J much larger than the charging energy E_C is commonly used for its immunity to charge noise. In this chapter, based on the theory we have developed so far, we will study the implementation of 1-qubit gates for a transmon. It turns out that it has similar results as the charge qubit. Finally, we will discuss the implementation of an entanglement gate, the **iSWAP gate**. Readers may refer [1] for deeper understanding after reading this chapter.

21.1.1 Learning Outcomes

Understand the differences and similarities between charge qubit and transmon and the implementation of their one-qubit gates; appreciate the concept of coupling and how it is used to implement the iSWAP gate; be able to use some critical equations for two-level systems.

21.1.2 Teaching Videos

- Search for Ch21 in this playlist
  - https://tinyurl.com/3yhze3jn
- Other videos
  - https://youtu.be/vXReRIsbQ90

21.2 One-Qubit Gate for Transmon

21.2.1 The Circuit

As discussed in the previous chapters, a transmon qubit is just a special case of Cooper pair box with $E_J \gg E_c$. Its structure and driving circuit are the same as that of a charge qubit ($E_J \ll E_c$). Therefore, we use the same circuit as in Fig. 20.3. It is redrawn and modified in Fig. 21.1 for convenience. The major difference is that there is no DC external voltage which is required to bias a charge qubit at the sweep spot. Moreover, an **envelope function**, $S(t)$, is used to modulate the sinusoidal function as shown in Fig. 21.1.

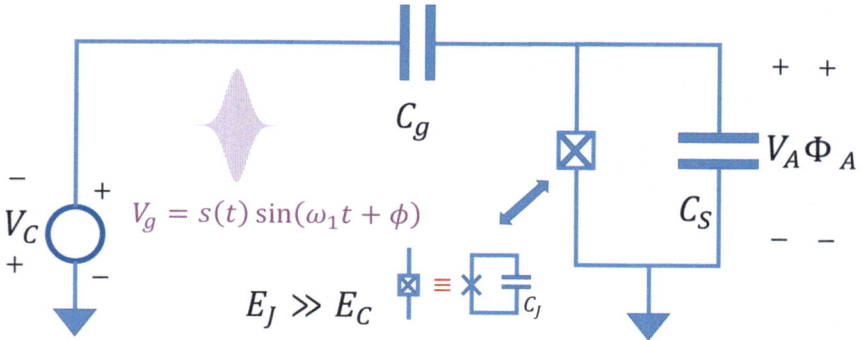

Fig. 21.1 A transmon with an AC external voltage for one-qubit gates implementation. The circuit is the same as that in Fig. 20.3 except that the external voltage only has an AC component with an envelope function, $s(t)$

21.2.2 The Hamiltonian

In Chap. 17, by following the quantization procedure in Fig. 17.2, we deduced the Hamiltonian of the circuit in Eq. (17.36) which is copied here for convenience:

$$\hat{\mathcal{H}} = \frac{(\hat{Q}_A + C_g V_g)^2}{2(C_\Sigma + C_g)} - E_J \cos \frac{\hat{\Phi}_A}{\Phi_0'}. \tag{21.1}$$

This is a rather complex Hamiltonian. We converted the operators into **phase operator**, $\hat{\varphi}$, and **Cooper pair number operator**, $\hat{n}$, in Eq. (17.38). Then, we used two-basis-state approximation (i.e., only use $|n=0\rangle$ and $|n=1\rangle$) to construct the Hamiltonian matrix for a charge qubit at the sweet spot to understand its dynamics and the implementation of one-qubit gates for charge qubit in Chap. 20. However, we have shown that the two-basis-state approximation is not accurate for a transmon qubit in Chap. 19 through numerical calculations. Therefore, we cannot use the same approach for a transmon qubit.

Luckily, since a transmon qubit has a small anharmonicity, we can just keep the fourth power of the Josephson junction energy and treat it as a perturbation. In this regime, a transmon qubit is just an LC tank with a perturbation of "potential energy" due to the fourth power of the Josephson junction energy. We thus showed that we could treat it as an LC tank in Sect. 19.4.2 and obtained Eq. (19.16) which is copied again for convenience:

$$\hat{\mathcal{H}} = \sqrt{8 E_c E_J} \left(\hat{b}^\dagger \hat{b} + \frac{1}{2} \right) - \frac{E_c}{12} \left(\hat{b} + \hat{b}^\dagger \right)^4. \tag{21.2}$$

We may drop the last term if we only work on two levels. So it becomes:

$$\hat{\mathcal{H}} = \sqrt{8 E_c E_J} \left(\hat{b}^\dagger \hat{b} + \frac{1}{2} \right),$$
$$= \sqrt{8 E_c E_J} \left(\hat{N} + \frac{1}{2} \right). \tag{21.3}$$

It is important to emphasize that a transmon qubit is essentially an SHO. Its eigenstates are **Fock states** with N runs from 0 to ∞. It is not the Cooper pair number operator eigenstates, $|n\rangle$, which runs from $-\infty$ to ∞. Its creation and annihilation operators increase and decrease the energy quanta N by one, respectively. As a qubit, we will limit the discussion to two levels, $N = 0$ and $N = 1$, we can express $\hat{N}$ in terms of $\hat{\sigma}_z$.

Example 21.1 Find the matrix representation of $\hat{N}$ in its eigenbasis for a two-level system. Then express $\hat{N}$ in terms of $\hat{\sigma}_z$.

In its eigenbasis, the diagonal has its eigenvalues (see Eq. (3.9)). For a two-level system, we have $N = 0$ and $N = 1$. Therefore,

$$N = \begin{pmatrix} 0 & 0 \\ 0 & 1 \end{pmatrix},$$

$$= -\begin{pmatrix} 0 & 0 \\ 0 & -1 \end{pmatrix},$$

$$= -\frac{1}{2}\begin{pmatrix} 0 & 0 \\ 0 & -2 \end{pmatrix},$$

$$= -\frac{1}{2}\left(\begin{pmatrix} 1 & 0 \\ 0 & -1 \end{pmatrix} + \begin{pmatrix} -1 & 0 \\ 0 & -1 \end{pmatrix}\right),$$

$$= -\frac{1}{2}(\sigma_z - I),$$

$$= -\frac{1}{2}\sigma_z + \frac{1}{2}I. \tag{21.4}$$

■

Therefore,

$$\hat{\mathcal{H}} = \sqrt{8E_c E_J}\left(-\frac{1}{2}\hat{\sigma}_z + 1\right),$$

$$= -\frac{\sqrt{8E_c E_J}}{2}\hat{\sigma}_z,$$

$$= -\frac{\hbar\omega_0}{2}\hat{\sigma}_z, \tag{21.5}$$

where we ignored the constant energy shift in the second line and used the energy quantum equation of a transmon (Eq. (19.17)) in the third line. ω_0 is the **Larmor frequency** or the **qubit freqency** and is also often donated as ω_q.

It should be noted that this equation is for a transmon *without* an external driving circuit. Therefore, we need to take the Hamiltonian due to the driving circuit, $\hat{\mathcal{H}}_d$, from Eq. (21.1) and add it to Eq. (21.5). Then, the total Hamiltonian describing a transmon qubit with an external drive voltage is

$$\hat{\mathcal{H}}' = \hat{\mathcal{H}} + \hat{\mathcal{H}}_d. \tag{21.6}$$

It is not difficult to find $\hat{\mathcal{H}}_d$ because without the external drive, $V_g = 0$. So $\hat{\mathcal{H}}_d$ must be due to all terms related to V_g in Eq. (21.1). Therefore,

21.2 One-Qubit Gate for Transmon

$$\hat{\mathcal{H}}_d = \frac{2\hat{Q}_A C_g V_g + (C_g V_g)^2}{2(C_\Sigma + C_g)}. \tag{21.7}$$

The second term does not contain any of the conjugate operators. It is just a diagonal matrix that will add an extra phase shift to the quantum gate. Therefore, we may ignore it. This is the same as what we did for Eq. (17.34) to Eq. (17.35). Therefore, it can be further simplified to

$$\hat{\mathcal{H}}_d = \frac{\hat{Q}_A C_g V_g}{C_\Sigma + C_g},$$

$$= \frac{V_g}{1 + \frac{C_\Sigma}{C_g}} \hat{Q}_A. \tag{21.8}$$

When we quantized the LC tank in Chap. 15, we already defined the creation and annihilation operators for the system and expressed the conjugate momentum, $\hat{Q}$, as a function of them (Eq. (15.32)). That is,

$$\hat{Q} = i\sqrt{\frac{C\omega_0\hbar}{2}}(-\hat{a} + \hat{a}^\dagger) = iQ_{zpf}(-\hat{a} + \hat{a}^\dagger), \tag{21.9}$$

where $C = C_\Sigma + C_g$ in our circuit. Therefore,

$$\hat{\mathcal{H}}_d = \frac{V_g}{1 + \frac{C_\Sigma}{C_g}} iQ_{zpf}(-\hat{b} + \hat{b}^\dagger), \tag{21.10}$$

where we replaced $\hat{a}$ and $\hat{a}^\dagger$ by $\hat{b}$ and $\hat{b}^\dagger$, respectively, because $\hat{b}$ and $\hat{b}^\dagger$ are the symbols we have been using for a transmon qubit.

We are working only on the two-level system (that is why we have dropped the last term in Eq. (21.2) to get Eq. (21.3)). Therefore, we also need to limit $\hat{\mathcal{H}}_a$ to a 2×2 matrix. We can then use this identity for a two-level system:

$$i(-\hat{b} + \hat{b}^\dagger) = \hat{\sigma}_y. \tag{21.11}$$

Or it can be written in its matrix form as

$$i(-\boldsymbol{b} + \boldsymbol{b}^\dagger) = \boldsymbol{\sigma}_y. \tag{21.12}$$

Example 21.2 Prove Eq. (21.11).

As mentioned earlier, the creation and annihilation operators increase and decrease the energy quanta N by one, respectively. Therefore, on the energy eigenstate basis, for a two-level system, they have the following form:

$$b = \begin{pmatrix} 0 & 1 \\ 0 & 0 \end{pmatrix},$$

$$b^\dagger = \begin{pmatrix} 0 & 0 \\ 1 & 0 \end{pmatrix}. \tag{21.13}$$

Thus,

$$i(-b + b^\dagger) = -ib + ib^\dagger,$$

$$= \begin{pmatrix} 0 & -i \\ 0 & 0 \end{pmatrix} + \begin{pmatrix} 0 & 0 \\ i & 0 \end{pmatrix},$$

$$= \sigma_y. \tag{21.14}$$

∎

Finally, substituting Eqs. (21.5), (21.10), and (21.11) into Eq. (21.6), we obtain the Hamiltonian for a *two-level* system with an external drive as

$$\hat{\mathcal{H}}' = \hat{\mathcal{H}} + \hat{\mathcal{H}}_d,$$

$$= -\frac{\hbar\omega_0}{2}\hat{\sigma}_z + \frac{V_g}{1 + \frac{C_\Sigma}{C_g}} Q_{zpf}\hat{\sigma}_y,$$

$$= -\frac{\hbar\omega_0}{2}\hat{\sigma}_z + \frac{s(t)\sin(\omega_1 t + \phi)}{1 + \frac{C_\Sigma}{C_g}} Q_{zpf}\hat{\sigma}_y,$$

$$= -\frac{\hbar\omega_0}{2}\hat{\sigma}_z + \hbar\Omega_R(t)\sin(\omega_1 t + \phi)\hat{\sigma}_y, \tag{21.15}$$

where we have substituted the AC voltage equation given in Fig. 21.1. We have also defined $\hbar\Omega_R(t) = \frac{s(t)Q_{zpf}}{1+\frac{C_\Sigma}{C_g}}$. This Hamiltonian only allows us to rotate the qubit about $\hat{z}$ and $\hat{y}$. As in Eq. (20.20) for a charge qubit, we will work on a rotating frame to enable us to have better control of the qubit.

21.2.3 Rotating Frame

21.2.3.1 Rotating Frame of Driving Frequency

We want to use the same trick as for the charge qubit in Sect. 20.3.1. Equation (21.15) is similar to Eq. (20.20) except that now it has $\hat{\sigma}_y$ instead of $\hat{\sigma}_x$. But when mapping back to the spin qubit case in Chap. 10, this is just the same as the

21.2 One-Qubit Gate for Transmon

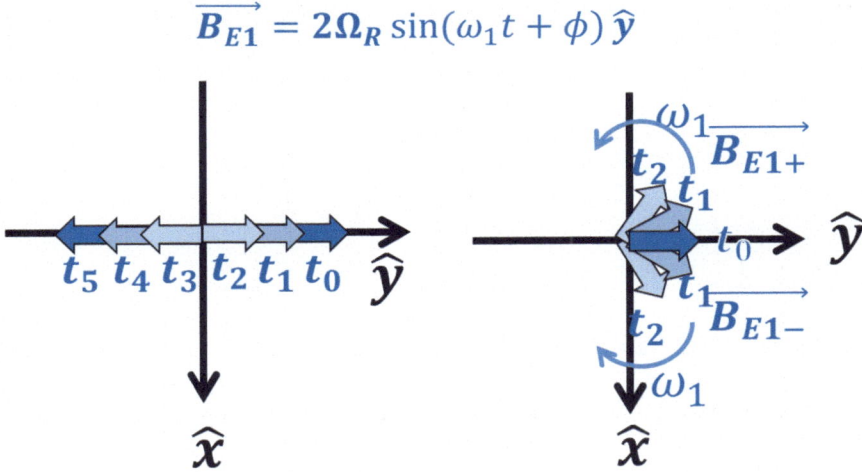

Fig. 21.2 The linearly oscillating fictitious magnetic field of a transmon (left) can be decomposed into two rotating magnetic fields rotating in opposite directions but with half of the original amplitude. Here $\phi = \frac{\pi}{2}$. This idea is similar to the spin qubit case in Fig. 9.4 and the charge qubit case in Fig. 20.6

Hamiltonian when there is a linearly oscillating magnetic field in the $\hat{y}$-direction (Fig. 21.2).

Therefore, we write:

$$\hat{\mathcal{H}}' = -\frac{\hbar\omega_0}{2}\hat{\sigma}_z + \hbar\Omega_R(t)\sin(\omega_1 t + \phi)\hat{\sigma}_y,$$

$$= -\frac{\hbar\omega_0}{2}\hat{\sigma}_z + \frac{\hbar\Omega_R(t)}{2}\left[(\sin(\omega_1 t + \phi)\hat{\sigma}_y - \cos(\omega_1 t + \phi)\hat{\sigma}_x)\right.$$
$$\left.+(\sin(\omega_1 t + \phi)\hat{\sigma}_y + \cos(\omega_1 t + \phi)\hat{\sigma}_x)\right],$$

$$= -\frac{\hbar\omega_0}{2}\hat{\sigma}_z + \frac{\hbar\Omega_R(t)}{2}(\hat{A}_- + \hat{A}_+). \tag{21.16}$$

We will first work on the rotating frame at the driving frequency, ω_1. We will follow $\vec{B}_{E1-}$ as this is the rotating vector in the clockwise direction which is the same as in the setup in Chap. 10. Again, $\vec{B}_{E1+}$ moves at $2\omega_1$ away in the rotating frame and thus has almost zero effect on average (**rotating wave approximation, RWA**). Therefore, it can be ignored. The Hamiltonian in the rotating frame becomes:

$$\hat{\mathcal{H}}' = -\frac{\hbar\omega_0}{2}\hat{\sigma}_z + \frac{\hbar\Omega_R(t)}{2}\left[\sin(\omega_1 t + \phi)\hat{\sigma}_y - \cos(\omega_1 t + \phi)\hat{\sigma}_x\right]. \tag{21.17}$$

Now we compare this to Eq. (10.3) and they are the same except the signs are different. If we perform the same transformation, the Hamiltonian in the rotating frame should be the same as in Eq. (10.27) except for the signs of $\hat{\sigma}_x$ and $\hat{\sigma}_y$:

$$\hat{\mathcal{H}}_{RF} = -\frac{\hbar \Delta}{2}\hat{\sigma}_z + \frac{\hbar \Omega_R(t)}{2}\left(-\cos\phi\hat{\sigma}_x + \sin\phi\hat{\sigma}_y\right),$$

$$= \frac{\hbar}{2}\begin{pmatrix}\hat{\sigma}_x & \hat{\sigma}_y & \hat{\sigma}_z\end{pmatrix} \cdot \begin{pmatrix} -\Omega_R(t)\cos\phi \\ \Omega_R(t)\sin\phi \\ -\Delta \end{pmatrix}. \quad (21.18)$$

We may set $I = \cos\phi$ and $Q = -\sin\phi$, then

$$\hat{\mathcal{H}}_{RF} = -\frac{\hbar \Delta}{2}\hat{\sigma}_z - \frac{\hbar \Omega_R(t)}{2}\left(I\hat{\sigma}_x + Q\hat{\sigma}_y\right). \quad (21.19)$$

I is the **in-phase** component and Q is the **out-of-phase** component.

21.2.3.2 Rotating Frame of Qubit Frequency

So far, in the treatments of spin qubits in Chap. 10, charge qubits in Chap. 20, and also the transmon qubits in the previous section, we have been using the driving frequency (ω_1 or also often called ω_d) of the magnetic field or electrical signal as the rotating frame. In the superconducting qubit case, we reuse the equations and derivations in the spin qubit case.

In practice, it is often to use the qubit frequency (ω_0, referring to the resonant frequency), which is sometimes also called ω_q (qubit frequency) or ω_L (Larmor frequency), as the rotating frame. To do so, we need to calculate from scratch (because we do not have equations to reuse from the spin qubit), but we can use Eqs. (10.32)–(10.36) in Sect. 10.7 to help us. In the new frame, we have $\hat{U}_{RF} = e^{-i\frac{\omega_0 t}{2}\hat{\sigma}_z}$ (Eq. (10.14)), and note that the frequency in the exponent is ω_0 instead of ω_1.

Using Eq. (10.20) and $\hat{\mathcal{H}}'$ in Eq. (21.15), therefore,

$$\hat{\mathcal{H}}_{RF,\omega_0} = i\hbar \frac{\partial \hat{U}_{RF}}{\partial t}\hat{U}^\dagger_{RF} + \hat{U}_{RF}\hat{\mathcal{H}}'\hat{U}^\dagger_{RF},$$

$$= \frac{\hbar \omega_0}{2}\hat{\sigma}_z + \hat{U}_{RF}\left[-\frac{\hbar \omega_0}{2}\hat{\sigma}_z + \hbar\Omega_R(t)\sin(\omega_1 t + \phi)\hat{\sigma}_y\right]\hat{U}^\dagger_{RF},$$

$$= \hbar\Omega_R(t)\sin(\omega_1 t + \phi)\hat{U}_{RF}\hat{\sigma}_y\hat{U}^\dagger_{RF}. \quad (21.20)$$

Note that $\hat{\sigma}_z$ terms are canceled because of Eq. (10.34). We then apply Eq. (10.33) but using ω_0 instead of ω_1 as we are using the qubit frequency as the rotating frame, or $\hat{U}_{RF} = e^{-i\frac{\omega_0 t}{2}\hat{\sigma}_z}$:

$$\hat{\mathcal{H}}_{RF,\omega_0} = \hbar\Omega_R(t)\sin(\omega_1 t + \phi)\hat{U}_{RF}\hat{\sigma}_y\hat{U}^\dagger_{RF},$$

$$= \hbar\Omega_R(t)\sin(\omega_1 t + \phi)\left(\cos\omega_0 t\,\hat{\sigma}_y - \sin\omega_0 t\,\hat{\sigma}_x\right). \quad (21.21)$$

We then perform multiplications and use product formulae in trigonometry. We again will use rotating wave approximation to ignore the high-frequency terms $(\omega_1 + \omega_0 + \phi)$; we finally get:

$$\hat{\mathcal{H}}_{RF,\omega_0} = \frac{\hbar\Omega_R(t)}{2}\left(\sin((\omega_1-\omega_0)t+\phi)\hat{\sigma}_y - \cos((\omega_1-\omega_0)t+\phi)\hat{\sigma}_x\right),$$

$$= -\frac{\hbar\Omega_R(t)}{2}\left(\sin(\Delta t - \phi)\hat{\sigma}_y + \cos(\Delta t - \phi)\hat{\sigma}_x\right), \quad (21.22)$$

where we continue to define the **detuning** as $\Delta = \omega_0 - \omega_1$. Then we apply trigonometric identities for the sum of angles and define $I = \cos\phi$ and $Q = -\sin\phi$ as before

$$\hat{\mathcal{H}}_{RF,\omega_0}$$

$$= -\frac{\hbar\Omega_R(t)}{2}\left[(\sin\Delta t\cos\phi - \cos\Delta t\sin\phi)\hat{\sigma}_y\right.$$
$$\left. + (\cos\Delta t\cos\phi + \sin\Delta t\sin\phi)\hat{\sigma}_x\right],$$

$$= -\frac{\hbar\Omega_R(t)}{2}\left[(I\sin\Delta t + Q\cos\Delta t)\hat{\sigma}_y + (I\cos\Delta t - Q\sin\Delta t)\hat{\sigma}_x\right]$$
$$(21.23)$$

When $\Delta = 0$, both Eq. (21.19) and Eq. (21.23) reduce to

$$\hat{\mathcal{H}}_{RF} = -\frac{\hbar\Omega_R(t)}{2}\left(I\hat{\sigma}_x + Q\hat{\sigma}_y\right). \quad (21.24)$$

This is, of course, expected because when $\omega_1 = \omega_0$, both the qubit and driving frequency frames are the same.

21.3 Two-Qubit Entanglement Gate for Transmon

As mentioned in Sect. 1.3, being able to implement a "universal" set of quantum gates is a necessary condition to build a quantum computer. In Sect. 5.5 we mentioned that we need a U_{XOR} gate (or the CNOT gate) which acts as the entanglement gate to complete a universal set of quantum gates. However, not every technology can implement U_{XOR} gate easily as it does not have the corresponding physics that can directly map to the U_{XOR} gate. That means U_{XOR} is not its **native entanglement gate**. For example, in Chap. 12, we mentioned that U_{Cphase} is the native entanglement gate for electron spin qubit. However, we can create a U_{XOR} by combining U_{Cphase} with one-qubit gates (see Eqs. (12.5), (12.9) and Fig. 12.3 without the first Hadamard gate).

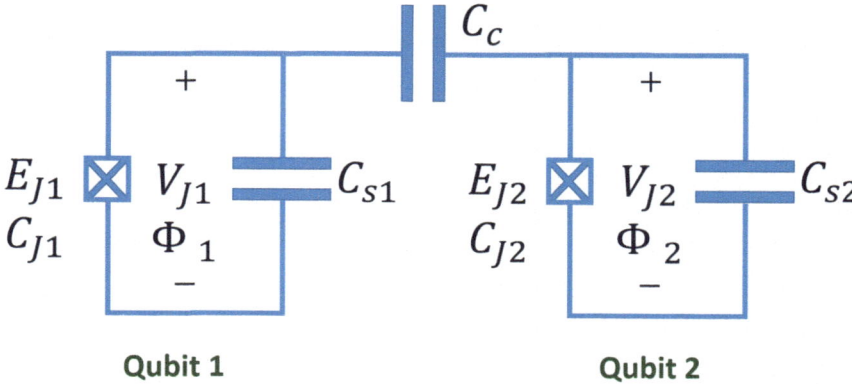

Fig. 21.3 A two-qubit entanglement gate is formed when two transmons are coupled to each other through capacitor C_c

This is the same for superconducting qubits made of transmon. Its native gate turns out to be the iSWAP gate which can be obtained easily by capacitively coupling two transmon qubits. Figure 21.3 shows the circuit.

21.3.1 iSWAP Gate as an Entanglement Gate

An iSWAP gate, U_{iSWAP}, is a two-qubit gate with the following matrix in the computational basis:

$$U_{iSWAP} = \begin{bmatrix} 1 & 0 & 0 & 0 \\ 0 & 0 & i & 0 \\ 0 & i & 0 & 0 \\ 0 & 0 & 0 & 1 \end{bmatrix}. \tag{21.25}$$

Let us first prove that we can construct U_{XOR} using iSWAP gates, U_{iSWAP}, and other one-qubit gates. A U_{XOR} gate (CNOT gate) can be decomposed into two U_{iSWAP} gates and some one-qubit gates as shown in Fig. 21.4 [2].

Readers are encouraged to prove the circuit equivalency in Fig. 21.4 using matrix multiplications. Here, we will use Python code to help us perform the verification. Readers can compare their calculations against the Python code. It should be noted that in the circuit in Fig. 21.4, the bottom qubit is the most significant bit (MSB). Also, the rotation matrix can be obtained from Eq. (5.23) and Eq. (5.25) by substituting the appropriate θ. Finally, a global phase shift of $\frac{\pi}{4}$ to the two-qubit system is applied. This is constructed from two one-qubit global phase shift gates of $\frac{\pi}{8}$ through a tensor product. A global phase shift gate, $Ph(\theta)$, is *not the phase*

21.3 Two-Qubit Entanglement Gate for Transmon

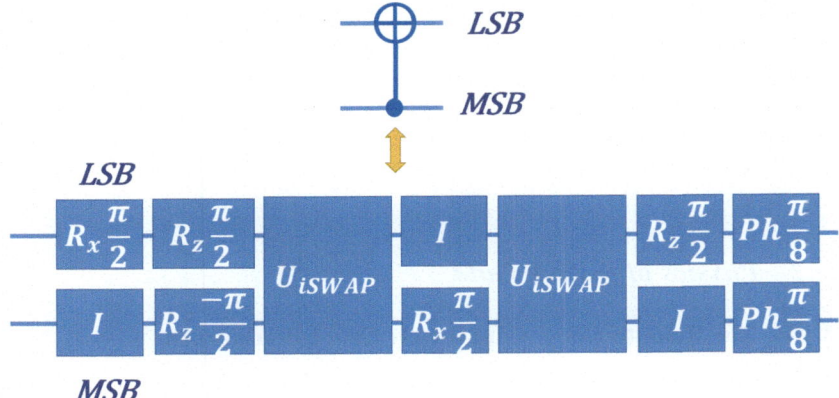

Fig. 21.4 Implementation of CNOT gate using two iSWAP gates and one-qubit gates

shift gate in Sect. 4.5.3. $Ph(\theta)$ has constant diagonal elements of $e^{i\theta}$. It introduces a global phase shift instead of a relative phase shift between $|0\rangle$ and $|1\rangle$.

The following shows the Python code with comments explaining how the code works. Note that in Python, the imaginary number i is coded as $1j$. The code can be downloaded from https://github.com/hywong2/Quantum_Computing_Architecture.

```
#Import library
import numpy as np
#Define individual gates
I = np.array([[1,0],[0,1]])
Rzpi_2 = np.array([[np.exp(-1j*np.pi/4),0],\
                   [0,np.exp(1j*np.pi/4)]])
Rxpi_2 = np.array([[np.cos(np.pi/4),-1j*np.sin(np.pi/4)],\
                   [-1j*np.sin(np.pi/4),np.cos(np.pi/4)]])
R_neg_zpi_2 = np.array([[np.exp(1j*np.pi/4),0],\
                        [0,np.exp(-1j*np.pi/4)]])
iSWAP = np.array([[1,0,0,0],[0,0,1j,0],[0,1j,0,0],[0,0,0,1]])
Phase_pi_8 = np.array([[np.exp(1j*np.pi/8),0],\
                       [0,np.exp(1j*np.pi/8)]])
#Perform tensor product and matrix multiplications
stage1 = np.kron(I,Rxpi_2)
stage2 = np.matmul(np.kron(R_neg_zpi_2,Rzpi_2),stage1)
stage3 = np.matmul(iSWAP,stage2)
stage4 = np.matmul(np.kron(Rxpi_2,I),stage3)
stage5 = np.matmul(iSWAP,stage4)
stage6 = np.matmul(np.kron(I,Rzpi_2),stage5)
stage7 = np.matmul(np.kron(Phase_pi_8,Phase_pi_8),stage6)
#print the final matrix
print(stage7)
```

The following shows the overall matrix of the circuit which is the same as that of a CNOT gate in Eq. (4.52):

```
[[1.+0.j 0.+0.j 0.+0.j 0.+0.j]
 [0.+0.j 1.+0.j 0.+0.j 0.+0.j]
 [0.+0.j 0.+0.j 0.+0.j 1.+0.j]
 [0.+0.j 0.+0.j 1.+0.j 0.+0.j]]
```

21.3.2 Physics of iSWAP Gate

21.3.2.1 The Hamiltonian of Capacitively Coupled Transmon

We need to derive the Hamiltonian for the circuit in Fig. 21.3. In principle, we need to follow the procedures in Fig. 17.2. Readers can do this as an exercise. Here we will only take the results from [1]. Firstly, since the Hamiltonian, $\hat{\mathcal{H}}$, is the total energy of a system, it is natural to partition it into three parts. It contains the Hamiltonians of the individual transmon when they are isolated ($\hat{\mathcal{H}}_1$ and $\hat{\mathcal{H}}_2$) and also the Hamiltonian due to their coupling through C_c, $\hat{\mathcal{H}}_{int}$. Based on Eq. (17.7), we have:

$$\hat{\mathcal{H}} = \hat{\mathcal{H}}_1 + \hat{\mathcal{H}}_2 + \hat{\mathcal{H}}_{int},$$
$$= 4E_{c1}\hat{n}_1^2 - E_{J1}\cos\hat{\varphi}_1 + 4E_{c2}\hat{n}_2^2 - E_{J2}\cos\hat{\varphi}_2 + \hat{\mathcal{H}}_{int}. \quad (21.26)$$

$\hat{\mathcal{H}}_{int}$ will be the term left over if we had found $\hat{\mathcal{H}}$ through the rigorous method in Fig. 17.2 after subtracting $\hat{\mathcal{H}}_1$ and $\hat{\mathcal{H}}_2$. When $C_c \ll C_1, C_2$, it is approximated as

$$\hat{\mathcal{H}}_{int} = 4e^2 \frac{C_c}{C_1 C_2} \hat{n}_1 \hat{n}_2, \quad (21.27)$$

where $C_1 = C_{J1} + C_{s1}$ and $C_2 = C_{J2} + C_{s2}$. Therefore,

$$\hat{\mathcal{H}} = 4E_{c1}\hat{n}_1^2 - E_{J1}\cos\hat{\varphi}_1 + 4E_{c2}\hat{n}_2^2$$
$$- E_{J2}\cos\hat{\varphi}_2 + 4e^2 \frac{C_c}{C_1 C_2} \hat{n}_1 \hat{n}_2. \quad (21.28)$$

Since they are transmons with anharmonicity, we now will limit them to two levels (the ground state and the lowest excited state). As proved in Sect. 19.4.2, the transmon qubit Hamiltonian can be also expressed as in Eq. (19.16), which is in the form of quantized SHO. By dropping the fourth-order term as we are only interested in the lowest two levels, we have:

$$\hat{\mathcal{H}} = \hbar\omega_{0,1} \hat{b}_1^\dagger \hat{b}_1 + \hbar\omega_{0,2} \hat{b}_2^\dagger \hat{b}_2 + 4e^2 \frac{C_c}{C_1 C_2} \hat{n}_1 \hat{n}_2, \quad (21.29)$$

21.3 Two-Qubit Entanglement Gate for Transmon

where $\hbar\omega_{0,1} = \sqrt{8E_{c1}E_{J1}}$ and $\hbar\omega_{0,2} = \sqrt{8E_{c2}E_{J2}}$. $\hat{b}_i$ and $\hat{b}_i^\dagger$ are the annihilation and creation operators for qubit-i, respectively. Note that we also dropped the zero-point energy, $\frac{1}{2}\hbar\omega_{0i}$, as it only provides a constant shift of energy level. For the interaction term, we use Eq. (19.20) to get

$$\begin{aligned}
\hat{\mathcal{H}} &= \hbar\omega_{0,1}\hat{b}_1^\dagger\hat{b}_1 + \hbar\omega_{0,2}\hat{b}_2^\dagger\hat{b}_2 \\
&\quad + \frac{4e^2}{\hbar}\frac{C_c}{C_1 C_2}\left(i\sqrt{\hbar}n_{zpf,1}\left(-\hat{b}_1 + \hat{b}_1^\dagger\right)\right)\left(i\sqrt{\hbar}n_{zpf,2}\left(-\hat{b}_2 + \hat{b}_2^\dagger\right)\right), \\
&= \hbar\omega_{0,1}\hat{b}_1^\dagger\hat{b}_1 + \hbar\omega_{0,2}\hat{b}_2^\dagger\hat{b}_2 \\
&\quad + \frac{4e^2 C_c}{C_1 C_2}\left(in_{zpf,1}\left(-\hat{b}_1 + \hat{b}_1^\dagger\right)\right)\left(in_{zpf,2}\left(-\hat{b}_2 + \hat{b}_2^\dagger\right)\right),
\end{aligned} \quad (21.30)$$

where zero-point fluctuation of the charge number of qubit-i, $n_{zpf,i}$, is defined as $\frac{1}{2}\left(\frac{E_{J_i}}{2E_{c_i}}\right)^{\frac{1}{4}}$.

Since we are working on two levels only, we can use the matrix to represent this equation and perform simplifications. Since $\hat{N}_i = \hat{b}_i^\dagger\hat{b}_i$ (Eq. (14.21)), and using Eq. (21.5), we have:

$$\begin{aligned}
H &= -\frac{\hbar\omega_{0,1}}{2}\sigma_{z,1}\otimes I - \frac{\hbar\omega_{0,2}}{2}I\otimes\sigma_{z,2} \\
&\quad + \frac{4e^2 C_c}{C_1 C_2}(in_{zpf,1}(-\boldsymbol{b_1}+\boldsymbol{b_1}^\dagger))\otimes(in_{zpf,2}(-\boldsymbol{b_2}+\boldsymbol{b_2}^\dagger)), \\
&= -\frac{\hbar\omega_{0,1}}{2}\sigma_{z,1}\otimes I - \frac{\hbar\omega_{0,2}}{2}I\otimes\sigma_{z,2} \\
&\quad - \frac{4e^2 C_c}{C_1 C_2}n_{zpf,1}n_{zpf,2}(-\boldsymbol{b_1}+\boldsymbol{b_1}^\dagger)\otimes(-\boldsymbol{b_2}+\boldsymbol{b_2}^\dagger), \\
&= -\frac{\hbar\omega_{0,1}}{2}\sigma_{z,1}\otimes I - \frac{\hbar\omega_{0,2}}{2}I\otimes\sigma_{z,2} \\
&\quad - g(-\boldsymbol{b_1}+\boldsymbol{b_1}^\dagger)\otimes(-\boldsymbol{b_2}+\boldsymbol{b_2}^\dagger),
\end{aligned} \quad (21.31)$$

where we have defined

$$g = \frac{4e^2 C_c}{C_1 C_2}n_{zpf,1}n_{zpf,2}. \quad (21.32)$$

g is the **coupling energy**. Also note that this is a two-qubit system; the tensor product is used to construct the larger space from the individual one-qubit systems.

The interaction Hamiltonian is thus

$$H_{int} = -g(-b_1 + b_1^\dagger) \otimes (-b_2 + b_2^\dagger). \tag{21.33}$$

Example 21.3 Express g in terms of the qubit frequencies, $\omega_{0,1}$ and $\omega_{0,2}$ and the capacitances, C_c, C_1, and C_2:

$$\begin{aligned}g &= \frac{4e^2 C_c}{C_1 C_2} n_{zpf,1} n_{zpf,2}, \\ &= \frac{4e^2 C_c}{C_1 C_2} \frac{1}{2}\left(\frac{E_{J_1}}{2E_{c_1}}\right)^{\frac{1}{4}} \frac{1}{2}\left(\frac{E_{J_2}}{2E_{c_2}}\right)^{\frac{1}{4}}, \\ &= \frac{e^2 C_c}{C_1 C_2}\left(\frac{E_{J_1}}{2E_{c_1}}\right)^{\frac{1}{4}}\left(\frac{E_{J_2}}{2E_{c_2}}\right)^{\frac{1}{4}}, \\ &= \frac{C_c}{\sqrt{C_1 C_2}}\left(\frac{e^4 E_{J_1}}{2E_{c_1} C_1^2}\right)^{\frac{1}{4}}\left(\frac{e^4 E_{J_2}}{2E_{c_2} C_2^2}\right)^{\frac{1}{4}}, \tag{21.34}\end{aligned}$$

where, in line 2, the definition of $n_{zpf,i}$ is used. Since $\hbar\omega_{0,i} = \sqrt{8E_{Ji}E_{ci}}$ (which is a good approximation if the anharmonicity is much smaller than the qubit frequency Eq. (19.18)), we have $E_{Ji} = \frac{(\hbar\omega_{0,i})^2}{8E_{ci}}$. Moreover, $E_{ci} = \frac{e^2}{2C_i}$. Therefore,

$$\begin{aligned}\left(\frac{e^4 E_{Ji}}{2E_{c_i} C_i^2}\right)^{\frac{1}{4}} &= \left(\frac{e^4 \frac{(\hbar\omega_{0,i})^2}{8E_{ci}}}{2E_{c_i} C_i^2}\right)^{\frac{1}{4}}, \\ &= \left(\frac{e^4(\hbar\omega_{0,i})^2}{16E_{c_i}^2 C_i^2}\right)^{\frac{1}{4}}, \\ &= \left(\frac{e^4(\hbar\omega_{0,i})^2}{16(\frac{e^2}{2C_i})^2 C_i^2}\right)^{\frac{1}{4}}, \\ &= \left(\frac{(\hbar\omega_{0,i})^2}{4}\right)^{\frac{1}{4}}, \\ &= \left(\frac{\hbar\omega_{0,i}}{2}\right)^{\frac{1}{2}}. \tag{21.35}\end{aligned}$$

21.3 Two-Qubit Entanglement Gate for Transmon

Therefore,

$$g = \frac{C_c}{\sqrt{C_1 C_2}} \left(\frac{\hbar \omega_{0,1}}{2}\right)^{\frac{1}{2}} \left(\frac{\hbar \omega_{0,2}}{2}\right)^{\frac{1}{2}},$$

$$= \frac{C_c}{2\sqrt{C_1 C_2}} \hbar \sqrt{\omega_{0,1} \omega_{0,2}}. \quad (21.36)$$

■

Before moving forward, we want to clarify a few possible confusions of the symbols. Here we use the 2×2 matrix forms of the operators. b_1 and $b_1^\dagger$ are the **annihilation** and **creation** operators, respectively. Using the matrices in Example 14.4, we have:

$$b_1 = \begin{pmatrix} 0 & 1 \\ 0 & 0 \end{pmatrix}, \quad (21.37)$$

$$b_1^\dagger = \begin{pmatrix} 0 & 0 \\ 1 & 0 \end{pmatrix}. \quad (21.38)$$

Sometimes, they are labeled as σ^- and σ^+, respectively. However, this is **NOT the convention in this book**. In this book, σ^- and σ^+ are the **lowering** and **raising** operators for the eigenvalues of σ_z, respectively. They are defined and given in Eq. (10.6) and Eq. (10.7). We are also **not** using the definitions in Example 10.1. Therefore,

$$b_1 = \begin{pmatrix} 0 & 1 \\ 0 & 0 \end{pmatrix} = \frac{1}{2}\sigma^+, \quad (21.39)$$

$$b_1^\dagger = \begin{pmatrix} 0 & 0 \\ 1 & 0 \end{pmatrix} = \frac{1}{2}\sigma^-. \quad (21.40)$$

The interaction Hamiltonian in Eq. (21.33) becomes:

$$H_{int} = -\frac{g}{4} \left(-\sigma_1^+ + \sigma_1^-\right) \otimes \left(-\sigma_2^+ + \sigma_2^-\right). \quad (21.41)$$

21.3.2.2 Rotating Frame and Rotating Wave Approximation

This Hamiltonian is not very useful. We will work in the **rotating frame** of the qubit frequencies. Note that this is a two-qubit system. So it is in the rotating frame of $\omega_{0,1}$ for qubit 1 and $\omega_{0,2}$ for qubit 2. The transformation matrix is $U_{RF} = e^{-i\frac{\omega_{0,1} t}{2}\sigma_z} \otimes e^{-i\frac{\omega_{0,2} t}{2}\sigma_z}$ (based on Eq. (10.14)). Like the one-qubit case in Eq. (21.20), for the total

Hamiltonian, we have:

$$H_{RF} = i\hbar \frac{\partial U_{RF}}{\partial t} U^{\dagger}_{RF} + U_{RF} H U^{\dagger}_{RF},$$
$$= i\hbar \frac{\partial U_{RF}}{\partial t} U^{\dagger}_{RF} + U_{RF} (H_1 + H_2 + H_{int}) U^{\dagger}_{RF}. \quad (21.42)$$

It can be shown that only $U_{RF} H_{int} U^{\dagger}_{RF}$ is left as there is no detuning like in Eq. (10.26). Therefore, in the rotating frame, we have:

$$H_{RF} = H_{RF,int} = U_{RF} H_{int} U^{\dagger}_{RF},$$
$$= -\frac{g}{4} U_{RF} \left(\sigma_1^+ \sigma_2^+ - \sigma_1^+ \sigma_2^- - \sigma_1^- \sigma_2^+ + \sigma_1^- \sigma_2^- \right) U^{\dagger}_{RF}, \quad (21.43)$$

where we dropped the tensor product symbols in line 2. We can evaluate each term using Eq. (10.35) and Eq. (10.36). For example,

$$U_{RF} \sigma_1^+ \sigma_2^+ U^{\dagger}_{RF} = U_{RF} \sigma_1^+ U_{RF}^{\dagger} U_{RF} \sigma_2^+ U^{\dagger}_{RF},$$
$$= e^{-i\omega_{0,1} t} \sigma_1^+ \otimes e^{-i\omega_{0,2} t} \sigma_2^+,$$
$$= e^{-i(\omega_{0,1} + \omega_{0,2}) t} \sigma_1^+ \otimes \sigma_2^+, \quad (21.44)$$

where we used $U_{RF}^{\dagger} U_{RF} = I$ in line 1. We also reintroduced the tensor product symbol in line 2 to emphasize that this is a two-qubit system, but it will be dropped again later. Similarly,

$$U_{RF} \sigma_1^+ \sigma_2^- U^{\dagger}_{RF} = e^{-i(\omega_{0,1} - \omega_{0,2}) t} \sigma_1^+ \sigma_2^-, \quad (21.45)$$

$$U_{RF} \sigma_1^- \sigma_2^+ U^{\dagger}_{RF} = e^{-i(-\omega_{0,1} + \omega_{0,2}) t} \sigma_1^- \sigma_2^+, \quad (21.46)$$

$$U_{RF} \sigma_1^- \sigma_2^- U^{\dagger}_{RF} = e^{i(\omega_{0,1} + \omega_{0,2}) t} \sigma_1^- \sigma_2^-. \quad (21.47)$$

Now we apply **rotating wave approximation** and thus drop the terms in Eq. (21.44) and Eq. (21.47) as their effects are averaged out due to their high frequencies, $\omega_{0,1} + \omega_{0,2}$. Equation (21.43) becomes:

$$H_{RF} = -\frac{g}{4} \left(-e^{-i(\omega_{0,1} - \omega_{0,2}) t} \sigma_1^+ \sigma_2^- - e^{-i(-\omega_{0,1} + \omega_{0,2}) t} \sigma_1^- \sigma_2^+ \right),$$
$$= \frac{g}{4} \left(e^{-i\delta\omega_{12} t} \sigma_1^+ \sigma_2^- + e^{i\delta\omega_{12} t} \sigma_1^- \sigma_2^+ \right), \quad (21.48)$$

where $\delta\omega_{12} = \omega_{0,1} - \omega_{0,2}$. To implement an **iSWAP gate**, we will tune the qubit frequency of qubit 2 so that it is the same as qubit 1 (resonant with $\delta\omega_{12} = 0$). This

21.3 Two-Qubit Entanglement Gate for Transmon

can be achieved using a flux tunable qubit (e.g., Sect. 16.4). Then,

$$H_{RF} = \frac{g}{4}\left(\sigma_1^+ \sigma_2^- + \sigma_1^- \sigma_2^+\right),$$
$$= \frac{g}{2}\left(\sigma_x \sigma_x + \sigma_y \sigma_y\right), \quad (21.49)$$

where the second line will be derived in the following example. Such a coupling is called **transverse coupling**.

Example 21.4 Show that $\frac{1}{4}(\sigma_1^+ \sigma_2^- + \sigma_1^- \sigma_2^+) = \frac{1}{2}(\sigma_x \sigma_x + \sigma_y \sigma_y)$.

Firstly, we find the matrix of $\frac{1}{4}(\sigma_1^+ \sigma_2^- + \sigma_1^- \sigma_2^+)$.

$$\frac{1}{4}(\sigma_1^+ \sigma_2^- + \sigma_1^- \sigma_2^+) = \frac{1}{4}\left(\begin{pmatrix} 0 & 2 \\ 0 & 0 \end{pmatrix} \otimes \begin{pmatrix} 0 & 0 \\ 2 & 0 \end{pmatrix} + \begin{pmatrix} 0 & 0 \\ 2 & 0 \end{pmatrix} \otimes \begin{pmatrix} 0 & 2 \\ 0 & 0 \end{pmatrix}\right),$$

$$= \frac{1}{4}\left(\begin{pmatrix} 0 & 0 & 0 & 0 \\ 0 & 0 & 4 & 0 \\ 0 & 0 & 0 & 0 \\ 0 & 0 & 0 & 0 \end{pmatrix} + \begin{pmatrix} 0 & 0 & 0 & 0 \\ 0 & 0 & 0 & 0 \\ 0 & 4 & 0 & 0 \\ 0 & 0 & 0 & 0 \end{pmatrix}\right),$$

$$= \frac{1}{4}\begin{pmatrix} 0 & 0 & 0 & 0 \\ 0 & 0 & 4 & 0 \\ 0 & 4 & 0 & 0 \\ 0 & 0 & 0 & 0 \end{pmatrix},$$

$$= \begin{pmatrix} 0 & 0 & 0 & 0 \\ 0 & 0 & 1 & 0 \\ 0 & 1 & 0 & 0 \\ 0 & 0 & 0 & 0 \end{pmatrix}. \quad (21.50)$$

Also,

$$\frac{1}{2}(\sigma_x \sigma_x + \sigma_y \sigma_y) = \frac{1}{2}\left(\begin{pmatrix} 0 & 1 \\ 1 & 0 \end{pmatrix} \otimes \begin{pmatrix} 0 & 1 \\ 1 & 0 \end{pmatrix} + \begin{pmatrix} 0 & -i \\ i & 0 \end{pmatrix} \otimes \begin{pmatrix} 0 & -i \\ i & 0 \end{pmatrix}\right),$$

$$= \frac{1}{2}\left(\begin{pmatrix} 0 & 0 & 0 & 1 \\ 0 & 0 & 1 & 0 \\ 0 & 1 & 0 & 0 \\ 1 & 0 & 0 & 0 \end{pmatrix} + \begin{pmatrix} 0 & 0 & 0 & -1 \\ 0 & 0 & 1 & 0 \\ 0 & 1 & 0 & 0 \\ -1 & 0 & 0 & 0 \end{pmatrix}\right),$$

$$= \frac{1}{2}\begin{pmatrix} 0 & 0 & 0 & 0 \\ 0 & 0 & 2 & 0 \\ 0 & 2 & 0 & 0 \\ 0 & 0 & 0 & 0 \end{pmatrix},$$

$$= \begin{pmatrix} 0 & 0 & 0 & 0 \\ 0 & 0 & 1 & 0 \\ 0 & 1 & 0 & 0 \\ 0 & 0 & 0 & 0 \end{pmatrix}. \tag{21.51}$$

Therefore, it is proved. ∎

Thus, when the two qubits are in resonance, the Hamiltonian in the rotating frame can be expressed as

$$H_{RF} = g \begin{pmatrix} 0 & 0 & 0 & 0 \\ 0 & 0 & 1 & 0 \\ 0 & 1 & 0 & 0 \\ 0 & 0 & 0 & 0 \end{pmatrix},$$

$$= \begin{pmatrix} 0 & 0 & 0 & 0 \\ 0 & 0 & g & 0 \\ 0 & g & 0 & 0 \\ 0 & 0 & 0 & 0 \end{pmatrix}. \tag{21.52}$$

21.3.2.3 iSWAP Gate

With the Hamiltonian, we can then solve the **Schrödinger equation** to get the evolution of a quantum state of the two-qubit system. The two-qubit quantum gate is related to the Hamiltonian as (Eq. (4.9))

$$U = e^{-i\frac{H_{RF}}{\hbar}t}. \tag{21.53}$$

The Hamiltonian is a non-diagonal matrix. To exponentiate it, we can diagonalize it or use Taylor expansion. We have already done that in Example 4.2. Therefore, based on Eq. (4.29)

$$U = e^{-i\frac{H_{RF}}{\hbar}t},$$

$$= \begin{pmatrix} 1 & 0 & 0 & 0 \\ 0 & \cos\frac{gt}{\hbar} & -i\sin\frac{gt}{\hbar} & 0 \\ 0 & -i\sin\frac{gt}{\hbar} & \cos\frac{gt}{\hbar} & 0 \\ 0 & 0 & 0 & 1 \end{pmatrix}. \tag{21.54}$$

21.4 Summary

Finally, if we only turn on the coupling for $t = \frac{3\pi\hbar}{2g}$ (i.e., setting $\delta\omega_{12} = 0$ for this amount of time), we get an **iSWAP gate**:

$$U_{iSWAP} = U\left(t = \frac{3\pi\hbar}{2g}\right) = \begin{pmatrix} 1 & 0 & 0 & 0 \\ 0 & 0 & i & 0 \\ 0 & i & 0 & 0 \\ 0 & 0 & 0 & 1 \end{pmatrix}. \quad (21.55)$$

21.3.2.4 Summary of iSWAP Gate

In summary, the two-qubit system has two qubits with different qubit frequencies. They are coupled to each other capacitively all the time. Since they have different frequencies, they would not affect each other as long as the time is long enough to let the effect be averaged to zero (Eq. (21.48)). When we want to apply the iSWAP gate, we will change the qubit frequency of qubit 2 to that of qubit 1. Therefore, at least qubit 2 needs to be flux tunable (Sect. 16.4). Then, we let it stay resonant with qubit 1 for $t = \frac{3\pi\hbar}{2g}$ and bring it back to the original frequency. This completes the action of an iSWAP gate. Note that, in the process, we work in the rotating frame of the qubit frequencies.

21.4 Summary

In this chapter, we studied the action of an external AC gate voltage on a transmon qubit. We again can use the rotating frame and rotating wave approximation to implement one-qubit gates for a transmon qubit. We emphasize that the rotating frame can be either at the driving voltage frequency or the qubit frequency. It is often to be at the qubit frequency. Regardless of the rotating frame used, we can reuse what we learned in spin qubits. More importantly, we concluded that the Hamiltonian has I and Q parts to control its rotation about $\hat{x}$ and $\hat{y}$ of the Bloch sphere, respectively. In the future chapters, we will demonstrate that they can be controlled easily by an **arbitrary waveform generator (AWG)**. We also studied the implementation of the iSWAP gate for two transmon qubits. It is an entanglement gate that is equivalent to a CNOT gate by using two iSWAP gates and other one-qubit gates. We derive that if we bring the two qubits into resonance for $t = \frac{3\pi\hbar}{2g}$ in the rotating frame of the qubit frequencies, we have effectively implemented an iSWAP gate.

Problems

21.1 Number Operator Matrix
N in Eq. (21.4) is a 2×2 Hermitian matrix. Therefore, it is natural that we can decompose it into a linear combination of I and σ_z. Use the inner product method to decompose it (Sect. 6.2). You should get the same result as in Eq. (21.4).

21.2 Rotating Frame in One-Qubit Gate
Draw the vector components at $\phi = 0$ and $t = 0$ for Fig. 21.2. When time increases, show that $\vec{B}_{E1+}$ is indeed moving counterclockwise.

21.3 CNOT and iSWAP Gates
Show the circuit equivalency in Fig. 21.4 using matrix multiplications and compare it to the Python code results. You can compare the intermediate results at each stage for debugging.

21.4 Rotating Frame in iSWAP Gate
Show that only $U_{RF} H_{int} U_{RF}^{\dagger}$ is left in Eq. (21.42) in the rotating frame of the qubit frequencies.

21.5 Transverse Coupling
Show that in a two-level two-qubit system, $(-b_1 + b_1^{\dagger}) \otimes (-b_2 + b_2^{\dagger}) = -\sigma_{y,1} \otimes \sigma_{y,2}$.

References

1. P. Krantz, M. Kjaergaard, F. Yan, T. P. Orlando, S. Gustavsson, and W. D. Oliver. A quantum engineer's guide to superconducting qubits. *Applied Physics Reviews*, 6(2):021318, 06 2019.
2. Norbert Schuch and Jens Siewert. Natural two-qubit gate for quantum computation using the XY interaction. *Phys. Rev. A*, 67:032301, Mar 2003.

Chapter 22
Superconducting Qubit: Readout and Initialization

22.1 Introduction

We studied 1-qubit and 2-qubit gates (last chapter) before the readout mechanism (this chapter) because they are mathematically easier. In that process, we only need to describe them as the interactions between the quantized artificial atom and classical voltage signal (although variables such as the charge are promoted to operators). One example of an artificial atom is the transmon qubit which has nonuniformly separated energy levels (anharmonicity). The readout scheme in a superconductor qubit is to interrogate the qubit through an electromagnetic wave. It thus requires the study of the interaction between the electromagnetic wave and the artificial atom. In this process, even the electromagnetic wave needs to be quantized which leads to second quantization. We can no longer treat the system semiclassically (classical electromagnetic wave and quantum mechanical artificial atoms). This system resembles cavity quantum electrodynamics (c-QED) [1]. Thus, it is called circuit quantum electrodynamics (circuit-QED) [2–4]. Since the relevant mathematics are profound, we will take electromagnetic wave quantization for granted. Interested readers can refer to [5] for a more rigorous treatment of second quantization.

22.1.1 Learning Outcomes

Understand how a qubit state pulls the resonant frequency of a resonator; be able to calculate the coupling factor using fundamental electrical circuit parameters; appreciate the meaning of Jaynes-Cumming Hamiltonian; be able to describe different schemes to initialize superconducting qubits.

22.1.2 Teaching Videos

- Search for Ch22 in this playlist
  - https://tinyurl.com/3yhze3jn
- Other videos
  - https://youtu.be/MS6eIhe7Hpg

22.2 Light-Matter Interaction

The **readout** process in a superconducting qubit resembles the interaction between a **photon** (light) in an optical cavity and an atom (matter). Therefore, it is instructive to have a brief overview of the light-matter interaction in an optical cavity first.

An optical cavity has certain resonant frequencies. It means that only light of certain modes (with certain frequencies) can exist in the cavity. This is just like a string on a violin that can only have standing waves of certain wavelengths and frequencies. We can quantize light, which is just an electromagnetic wave, through **second quantization**. The name second quantization can be confusing. It does *not* imply that we perform another quantization over the quantization we have been doing in quantum mechanics. In quantum mechanics, we have been quantizing physical quantities such as momentum (due to boundary conditions in wavefunctions). In second quantization, we quantize also fields such as an electromagnetic field. The result is that light (i.e., EM wave) has quantized photon **Fock states** $|N\rangle$, and **creation**, $\hat{a}^\dagger$, and **annihilation**, $\hat{a}$, operators are introduced which can be used to increase and decrease N, respectively. The result is similar to what we saw when we quantized a simple harmonic oscillator. We do not need to study the details of EM wave quantization (e.g., [5]). We just need to agree that, in a cavity, only certain modes can exist. Each mode can have different numbers of photons ($|0\rangle$, $|1\rangle$, $|2\rangle$, $|3\rangle$, etc.) as its Fock states. Here we will only consider one mode.

Now if we put an atom with two different levels (ground state $|g\rangle$ and excited state $|e\rangle$) in the cavity, the EM field will interact with the atom through **coupling energy**, g (Fig. 22.1). The coupling is a result of the interaction between the oscillating electric field in the EM wave and the atom dipole moment (the negative electron cloud center of the atom being not aligned with the positive nucleus charge center). We do not need to understand the details [8]. The result is that the atom and the cavity will exchange energy and this is easy to understand. Starting with $n - 1$ photons (state $|n - 1\rangle$) and an excited atom ($|e\rangle$), the atom can relax to its ground state $|g\rangle$ by emitting a photon, resulting in $N = n$. Similarly, the atom can also absorb a photon from the cavity (N is decreased to $n - 1$ from n) and get excited (becomes $|e\rangle$). Therefore, the system will be switching between state $|g\rangle |N = n\rangle$

22.2 Light-Matter Interaction

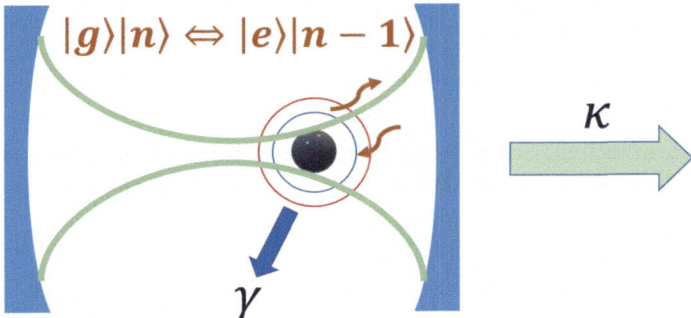

Fig. 22.1 Illustration of light-matter interaction. Two mirrors (blue) form an optical cavity in which a certain mode of light is trapped (green). Here we will only consider one mode. After the second quantization, it can have different numbers of photons ($|N\rangle$) of a certain mode. A two-level atom with ground state $|g\rangle$ and excited state $|e\rangle$ is placed inside the cavity. The cavity and the atom form a new system and can change between state $|g\rangle|n\rangle$ and state $|e\rangle|n-1\rangle$ through light-matter interaction. The cavity has "leakage" and the photon can leak at a rate of κ. The atom's excited state can also relax through non-radiative processes or radiative processes not coupled to the cavity mode through a rate of γ

and state $|e\rangle|N = n-1\rangle$ as a whole. This is the **Rabi oscillation** of the atom. When $n = 1$, it is called the **vacuum Rabi oscillation**.

22.2.1 Vacuum Rabi Oscillation

It is instructive to discuss a little bit more about vacuum Rabi oscillation. In quantum mechanics, there is no reason why an excited atom should decay from $|e\rangle$ to $|g\rangle$. An atom will only decay if it interacts with an electric field. For example, if there is a photon of the same mode existing in the cavity, its electric field will interact with the atom causing it to decay and emit another photon. This is the famous **stimulated emission** which is the working principle of a LASER. When $N = 0$ (i.e., in vacuum), there should be no electric field and the atom should stay at $|e\rangle$ forever. However, due to the quantization of EM wave, just like the quantization of SHO in Chap. 14, it has finite energy (**vacuum energy**) even when $N = 0$ (Eq. (14.26)). This can also be understood as the result of the uncertainty principle when the field is quantized. Therefore, even in a vacuum, there is a fluctuation of the electric field due to uncertainty. The fluctuating electric field will interact with the atom resulting in **spontaneous emission** after which a photon is emitted ($N = 1$) and the atom relaxes to ground state $|g\rangle$.

22.3 Jaynes-Cummings Hamiltonian

22.3.1 Overview and Symbols

The formalism of light-matter interaction in an optical cavity is called the **cavity quantum electrodynamics (c-QED)**. For the simplest case where the matter only has two levels and there is only one *single* mode in the optical cavity, it is described by the famous **Jaynes-Cummings Hamiltonian**, $\hat{\mathcal{H}}_{JC}$, [9], which is written as

$$\hat{\mathcal{H}}_{JC} = \hat{\mathcal{H}}_{field} + \hat{\mathcal{H}}_{atom} + \hat{\mathcal{H}}_{int}, \tag{22.1}$$

where $\hat{\mathcal{H}}_{field}$, $\hat{\mathcal{H}}_{atom}$, and $\hat{\mathcal{H}}_{int}$ are the Hamiltonians of the EM field, atom, and their interaction, respectively. $\hat{\mathcal{H}}_{int}$ is the term that results in interesting dynamics in the light-matter interaction such as Rabi oscillation.

The components in the Jaynes-Cummings Hamiltonian can be expressed as

$$\hat{\mathcal{H}}_{field} = \hbar\omega_{EM}\hat{a}^\dagger\hat{a}, \tag{22.2}$$

$$\hat{\mathcal{H}}_{atom} = \hbar\omega_{atom}\frac{\hat{\sigma}_z'}{2}, \tag{22.3}$$

$$\hat{\mathcal{H}}_{int} = g(\hat{a}\hat{\sigma}_+ + \hat{a}^\dagger\hat{\sigma}_-). \tag{22.4}$$

Let us first discuss Eq. (22.2). ω_{EM} is the frequency of the photon of the cavity mode. $\hat{a}^\dagger$ and $\hat{a}$ are the creation and annihilation operators of the photon number state (Fock state), respectively. Just like in the case of SHO, the photon number operator, $\hat{N}$, is $\hat{a}^\dagger\hat{a}$ (e.g., Eq. (14.21)). Note that the energy offset due to vacuum energy (the $\frac{1}{2}$ term) is not included for simplicity. Therefore, Eq. (22.2) is easy to understand as the total energy due to the cavity is just the number of photons multiplied by the energy of one photon.

In Eq. (22.3), $\hbar\omega_{atom}$ is the energy difference between state $|e\rangle$ and state $|g\rangle$ of the atom. We will derive Eq. (22.3) in the following section. But we want to discuss the meaning of $\hat{\sigma}_z'$ first. This is *not* the Pauli matrix $\hat{\sigma}_z = \begin{pmatrix} 1 & 0 \\ 0 & -1 \end{pmatrix}$. But very often, in the literature, $\hat{\sigma}_z'$ is written as $\hat{\sigma}_z$ and can cause confusion. $\hat{\sigma}_z'$ is defined as

$$\hat{\sigma}_z' = |e\rangle\langle e| - |g\rangle\langle g|,$$

$$= \begin{pmatrix} 0 \\ 1 \end{pmatrix}(0\ 1) - \begin{pmatrix} 1 \\ 0 \end{pmatrix}(1\ 0),$$

$$= \begin{pmatrix} 0 & 0 \\ 0 & 1 \end{pmatrix} - \begin{pmatrix} 1 & 0 \\ 0 & 0 \end{pmatrix},$$

22.3 Jaynes-Cummings Hamiltonian

$$= \begin{pmatrix} -1 & 0 \\ 0 & 1 \end{pmatrix},$$

$$= -\hat{\sigma}_z. \tag{22.5}$$

Of course, the matrix forms of $\hat{\sigma}_z'$ and $\hat{\sigma}_z$ depend on how we order the atom states. In this book, we have been ordering them with the ground state first. This results in the matrix in Eq. (22.5).

In Eq. (22.4), we see that the interaction Hamiltonian involves the creation and annihilation operators of the photon number states and also two new operators, $\hat{\sigma}_+$ and $\hat{\sigma}_-$. They are defined as

$$\hat{\sigma}_+ = |e\rangle \langle g|,$$

$$= \begin{pmatrix} 0 \\ 1 \end{pmatrix} (1\ 0),$$

$$= \begin{pmatrix} 0 & 0 \\ 1 & 0 \end{pmatrix}, \tag{22.6}$$

and

$$\hat{\sigma}_- = |g\rangle \langle e|,$$

$$= \begin{pmatrix} 1 \\ 0 \end{pmatrix} (0\ 1),$$

$$= \begin{pmatrix} 0 & 1 \\ 0 & 0 \end{pmatrix}. \tag{22.7}$$

$\hat{\sigma}_+$ and $\hat{\sigma}_-$ are the **raising** and **lowering** operators of *a two-level system*, respectively. It means that when they are applied to $|g\rangle$ and $|e\rangle$, they will change the atom state to $|e\rangle$ and $|g\rangle$, respectively (see Problem 22.1). Note that they are different from the raising operator $\hat{\sigma}^+$ and lowering operator $\hat{\sigma}^-$ *of* σ_z (Problem 10.1), which are to raise to a state with *a higher eigenvalue in* σ_z or to lower to a state with *a lower eigenvalue in* σ_z, respectively. Moreover, they also have different notations (note that one has a superscript and one has a subscript). We also note that sometimes $\hat{\sigma}^+$ and $\hat{\sigma}^-$ are defined slightly differently (see Eq. (10.5)).

We may have a hand-waiving understanding of Eq. (22.4) by reading its terms one by one. Firstly, the interaction Hamiltonian strength depends on the coupling energy g. When $g = 0$, there is no coupling, and the cavity and the atom will stay as two independent systems, and there will be no Rabi oscillation. The term $\hat{a}\hat{\sigma}_+$ says that the interaction can be due to the raising operator of the atom (i.e., bringing the atom to the excited state) and the annihilation operator of the photon which will reduce one photon. This represents the photon absorption process. The term $\hat{a}^\dagger \hat{\sigma}_-$ says that the interaction can be due to the lowering operator of the atom (i.e., bringing the atom to the ground state) and the creation operator of the photon. This represents the photon emission process.

22.3.2 Circuit Quantum Electrodynamics

When a transmon is coupled to an LC tank, mathematically, we have the same system as the one in Fig. 22.1. And the formalism becomes **circuit quantum electrodynamics (circuit-QED)**. Of course, circuit-QED is more profound than just coupling a transmon to an LC tank. One may refer to [2] and [3] for more details. In the case of a superconducting qubit interacting with a resonator (which can be modeled as an LC tank), if we limit the superconducting qubit to two levels ($|0\rangle / |1\rangle$ or $|g\rangle / |e\rangle$), it can be described by the Jaynes-Cummings Hamiltonian as well. We can limit it to two levels if it has sufficient **anharmonicity**. A transmon qubit satisfies this requirement. Figure 22.2 shows the circuit. From the c-QED point of view, the LC tank acts as the cavity. We already mentioned that a single-mode optical cavity, with angular frequency ω_{EM}, can have different numbers of photons (N), and its energy is given by Eq. (22.2). In a quantized LC tank, it also can have different numbers of photons (as photons are just electromagnetic waves), and each photon has an energy of $\hbar\omega_r$, where $\omega_r = \frac{1}{\sqrt{LC}}$ is the resonant frequency of the tank. Here we rename ω_0 in Eq. (15.5) as ω_r. The transmon acts as an **artificial atom** which has two levels and anharmonicity (although we ignore higher levels in the treatment here). Moreover, unlike a natural atom, we can control its "atomic properties" such as its energy separation by changing the Josephson energy and capacitive energy (Eq. (19.13)). Since it is a qubit, we rename ω_{atom} as ω_q.

Therefore, Jaynes-Cummings Hamiltonian for the qubit-resonator system is

$$\hat{\mathcal{H}}_{JC} = \hbar\omega_r \hat{a}^\dagger \hat{a} + \hbar\omega_q \frac{\hat{\sigma}_z'}{2} + g(\hat{a}\hat{\sigma}_+ + \hat{a}^\dagger \hat{\sigma}_-). \tag{22.8}$$

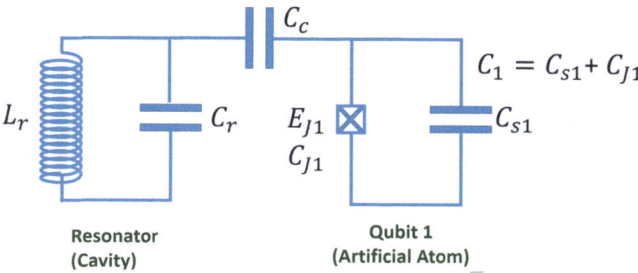

Fig. 22.2 A transmon qubit coupled to an LC tank. This resembles the coupling between an optical cavity and a two-level atom, where the LC tank acts as the cavity and the transmon qubit acts as an artificial atom

22.3.3 Proof of Jaynes-Cummings Hamiltonian

One can go through the procedure in Fig. 17.2 to quantize the circuit to obtain the Jaynes-Cummings Hamiltonian in circuit-QED. For example, one may refer to [2].

Here, we will use another approach by reusing the equations for a two-qubit system in which two transmons are coupled through a coupling capacitor, C_c. The two transmons are just two simple harmonic oscillators when both are limited to two levels. As long as we also limit the LC tank to two levels, we can reuse the equations in the two-qubit system in Fig. 21.3. Of course, the LC tank has multiple levels. Therefore, *the derivation below is incomplete although the result is correct for multiple levels.*

We start with Eq. (21.29), which describes the total Hamiltonian of the two-transmon system and is copied here for convenience:

$$\hat{\mathcal{H}} = \hbar\omega_{0,1}\hat{b}_1^\dagger \hat{b}_1 + \hbar\omega_{0,2}\hat{b}_2^\dagger \hat{b}_2 + 4e^2\frac{C_c}{C_1 C_2}\hat{n}_1\hat{n}_2. \tag{22.9}$$

We need to remind ourselves that $\hat{b}_1$ and $\hat{b}_1^\dagger$ are the annihilation and creation operators for qubit 1's eigenenergy number $|N_1\rangle$ instead of Cooper pair number $|n_1\rangle$. This is the same for $\hat{b}_2$ and $\hat{b}_2^\dagger$. Moreover, once we limit ourselves to studying the first two levels of a transmon qubit, a transmon qubit is just equivalent to an SHO. Since an $L - C$ tank (resonator) also has its own creation and annihilation operators, $\hat{a}_1$ and $\hat{a}_1^\dagger$ (see Eq. (15.30)), we can just replace the Hamiltonian of qubit 2, $\hbar\omega_{0,2}\hat{b}_2^\dagger\hat{b}_2$, by $\hbar\omega_r \hat{a}^\dagger \hat{a}$.

There is no direct physical meaning of $|n\rangle$ in an $L - C$ tank because n refers to the number of excess Cooper pairs in a Cooper pair box. However, this is just the conjugate momentum of the system. Therefore, we can rewrite it using Eq. (15.32):

$$2e\hat{n} = \hat{Q},$$
$$= i\sqrt{\frac{C_r\omega_r\hbar}{2}}(-\hat{a} + \hat{a}^\dagger),$$
$$= iQ_{zpf}(-\hat{a} + \hat{a}^\dagger). \tag{22.10}$$

where $\omega_r = \omega_0$ is the resonant frequency of the resonator and C_r is the effective capacitance of the resonator. Substituting Eq. (22.10) into Eq. (22.9) and using Eq. (19.20), we finally get:

$$\hat{\mathcal{H}} = \hbar\omega_q \hat{b}^\dagger \hat{b} + \hbar\omega_r \hat{a}^\dagger \hat{a} + 2e\frac{C_c}{C_1 C_r}\hat{n}_1(iQ_{zpf}(-\hat{a} + \hat{a}^\dagger)),$$
$$= \hbar\omega_q \hat{b}^\dagger \hat{b} + \hbar\omega_r \hat{a}^\dagger \hat{a}$$

$$+ \frac{2eC_c}{C_1C_r}(in_{zpf}(-\hat{b}+\hat{b}^\dagger))(iQ_{zpf}(-\hat{a}+\hat{a}^\dagger)),$$

$$= \hbar\omega_q \hat{b}^\dagger\hat{b} + \hbar\omega_r \hat{a}^\dagger\hat{a}$$

$$- \frac{2eC_c}{C_1C_r}(n_{zpf}(-\hat{b}+\hat{b}^\dagger))(Q_{zpf}(-\hat{a}+\hat{a}^\dagger)),$$

$$= \hbar\omega_q \hat{b}^\dagger\hat{b} + \hbar\omega_r \hat{a}^\dagger\hat{a} - g(-\hat{b}+\hat{b}^\dagger)(-\hat{a}+\hat{a}^\dagger), \quad (22.11)$$

where we defined the coupling energy of this system, g, as

$$g = \frac{2eC_c}{C_1C_r} n_{zpf} Q_{zpf},$$

$$= \frac{2eC_c}{C_1C_r} \frac{1}{2}\left(\frac{E_J}{2E_c}\right)^{\frac{1}{4}} \sqrt{\frac{C_r\omega_r\hbar}{2}},$$

$$= e\frac{C_c}{C_1C_r}\left(\frac{E_J}{2E_c}\right)^{\frac{1}{4}} \sqrt{\frac{C_r\omega_r\hbar}{2}}. \quad (22.12)$$

It is also instructive to further show that

$$g = \hbar\omega_r \frac{C_c}{C_1}\left(\frac{E_J}{2E_c}\right)^{\frac{1}{4}} \sqrt{\frac{\pi Z_r}{R_K}}, \quad (22.13)$$

where Z_r is the resonator's characteristic impedance and $R_k = \frac{h}{e^2}$, which is the resistance quantum (see Eq. (33) in [2] and Problem 22.2).

Similar to Eq. (21.31) for a two-qubit interaction, we can represent the Hamiltonian in its matrix form in the following way since we are dealing with a two-level system:

$$H = -\frac{\hbar\omega_q}{2}\sigma_z \otimes I - \frac{\hbar\omega_r}{2}I \otimes \sigma_{za} - g(-b+b^\dagger) \otimes (-a+a^\dagger), \quad (22.14)$$

where we *temporarily* introduced σ_{za} for the LC tank which has the same matrix form as $\sigma_z = \begin{pmatrix} 1 & 0 \\ 0 & -1 \end{pmatrix}$. This is valid because we only study two levels of the LC tank, and therefore, it behaves like a qubit. We can thus repeat what we did for the two-qubit system and use Eq. (21.37) to (21.41).

Then, working in a rotating frame using the qubit frame at ω_q and the LC tank frame at ω_R and applying **rotating wave approximation** (see Sect. 21.3.2.2), we obtain the Hamiltonian in the rotating frame, $H_{JC,RF}$, as in Eq. (21.48), in which the high-frequency terms are ignored.

Now, we transform it back to the original laboratory frame. Since the high-frequency terms (Eqs. (21.44) and (21.47)) have been eliminated during the rotating wave approximation, we obtain the final Jaynes-Cummings Hamiltonian as

$$H_{JC} = -\frac{\hbar\omega_q}{2}\sigma_z \otimes I - \frac{\hbar\omega_r}{2}I \otimes \sigma_{za} + g(b \otimes a^\dagger + b^\dagger \otimes a),$$

$$= \frac{\hbar\omega_q}{2}\sigma_z' \otimes I + \hbar\omega_r I \otimes a^\dagger a + g(b \otimes a^\dagger + b^\dagger \otimes a), \quad (22.15)$$

where we have substituted $\sigma_z = -\sigma_z'$ from Eq. (22.5). This is the same as Eq. (22.14) with the high-frequency terms removed ($b^\dagger a^\dagger$ and ba). We have also rewritten the resonator Hamiltonian with the creation and annihilation operators using Eq. (21.4). Again, constant energy terms are ignored.

To compare to Eq. (22.8), we swap the order of the resonator state and the qubit state. We then use the fact that $b = \sigma_-$ and $b^\dagger = \sigma_+$ (comparing Eqs. (21.37) and Eq. (21.38) to Eqs. (22.7) and (22.6), respectively), we get:

$$H_{JC} = \hbar\omega_r a^\dagger a \otimes I + \frac{\hbar\omega_q}{2}I \otimes \sigma_z' + g(a^\dagger \otimes \sigma_- + a \otimes \sigma_+), \quad (22.16)$$

which is the 2 × 2 matrix representation of Eq. (22.8).

22.4 Dispersive Readout

The resonator in the resonator-qubit system can be used to measure the state of the qubit (**qubit readout**) by interrogating the resonant frequency of the resonator with a microwave pulse. This is because the state of the qubit will shift the resonant frequency differently for different qubit states ($|e\rangle$ or $|g\rangle$). However, although it is beautiful, the Jaynes-Cummings Hamiltonian in Eq. (22.8) is only valid when we consider the transmon as a two-level artificial atom, and it cannot show us how the resonant frequency will be shifted. In order to understand the readout process, higher levels in the transmon need to be considered. Moreover, we do not want the qubit to be strongly coupled to the resonator. This is to avoid actual photons being absorbed by the qubit during the interrogation. This is achieved by setting the resonator frequency, ω_r, differently from the qubit frequency, ω_q. That is to have nonzero detuning with **detuning frequency**, $\Delta = \omega_q - \omega_r \neq 0$. Note that in the literature, detuning is often defined as $\Delta = \omega_r - \omega_q$, but we want to continue to use the same definition we have defined earlier throughout the book. When they have different frequencies, of course, they cannot exchange energies through direct photon emission and absorption. The interaction needs to involve virtual photons and that is why higher levels need to be considered. The detuning needs to be large enough such that $|\hbar\Delta| \gg g$. This is called the **dispersive regime**, and the readout process to read the qubit state through a resonator coupled to a qubit in the dispersive regime is called **dispersive readout**. Dispersive readout also means the signal obtained during interrogation depends on the effective frequency of the resonator.

22.4.1 Dispersive Shift

By taking higher levels into consideration and working in the dispersive regime ($\hbar\Delta \gg g$), the resulting Hamiltonian of the resonator-qubit system is [7]

$$\hat{\mathcal{H}} = \hbar\omega_q' \frac{\hat{\sigma}_z'}{2} + (\hbar\omega_r' + \hbar\chi\hat{\sigma}_z')\hat{a}^\dagger\hat{a}. \quad (22.17)$$

Before explaining the newly introduced variables, let us first compare this equation to the Jaynes-Cummings Hamiltonian in Eq. (22.8). Firstly, we see that both of them have a resonator energy term due to the photon number ($\hat{a}^\dagger\hat{a}$) and a qubit frequency term. But there is no explicit interaction Hamiltonian in Eq. (22.17). This is expected because now we are operating in the dispersive regime with weak coupling. So we should not see Rabi oscillation due to the interaction with the resonator. Secondly, the effective qubit frequency and the effect resonator frequency are shifted. For the qubit, it changes from ω_q to ω_q'. ω_q is the so-called **bare qubit frequency** which is the frequency of the qubit when it is isolated. ω_q' is the qubit frequency when the qubit is coupled to the resonator and is given by $\omega_q' = \omega_q + \chi_{01}$ where χ_{01} will be defined later. More importantly, the resonator frequency is shifted from ω_r to $\omega_r' + \chi\hat{\sigma}_z'$ with $\omega_r' = \omega_r - \chi_{12}/2$, where χ_{12} will be defined later, too. Let us now examine how a qubit state will change the resonator frequency.

Example 22.1 Find the resonator frequency difference between when the qubit is at state $|g\rangle$ and when the qubit is at state $|e\rangle$.

We can find the frequency of the resonator when the qubit is at $|g\rangle$, $\omega_{r,g}$, by performing $\langle g|\hbar\omega_r' + \hbar\chi\hat{\sigma}_z'|g\rangle/\hbar$. This is equivalent to finding the expectation value of the photon energy divided by $\hbar$:

$$\begin{aligned}
\omega_{r,g} &= \langle g|\omega_r' + \chi\hat{\sigma}_z'|g\rangle, \\
&= \langle g|\omega_r'|g\rangle + \langle g|\chi\hat{\sigma}_z'|g\rangle, \\
&= \omega_r' + \langle g|\chi(|e\rangle\langle e| - |g\rangle\langle g|)|g\rangle, \\
&= \omega_r' - \langle g|\chi|g\rangle\langle g|g\rangle, \\
&= \omega_r' - \chi, \quad (22.18)
\end{aligned}$$

where in the first line we have canceled $\hbar$. In the third line, the definition of $\hat{\sigma}_z'$ is used (Eq. (22.5)). In the fourth line, the orthonormal property of the atom states is used.

Similarly, the resonator frequency becomes $\omega_r' + \chi$ when the atom is at the excited state $|e\rangle$. Therefore, the frequency difference is $|2\chi|$. ∎

22.4 Dispersive Readout

χ is the **cross Kerr** or **effective dispersive shift** of the system. It is defined as

$$\chi = \chi_{01} - \chi_{12}/2, \qquad (22.19)$$

where

$$\chi_{01} = \frac{|g_{01}/\hbar|^2}{\omega_{10} - \omega_r},$$

$$\chi_{12} = \frac{|g_{12}/\hbar|^2}{\omega_{21} - \omega_r}, \qquad (22.20)$$

where $\omega_{10} = \omega_q = \frac{E_{10}}{\hbar}$ and $\omega_{21} = \frac{E_{21}}{\hbar}$ (which is different from [7] because we want to retain the same definition of energy difference that we have been using so far, e.g., in Figs. 15.3 and 19.1). Also, g_{01}, g_{12}, and g (in the following equations) have the *unit of energy* as they are the coupling energies while in [7], they are angular frequencies. Therefore, it is divided by the reduced Planck constant to preserve the form in [7]. It is also given that

$$|g_{01}| = g,$$
$$|g_{12}| = \sqrt{2}g, \qquad (22.21)$$

Therefore, cross-Kerr can also be expressed as

$$\chi = \chi_{01} - \frac{\chi_{12}}{2},$$

$$= \frac{|g_{01}/\hbar|^2}{\omega_{10} - \omega_r} - \frac{|g_{12}/\hbar|^2}{2(\omega_{21} - \omega_r)},$$

$$= \frac{|g_{01}/\hbar|^2}{\omega_{10} - \omega_r} - 2\frac{|g_{01}/\hbar|^2}{2(\omega_{21} - \omega_r)},$$

$$= \frac{|g_{01}/\hbar|^2}{\omega_{10} - \omega_r} - \frac{|g_{01}/\hbar|^2}{\omega_{10} + \frac{\alpha}{\hbar} - \omega_r},$$

$$= \frac{|g_{01}/\hbar|^2}{\Delta} - \frac{|g_{01}/\hbar|^2}{\Delta + \frac{\alpha}{\hbar}},$$

$$= |g_{01}/\hbar|^2 \frac{\frac{\alpha}{\hbar}}{\Delta(\Delta + \frac{\alpha}{\hbar})},$$

$$= -|g_{01}/\hbar|^2 \frac{E_c/\hbar}{\Delta(\Delta - E_c/\hbar)},$$

$$= -|g/\hbar|^2 \frac{E_c/\hbar}{\Delta(\Delta - E_c/\hbar)}, \qquad (22.22)$$

where $\Delta = \omega_{10} - \omega_r = \omega_q - \omega_r$. In line one, we have used Eq. (22.19). In line two and the last line, Eq. (22.20) is used. Then, the definition of anharmonicity, $\alpha = -E_c$ in Eq. (19.19), is used in line four.

Usually, $\Delta < 0$. This means that the resonator is designed at a higher frequency than the qubit frequency. In this case, $\chi < 0$. Therefore, when a qubit is at $|g\rangle$, the resonator frequency will shift to a higher frequency (Eq. (22.18)). When the qubit is at $|e\rangle$, the resonator frequency will shift to a lower frequency. Figure 22.4 illustrates the situation.

22.5 Microwave Readout Circuit Example

The readout process in superconducting qubits usually relies on the dispersion shift mechanism. One of the advantages is that it allows **quantum non-demolition (QND) measurement**. QND means that after the measurement, the qubit will stay in the same state it collapsed to. This will allow multiple measurements to increase the accuracy. Figure 22.3 shows a commonly used setup [4, 6]. A microwave pulse in the GHz regime is applied from the left of the transmission line. The transmission line should have the same characteristic impedance (e.g., 50Ω) as the source. If the transmission line is not interrupted, the signal should not be distorted and will reach the output with the same frequency and magnitude. This can be measured by the **scattering matrix** [10]. In particular, we are interested in the S_{21} component which

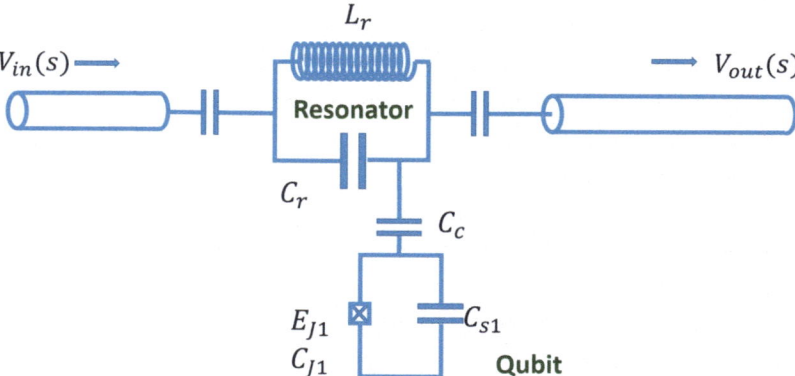

Fig. 22.3 A microwave readout circuit using the dispersive shift mechanism. Microwave pulses go into a transmission line from the left which is interrupted by a resonator (modeled as an LC tank here). The signal can only pass the resonator if it has the same frequency as the resonator's resonant frequency. The resonator is capacitively coupled to a transmon qubit which will modify the resonant frequency of the resonator based on its state. By measuring the output signal on the right, one can distinguish the state of the qubit based on the phase shift

22.5 Microwave Readout Circuit Example

is defined as

$$S_{21} = \frac{V_{out}}{V_{in}},$$
$$= Re(S_{21}) + i Im(S_{21}),$$
$$= |S_{21}|e^{i\phi}. \quad (22.23)$$

S_{21} is a complex number with a real part ($Re(S_{21})$) and an imaginary part ($Im(S_{21})$). Its magnitude is denoted as $|S_{21}|$ and its phase is ϕ with $\phi = \arctan \frac{Im(S_{21})}{Re(S_{21})}$.

If the transmission line is interrupted by a resonator, the signal cannot propagate from left to right if its frequency is far from the resonator's resonant frequency (ω_r). As a result, $|S_{21}|$ is small (Fig. 22.4). Since the resonator is coupled to a transmon qubit, due to dispersive shift (Example 22.1 and Eq. (22.22)), ω_r will be increased by χ when the qubit is at $|e\rangle$ and decreased by χ when the qubit is at $|g\rangle$. If $\omega_r > \omega_q$, we have $\chi < 0$. Then, the peak transmission is shifted to a lower frequency at $|e\rangle$ and a higher frequency at $|g\rangle$. If we set to measure $|S_{21}|$ at ω_r, we cannot distinguish the two states as both give a small $|S_{21}|$ (left of Fig. 22.4). On the other hand, if we measure the real and imaginary parts (which is equivalent to measuring the *phase*), we will be able to distinguish them. For example, on the right of Fig. 22.4, the imaginary parts of S_{21} are both near zero for $|g\rangle$ and $|e\rangle$ but it is positive for $|g\rangle$ and negative for $|e\rangle$. Therefore, *under this setup*, if one measures positive $Re(S_{21})$, we know that the qubit is at $|g\rangle$. If one measures negative $Re(S_{21})$, we know that the qubit is at $|e\rangle$.

We can plot the real and imaginary parts of S_{21} on the complex plane as shown in the left of Fig. 22.5. In general, due to parasitic components, non-ideality, and the choice of measurement frequency, the measurement might not lie exactly along the real axis. However, one can always transform them. It is common to transform it so

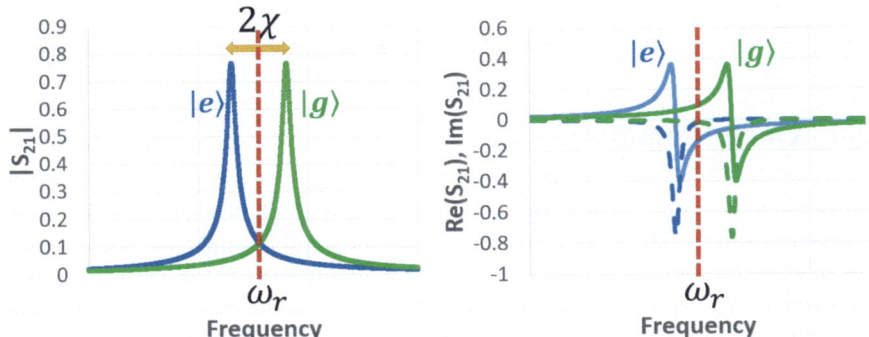

Fig. 22.4 S_{21} as a function of input pulse frequency with the qubit in Fig. 22.3 in different states. Left: Magnitude of S_{21}. Right: The real part (solid lines) and imaginary part (dashed lines) of S_{21}. The vertical red dashed line shows the measurement frequency to obtain S_{21} to distinguish the two states. Here it is assumed that ω_r is larger than ω_q

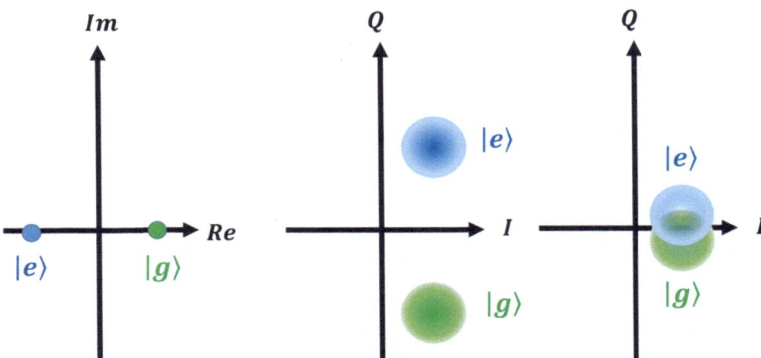

Fig. 22.5 Left: Plot of the real and imaginary parts of S_{21} using the measurement setup in Fig. 22.4. Middle: Measurement results of a general measurement setup when there is noise and after transformation. Right: When the noise is too large or the cross-Kerr is not sufficiently large, the two blobs may overlap rendering significant measurement error

that the $|g\rangle$ and $|e\rangle$ measurement results are symmetrical about the horizontal axis as shown in the middle of Fig. 22.5. The plane on which this is plotted is called the *IQ*-**plane**. Moreover, due to noise, measurement results of the same qubit state will not align on one single point [6]. Instead, they will reside within a certain radius of the ideal point. These circles are called the **IQ blobs**.

If the noise is large (large blob size) or the centers of the two blobs are not well separated (small cross-Kerr magnitude, $|\chi|$), then one will not be able to distinguish $|e\rangle$ from $|g\rangle$. The right of Fig. 22.5 shows that in the overlapping area, both states can give the same Q-value. As a result, one cannot distinguish the two states resulting in measurement error. Readers can also refer to Sect. 23.4.3 on how $I - Q$ blobs can be constructed from the readout signal.

Interested users may refer to [11] for more details on superconducting qubit measurements.

22.6 Qubit Initialization

Superconducting qubit initialization is usually done through two approaches. One is by **thermalization**. This is to let the qubit wait for about 10 times the decoherence time of the qubit (typically $100\,\mu s$) so that it will thermalize to drop from $|e\rangle$ to $|g\rangle$. The issue of this approach is that it takes a long time (in the order of ms). Another approach is to use **active reset**. Firstly, the qubit will be measured. If it is at $|g\rangle$, then nothing needs to be performed as it is initialized to $|g\rangle$ already. If it is measured to be $|e\rangle$, then a π-pulse is applied to rotate the qubit to $|g\rangle$ from $|e\rangle$. On the Bloch sphere, this is equivalent to rotating the state by an angle π about $\hat{y}$. This is fast as the gate time and measurement time are $< 1\mu s$. However, measurement error can result in a wrong operation, and thus the initialization error will be larger (see also Chap. 25).

22.7 Summary

In this chapter, we discuss the last two important DiVincenzo's criteria to make superconducting qubits useful to build a quantum computer, namely, the ability to perform readout and state initialization (state preparation). We first discuss c-QED in which the interaction between a two-level atom and an optical cavity is studied, which can be explained by Jaynes-Cummings Hamiltonian. We then point out that mathematically, a transmon coupling to a microwave resonator is the same as an atom in a single mode cavity. The state of the transmon will pull the resonator frequency of the resonator. The difference in resonator's resonant frequencies due to $|e\rangle$ and $|g\rangle$ is 2χ, where χ is the cross-Kerr. Finally, we discuss how to perform measurement (measuring phase shift) to determine the state of a qubit.

Problems

22.1 Raising and Lowering Operators of a Two-Level System
By applying Eq. (22.6) to $|g\rangle$ and Eq. (22.7) to $|e\rangle$, show that they are the raising and lowering operators of a two-level system, respectively.

22.2 Coupling Energy Between a Resonator and a Transmon
Prove Eq. (22.13). Note that $Z = \sqrt{\frac{L_r}{C_r}}$. You may refer to Eq. (33) in [2]. Note that in that paper, the coupling energy is given by $\hbar g$ instead of g.

22.3 Cross-Kerr
Show that the resonator frequency becomes $\omega'_r + \chi$ when the atom is at the excited state $|e\rangle$.

References

1. Herbert Walther, Benjamin T H Varcoe, Berthold-Georg Englert, and Thomas Becker. Cavity quantum electrodynamics. *Reports on Progress in Physics*, 69(5):1325, apr 2006.
2. Alexandre Blais, Arne L. Grimsmo, S. M. Girvin, and Andreas Wallraff. Circuit quantum electrodynamics. *Rev. Mod. Phys.*, 93:025005, May 2021.
3. David Isaac Schuster. *Circuit Quantum Electrodynamics*. PhD thesis, Yale University, 2006.
4. Alexandre Blais, Ren-Shou Huang, Andreas Wallraff, S. M. Girvin, and R. J. Schoelkopf. Cavity quantum electrodynamics for superconducting electrical circuits: An architecture for quantum computation. *Phys. Rev. A*, 69:062320, Jun 2004.

5. L.D. Landau, V. B. Berestetski, and E.M. Lifshitz. *Quantum Electrodynamics (Course of Theoretical Physics, Vol. 4)*. Pergamon, 1982.
6. Hiu Yung Wong, Prabjot Dhillon, Kristin M. Beck, and Yaniv J. Rosen. A simulation methodology for superconducting qubit readout fidelity. *Solid-State Electronics*, 201:108582, 2023.
7. Jens Koch, Terri M. Yu, Jay Gambetta, A. A. Houck, D. I. Schuster, J. Majer, Alexandre Blais, M. H. Devoret, S. M. Girvin, and R. J. Schoelkopf. Charge-insensitive qubit design derived from the cooper pair box. *Phys. Rev. A*, 76:042319, Oct 2007.
8. Christopher J. Foot. *Atomic Physics*. Oxford University Press, 2005.
9. E. T. Jaynes and F. W. Cummings. Comparison of quantum and semiclassical radiation theories with application to the beam maser. *IEEE Proc.*, 51:89–109, 1963.
10. David M. Pozar. *Microwave Engineering*. Wiley; 4th edition, 201.
11. Daniel Thomas Sank. *Fast, Accurate State Measurement in Superconducting Qubitss*. PhD thesis, University of California, Santa Barbara, 2014.

Part IV
Quantum Computer Design and Implementation

Chapter 23
Microwave Electronics in Quantum Computers

23.1 Introduction

We have demonstrated how to build a quantum computer using spin qubits and superconducting qubits. Particularly, we showed the mechanisms to meet four of the five DiVincenzo's criteria (Sect. 1.3), namely, well-characterized qubits, efficient initialization, efficient readout, and the availability of a "universal set" of quantum gates in both types of qubits. We also have made two assumptions. Firstly, they have long enough coherence times (the fifth DiVincenzo's criterion). This is an active research topic. We will discuss their definitions, measurements, and equations in Chap. 25. The second is that we assume we are able to generate the required signals to initialize, manipulate, and read out qubits. These involve high-speed electronics and microwave circuits. In this chapter, we will study a typical superconducting qubit quantum computer and follow the propagation paths of the readout and manipulation microwave pulses to understand how microwave circuits play an important role in quantum computers. Although a superconducting quantum computer is used as an example, most of the theory and components are applicable directly to other types of quantum computers.

23.1.1 Learning Outcomes

Appreciate the critical roles of microwave electronics in the realization of quantum computers; be able to relate the $I - Q$ signal to microwave's $I - Q$ mixing; be able to describe how the manipulation and readout signals propagate in a quantum computer.

23.1.2 Teaching Videos

- Search for Ch23 in this playlist
    - https://tinyurl.com/3yhze3jn
- Other videos
    - https://youtu.be/kg80aJNY3jw
    - https://youtu.be/DtjncNDGToI
    - https://youtu.be/X9Y4q6fNnwo

23.2 Overview

We have shown the schematic of a typical superconducting qubit quantum computer in Fig. 1.3. Now, we want to follow the paths of the manipulation pulses and readout pulses. Note that for initialization, since either thermalization is used (with no pulses) or active reset is used (a combination of readout pulse and manipulation pulse), we do not need to discuss it separately (Sect. 22.6).

For convenience, the schematic of a typical superconducting qubit quantum computer is modified and shown again in Fig. 23.1. We see it has the following major microwave components. It has **mixers**, **local oscillators (LOs)**, **attenuators**, **isolators**, and **amplifiers**.

We also should not overlook the signal generator at the top. This component is a combination of **high-speed digital and analog electronics** such as **field**

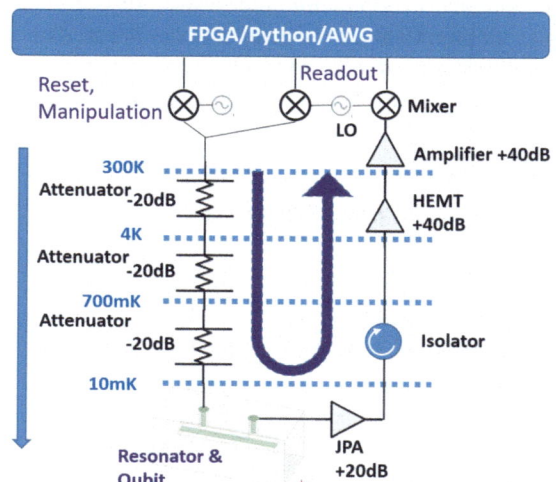

Fig. 23.1 Schematic of a typical superconducting qubit quantum computer. The qubit manipulation and initialization paths (blue vertical arrow) and the readout path (purple U-shaped arrow) are indicated

programmable gate arrays (FPGAs), microprocessors, and arbitrary waveform generators (AWGs), controlled by classical software programs such as Python. *The process of quantum computing is to program the high-speed circuits so that appropriate waveforms are generated by AWGs and sent to the qubits for the desired operations. During the readout process, again, appropriate signals are sent to the qubit, and the reflected or transmitted pulses are collected and analyzed.*

In Fig. 23.1, the same readout scheme as the one in Sect. 22.5 is used. A resonator (which is in a 3D cavity) is coupled to a transmon qubit. The readout pulse interrogates the resonator which has its resonant frequency pulled by the qubit state [1].

23.3 Qubit Manipulation Path

23.3.1 Mixer

In Fig. 21.1, we showed that a pulse of $V_g = s(t)\sin(\omega_1 t + \phi)$ is sent to the transmon. After a lengthy derivation and working in the rotating frame of the qubit frequency, $\omega_0 = \omega_q$, we obtained the effective Hamiltonian in Eq. (21.24) which is repeated here for convenience. We will also use its 2×2 matrix form:

$$H_{RF} = -\frac{\hbar \Omega_R(t)}{2} \left(I\sigma_x + Q\sigma_y \right), \qquad (23.1)$$

where

$$\hbar \Omega_R(t) = \frac{s(t) Q_{zpf}}{1 + \frac{C_\Sigma}{C_g}}, \qquad (23.2)$$

$$I = \cos\phi, \qquad (23.3)$$

$$Q = -\sin\phi. \qquad (23.4)$$

How do we obtain $V_g = s(t)\sin(\omega_1 t + \phi)$? We use an $I - Q$ mixer to achieve this goal.

Firstly, an "in-phase" (or I) component and an "out-of-phase" (or Q or **quadrature** component) are generated by AWGs (Fig. 23.2). AWGs can generate any (arbitrary) waveform as long as the required frequency is not too high. The I component is

$$2s(t)I \sin\omega_{AWG}t = 2s(t)\cos\phi \sin\omega_{AWG}t, \qquad (23.5)$$

and the Q component is

$$2s(t)Q \sin\omega_{AWG}t = -2s(t)\sin\phi \sin\omega_{AWG}t, \qquad (23.6)$$

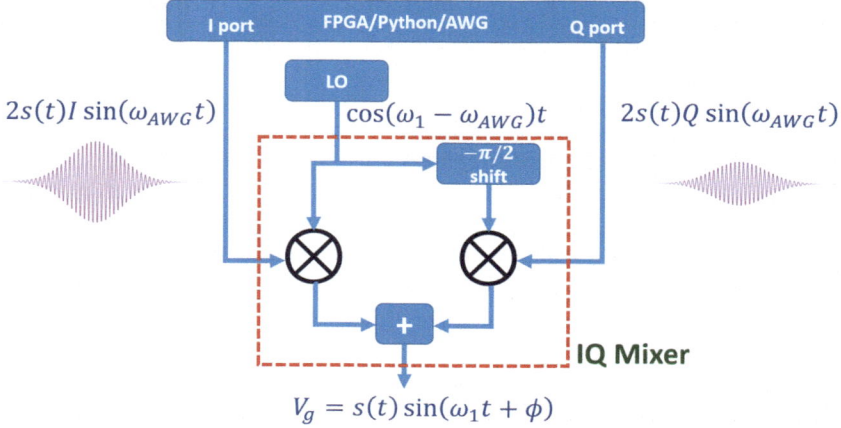

Fig. 23.2 Schematic showing how the I and Q signals generated by AWG are mixed with the LO signal in an IQ mixer to achieve the desired wavefunction for qubit manipulations

where ω_{AWG} is the frequency of the pulses and $s(t)$ is the envelope function. Note that $s(t)$ is scaled by $2I$ and $2Q$, respectively, as the final envelope functions of the pulses generated by the AWG in this particular example. Since ω_{AWG} is in the order of tens of MHz (e.g., 50 MHz), it can be generated digitally by the AWGs easily. Note that ϕ is embedded as the amplitude of the I and Q signals.

However, the pulses we need to interact with the qubits are in GHz. An $I - Q$ mixer is then used to multiply the AWG signals with the signal from a local oscillator to bring it to a high enough frequency. This is called **up-conversion**. The LO generates a high-frequency sinusoidal wave at the GHz range, $\cos(\omega_1 - \omega_{AWG})t$ (e.g., 5 GHz). We assume the amplitude is one for simplicity. Its frequency is chosen to be $\omega_1 - \omega_{AWG}$ because the goal is to achieve a signal with ω_1 at the mixer output. Note that, in an $I - Q$ mixer, the I part is multiplied by the LO signal (in-phase component) directly. The Q component is multiplied by the LO signal phase-shifted by $-\pi/2$ (quadrature component). Note that we make it $-\pi/2$ instead of the commonly used $\pi/2$ because we want to make the final equation in the desired form for instructional purposes. They are then added together as the output.

Therefore, the signal after the $I - Q$ mixer, V_g, is given by

$$\begin{aligned}
V_g &= 2s(t)I \sin\omega_{AWG}t \cos(\omega_1 - \omega_{AWG})t \\
&\quad + 2s(t)Q \sin\omega_{AWG}t \cos\left((\omega_1 - \omega_{AWG})t - \frac{\pi}{2}\right), \\
&= 2s(t)\cos\phi \sin\omega_{AWG}t \cos(\omega_1 - \omega_{AWG})t \\
&\quad - 2s(t)\sin\phi \sin\omega_{AWG}t \sin(\omega_1 - \omega_{AWG})t,
\end{aligned} \quad (23.7)$$

where we have used the identity, $\cos(\theta - \pi/2) = \sin\theta$ in the second line. Then, we will use the product formulae, $\sin\alpha\cos\beta = \frac{1}{2}(\sin(\alpha+\beta) + \sin(\alpha-\beta))$ for the first

23.3 Qubit Manipulation Path

term and $\sin\alpha \sin\beta = \frac{1}{2}(\cos(\alpha - \beta) - \cos(\alpha + \beta))$ for the second term. We get:

$$V_g = 2s(t)\cos\phi\frac{1}{2}(\sin(\omega_1 t) + \sin(2\omega_{AWG}t - \omega_1 t))$$
$$-2s(t)\sin\phi\frac{1}{2}(\cos(2\omega_{AWG}t - \omega_1 t) - \cos\omega_1 t), \quad (23.8)$$

where the $(2\omega_{AWG}t - \omega_1 t)$ terms will be filtered by a **low-pass filter (LPF)** and the signal becomes:

$$V_g = s(t)\cos\phi\sin\omega_1 t + s(t)\sin\phi\cos\omega_1 t,$$
$$= s(t)\sin(\omega_1 t + \phi). \quad (23.9)$$

Therefore, we have achieved the pulse in Fig. 21.1 for qubit manipulation. More specifically, we only need to set the amplitude of the pulse from the I (Q) port to control the amount of rotation about the x-axis (y-axis) on the Bloch sphere. We can also control the rotation speed ($\Omega_R(t)$) by applying an appropriate envelope function $s(t)$ using AWG.

It should be noted that if ω_{AWG} is not large enough, filtering of the $(2\omega_{AWG}t - \omega_1 t)$ terms can be difficult as it is approximately the same as $-\omega_1 t$. Then a **single-side band (SSB)** mixer is required, which can achieve the same purpose of keeping only one of the components.

23.3.2 Attenuators and Noise

The signals are generated at room temperature. They contain thermal noise. If they are sent to the qubit, which is cooled to tens of mK, the noise will be too large for any useful quantum operations. Therefore, the signals (including the thermal noise) go through a series of attenuators. In Fig. 23.1, they are attenuated by 60 dB. In this section, we will not discuss the mechanism of attenuation. We will spend time discussing some aspects of thermal noise and power units.

Noise is unpredictable. For example, a voltage noise ($V(t)$) fluctuates as a function of time randomly (Fig. 23.3). It also has a zero average value. Therefore, to describe the noise, one would use the average of its square. This is the **mean square**. However, a more useful way to describe its property is to find the mean square of each frequency component. This is called the **noise power spectral density (NSD)**, $\overline{v_n^2}(f)$, with a unit of V^2/Hz. NSD does *not* have the unit of power per hertz because the resistance is omitted (power has a unit of V^2/Ω). This is the convention, and it is used for convenience because it is expected that R should be the characteristic impedance of the circuit (e.g., 50 Ω). Therefore, one should perform the appropriate conversion when needed.

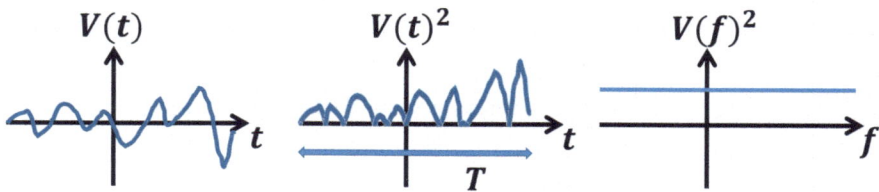

Fig. 23.3 Left: Noise voltage as a function of time. Middle: Noise voltage squared as a function of time. Right: The noise power spectral density of white noise

Thermal noise is white noise. This means that it has a constant NSD (Fig. 23.3). This resembles the fact that white light is a combination of various colors (frequencies) of equal power. The NSD of thermal noise of a resistor with resistance R is given as [2]

$$\overline{v_n^2} = 4\,\text{kTR}, \tag{23.10}$$

which is frequency-independent as expected.

Example 23.1 Find the noise voltage of a $50\,\Omega$ resistor at $T = 300\,K$.
Firstly, the NSD is

$$\overline{v_n^2} = 4\,\text{kTR},$$
$$= 4 \times 1.38 \times 10^{-23}\,\text{J/K} \times 300K \times 50\,\Omega,$$
$$= 8.29 \times 10^{-19} V^2 \cdot \text{Hz}^{-1} \tag{23.11}$$

Therefore, its noise voltage is $\sqrt{8.29 \times 10^{-19}} V \cdot \text{Hz}^{\frac{-1}{2}} = 0.91\,\text{nV} \cdot \text{Hz}^{\frac{-1}{2}}$. ∎

After the attenuator, the thermal noise is attenuated by 60 dB.

Example 23.2 Find the noise power, $\overline{v_{n,a}^2}$, when it reaches the resonator.
The equation relating power ratio to decibels is given by

$$60\,\text{dB} = 10 \log \frac{\overline{v_n^2}(300K)}{\overline{v_{n,a}^2}}. \tag{23.12}$$

Therefore,

$$\overline{v_{n,a}^2} = \overline{v_n^2}(300K)/10^{\frac{60}{10}},$$
$$= 8.29 \times 10^{-19} V^2 \cdot \text{Hz}^{-1} \times 10^{-6},$$
$$= 8.29 \times 10^{-25} V^2 \cdot \text{Hz}^{-1}. \tag{23.13}$$

It should also be noted that the equation relating voltage ratio to decibels is given by

$$dB = 20 \log \frac{v_1}{v_2}.\qquad(23.14)$$

∎

23.4 Readout Path

While the qubit manipulation path stops at the qubit, the readout path goes from AWG, reaches the qubit, and then goes through a chain of amplifiers. Eventually, the signal will be analyzed in the classical computing unit after the signal is **down-converted** to a lower frequency.

23.4.1 Circulator and Isolator

In Fig. 23.1, there is an **isolator** highlighted. The purpose of an isolator is to limit the microwave signal to propagating in one direction instead of the other. For example, the isolator at the right part of the figure in the readout path ensures that the signal only propagates from the low-temperature zone to the high-temperature zone (upward in the figure). This is to avoid thermal noise propagating from the room temperature region to the qubit (downward in the figure).

23.4.1.1 3-Port Network

An isolator can be implemented by a **circulator**, which is a special **three-port network**. A three-port network has three ports (Fig. 23.4). Therefore, it can be represented as a 3×3 **scattering matrix**, S:

$$S = \begin{pmatrix} S_{11} & S_{12} & S_{13} \\ S_{21} & S_{22} & S_{23} \\ S_{31} & S_{32} & S_{33} \end{pmatrix}.\qquad(23.15)$$

It is instructive to review three properties of the scattering matrix. If the network is **lossless** (does not lose energy), then it must be **unitary**. That means the i-th and j-th columns in the scattering matrix are orthonormal (see also Sect. 3.3.4 and Eq. (3.22)):

$$\langle v_i | v_j \rangle = \delta_{i,j}.\qquad(23.16)$$

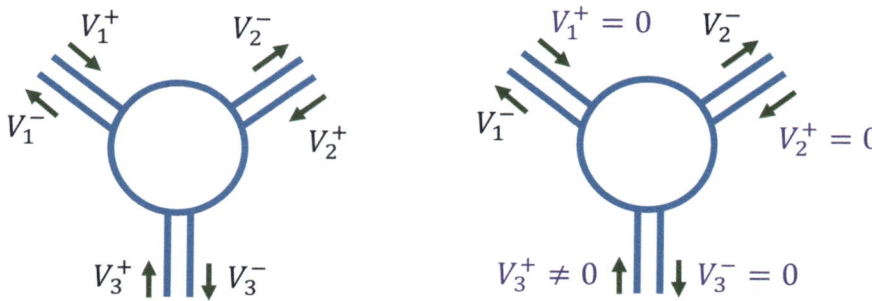

Fig. 23.4 Left: Schematic of a three-port network. Right: An example of when port 3 is matched

Another possible property is that the network is **reciprocal**, which means that it can only have passive components. In this case, the matrix is **symmetric** with

$$S_{ij} = S_{ji}. \qquad (23.17)$$

Finally, the network can be **matched**. This means that there is no reflection from any of the ports when a signal is applied to that port. This is equivalent to having ($V_i^- = 0$) when only that port is applied with a signal and other ports are not applied with any signals. For example, in Fig. 23.4, there are no signals applied to port 1 and port 2 ($V_1^+ = V_2^+ = 0$). When a signal is applied to port 3 ($V_3^+ \neq 0$), there is no reflection ($V_3^- = 0$). Then port 3 is *matched*. Therefore, when the i-th port is matched, we have

$$S_{ii} = \frac{V_i^-}{V_i^+}\Big|_{V_{j \neq i}^+ = 0} = 0. \qquad (23.18)$$

However, it can be proved that a three-port network cannot be lossless, reciprocal, and matched simultaneously (see Problem 23.2).

23.4.1.2 Circulator

A circulator is a three-port network that is matched and lossless (but not reciprocal). Therefore, it has the form of

$$S = \begin{pmatrix} 0 & S_{12} & S_{13} \\ S_{21} & 0 & S_{23} \\ S_{31} & S_{32} & 0 \end{pmatrix}, \qquad (23.19)$$

23.4 Readout Path

which is subjected to the following six unitary conditions

$$|S_{21}|^2 + |S_{31}|^2 = 1, \tag{23.20}$$

$$|S_{12}|^2 + |S_{32}|^2 = 1, \tag{23.21}$$

$$|S_{13}|^2 + |S_{23}|^2 = 1, \tag{23.22}$$

$$S_{31}^* S_{32} = 0, \tag{23.23}$$

$$S_{12}^* S_{13} = 0, \tag{23.24}$$

$$S_{21}^* S_{23} = 0. \tag{23.25}$$

Two of the possible solutions are the following. Firstly, we choose to set $S_{12} = 0$ from Eq. (23.24). This means that we do not allow the input signal at port 2 (V_2^+) to propagate to the output at port 1 (V_1^-). Then, we obtain $|S_{32}|^2 = 1$ based on Eq. (23.21). We may choose $S_{32} = 1$. Then $S_{31} = 0$ based on Eq. (23.23) and thus $|S_{21}|^2 = 1$ based on Eq. (23.20). We choose $S_{21} = 1$ and thus $S_{23} = 0$ based on Eq. (23.25). Using Eq. (23.22), $|S_{13}|^2 = 1$ and we choose $S_{13} = 1$. Therefore,

$$S = \begin{pmatrix} 0 & 0 & 1 \\ 1 & 0 & 0 \\ 0 & 1 & 0 \end{pmatrix} \tag{23.26}$$

To see the effect of this circulator, let us observe how the input signal at each port propagates:

$$\begin{pmatrix} V_1^- \\ V_2^- \\ V_3^- \end{pmatrix} = \begin{pmatrix} 0 & 0 & 1 \\ 1 & 0 & 0 \\ 0 & 1 & 0 \end{pmatrix} \begin{pmatrix} V_1^+ \\ V_2^+ \\ V_3^+ \end{pmatrix},$$

$$\begin{pmatrix} V_1^- \\ V_2^- \\ V_3^- \end{pmatrix} = \begin{pmatrix} V_3^+ \\ V_1^+ \\ V_2^+ \end{pmatrix}. \tag{23.27}$$

Therefore, this circulator propagates signal from ports 3, 1, and 2 to ports 1, 2, and 3, respectively (clockwise in Fig. 23.5).

We can also set up another condition so that the signal propagates anticlockwise which has the following matrix (see the middle of Fig. 23.5):

$$S = \begin{pmatrix} 0 & 1 & 0 \\ 0 & 0 & 1 \\ 1 & 0 & 0 \end{pmatrix}. \tag{23.28}$$

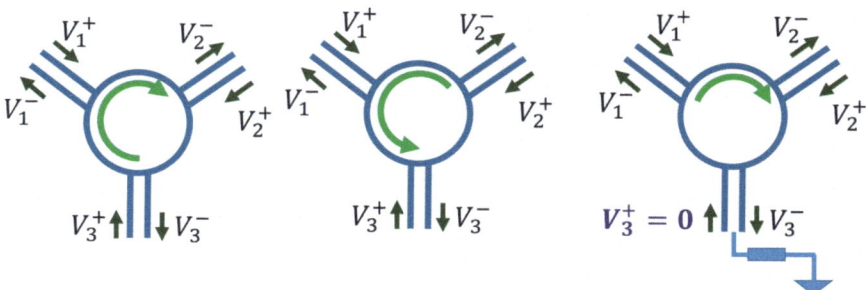

Fig. 23.5 Left: A circulator that only allows the signal to propagate clockwise from one port to the adjacent port. Middle: A circulator that only allows the signal to propagate anticlockwise from one port to the adjacent port. Right: An isolator that allows only signal to propagate from port 1 to port 2 but not vice versa

23.4.1.3 Isolator

To implement an isolator, we can use a circulator with one of the ports properly terminated. For example, in the right of Fig. 23.5, port 3 of the circulator is terminated with an impedance that matches its characteristic impedance (e.g., 50 Ω). Therefore, V_3^- will not be reflected and, thus, $V_3^+ = 0$. Since it is a circulator, the signal will only propagate from port 1 to port 2 but not from port 2 to port 1. While the signal from port 2 can propagate to port 3, due to the termination, it would not be further propagated to port 1. Therefore, it works well as an isolator between port 2 and port 1. In our superconducting qubit quantum computer, port 1 is connected to the lower-temperature region, and port 2 is connected to the higher-temperature region.

23.4.2 Amplification Chain

After interrogating the qubit (e.g., Sect. 22.5), the microwave pulse will be amplified. It is usually amplified by three stages. The first stage is a **quantum-limited amplifier (QLA)** [3, 4]. The second stage is a low-noise amplifier such as those built with **high electron mobility transistor (HEMT)** at a cryogenic temperature [5]. Finally, it is amplified using another amplifier at room temperature.

Why do we need an amplification chain? Is it possible to just use one single amplifier with a large amplification? To answer this question, we need to introduce the concept of **noise factor (NF)** [2].

We have discussed the concept of thermal noise in Sect. 23.3.2. While we want to minimize the noise, what matters is the **signal-to-noise ratio (SNR)**, which is the ratio of signal *power* to noise *power*, and in the calculation, we usually use power spectral density (i.e., power per unit bandwidth). This is just like even though

23.4 Readout Path

I usually speak much louder than other people, when we are in a noisy restaurant, I need to speak even louder (larger signal) to maintain the SNR so that you can hear me clearly. When an amplifier with a gain of A is used to amplify a signal, it inevitably will amplify the noise by the same gain, too. And their powers (V_{in}^2 and $\overline{v_n^2}$) will be amplified by A^2. At the same time, as a physical object, the amplifier will introduce an additional noise power of $\overline{v_{n,amp}^2}$ to its output. As a result, the SNR will degrade even though the signal is amplified. Therefore, the amplified signal SNR (SNR_{out}) is lower than the input signal SNR (SNR_{in}). This can be understood by this equation:

$$SNR_{out} = \frac{V_{in}^2 \times A^2}{\overline{v_n^2} \times A^2 + \overline{v_{n,amp}^2}},$$

$$< SNR_{in} = \frac{V_{in}^2}{\overline{v_n^2}}. \tag{23.29}$$

The ratio between the input and output SNR signifies how much an amplifier degrades the SNR of the signal it receives. This ratio is called the noise factor, NF:

$$NF = \frac{SNR_{in}}{SNR_{out}},$$

$$= 1 + \frac{\overline{V_{v,amp}^2}}{\overline{v_n^2} A^2}, \tag{23.30}$$

which is always larger than 1.

There is a trade-off between the gain and the NF of an amplifier. In general, an amplifier with a larger gain also has a larger NF because more physical components are needed. This is why we do not use a single amplifier even if we can make an amplifier with a large gain. On the other hand, we can achieve a large gain by cascading a chain of amplifiers like what is shown in Fig. 23.1 ($A_{total} = A_1 A_2 A_3 \cdots$), but the *overall* NF is only approximately that of the first stage. This is due to the **Friis equation**, which states that the overall NF of an amplification chain, NF_{total}, is given by

$$NF_{total} = 1 + (NF_1 - 1) + \frac{NF_2 - 1}{A_{P1}} + \frac{NF_3 - 1}{A_{P1} A_{P2}} \cdots,$$

$$\approx NF_1, \tag{23.31}$$

where NF_i and A_{Pi} are the NF and **available power gain** of the i-th stage, respectively. We will not discuss available power gain here. However, it is related to the amplifier's gain and is supposed to be large. We see that the NF of a later stage is scaled down by the available power gains of the previous stages. Therefore, it is very important to design an amplifier with minimal NF in the first stage and then add amplifiers with large gains in the later stages.

23.4.2.1 Parametric Amplification

In the amplification chain, a quantum-limited amplifier (QLA) is used as the first stage. QLA has the smallest possible NF although its gain is not high (usually less than 20 dB). The source of noise in a QLA is due to **uncertainty principle**. There are two commonly used QLA. One is the **Josephson parametric amplifier (JPA)** [3] and one is the **traveling wave parametric amplifier (TWPA)** [4]. The one shown in Fig. 23.1 is a JPA. We will not discuss the working principle of QLA. Figure 23.6 shows the schematic of a JPA. The JPA is composed of a SQUID formed by two Josephson junctions (see also Sect. 16.4). The reason for using Josephson junctions is to provide nonlinearity to mix a pump signal with the signal to be amplified. The pump source (ω_{pump}) is twice the frequency of the signal to be amplified (ω_{signal}). The input signal is directed to JPA through the circulator which is then mixed with the pump signal, after which it is directed to the output through the circulator. Examples of pump signals are voltage and flux pumps. For example, the SQUID can be put in parallel with a capacitor to form an LC tank, and its resonant frequency is modulated by the varying flux when the pump is a flux.

It is instructive to understand the concept of parametric amplification. The following shows a *classical example. Note that this is NOT how JPA works.*

Figure 23.7 shows an example of parametric amplification. Assume there is an LC tank oscillating at a frequency of ω_{signal}. We know that the voltage across the capacitor, V_C, will reach the peak twice in each cycle (one positive and one negative) as shown in Eq. 15.7. When $|V_C|$ is maximum, it has maximum capacitive charge, Q, and electric field, $\vec{E}$, across the capacitor. If we quickly move the capacitor plates

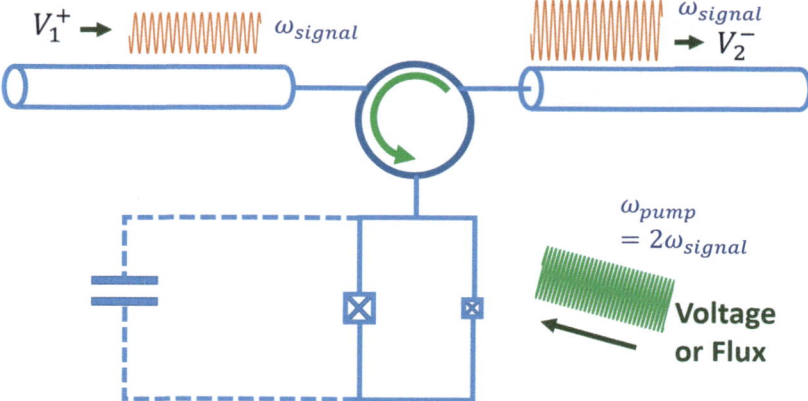

Fig. 23.6 An example of using Josephson parametric amplifier to amplify a signal. Here, a SQUID formed by Josephson junctions is used

23.4 Readout Path

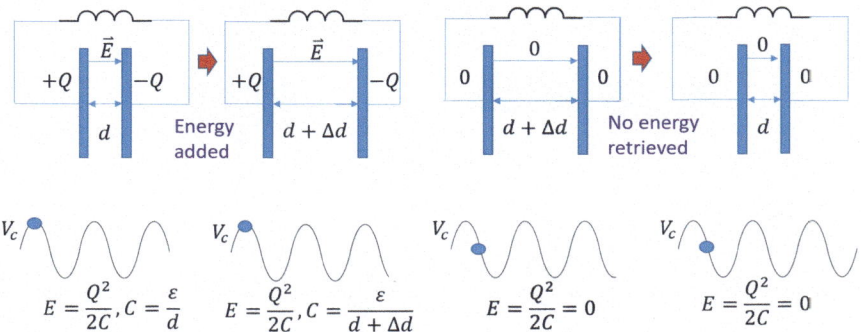

Fig. 23.7 An example of parametric amplification. E is the capacitive energy. Note that this is *not* the mechanism of a QLA

apart mechanically by Δd, the charges cannot be changed instantaneously (i.e , Q is unchanged) because they are matters. In this process, the mechanical energy will be changed to electrical potential energy when the plates are pulled apart because it has to perform work against the attractive force between the capacitive plates. Therefore, we have increased the capacitive energy by applying the external mechanical energy.

Then, we wait unit it completes a quarter of the oscillation cycle, at which all capacitive energy is converted to inductive energy in the inductor with zero charges across the capacitive plate (Fig. 15.2). Now the plate is pushed back to the original distance (d). Since there is no charge and no electrical field between the capacitive plates, no work will be done and, therefore, no energy is extracted or added to the LC tank.

This process repeats twice in each period, and the mechanical energy is pumped into the electrical energy of the LC tank to amplify its signal. We see that the pump mechanical signal has a frequency of $2\omega_{signal}$. We also see that this is a parametric amplification in which we vary a parameter of the capacitor (distance between the capacitor plates) to achieve the amplification.

23.4.3 $I - Q$ Demodulation

Finally, the amplified signal needs to be detected. As discussed in Fig. 22.4, we can distinguish qubit states $|g\rangle$ and $|e\rangle$ by detecting the phase, ϕ, of the signal (which is equivalent to detecting the real and imaginary parts of the signal). Assume that the signal has the following form:

$$s(t) = A(t) \sin(\omega_1 t + \phi). \tag{23.32}$$

In order to extract ϕ, we will perform $I - Q$ demodulation to the signal (Fig. 23.8). This is similar to a reverse process of input pulse creation (Fig. 23.2).

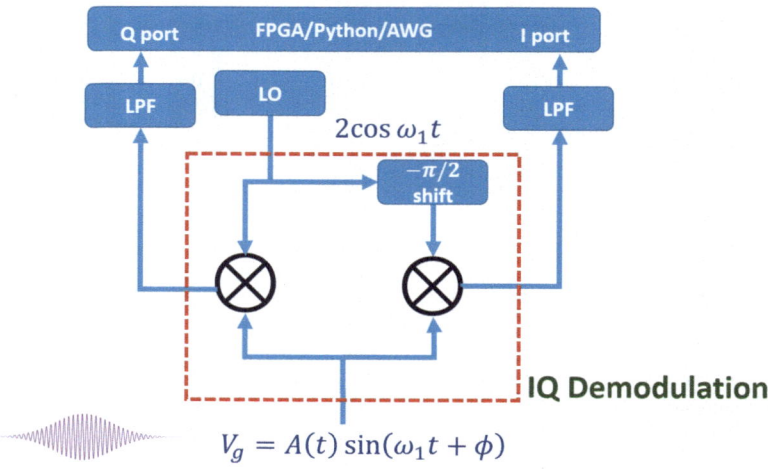

Fig. 23.8 $I - Q$ demodulation of readout signals. Note that its I and Q ports are not at the same positions as in Fig. 23.2

Firstly, the signal will go through an $I - Q$ mixer in the reverse direction. We get the out-of-phase (quadrature) part by mixing it with LO signal, $2\cos\omega_1 t$, from the local oscillator followed by a **low-pass filter (LPF)**. For simplicity, we assume the magnitude to be two. The local oscillator this time has the same frequency as the readout signal. After the mixer, we get:

$$V_Q(t) = A(t)\sin(\omega_1 t + \phi)2\cos\omega_1 t,$$
$$= 2A(t)\frac{1}{2}(\sin(\omega_1 t + \phi + \omega_1 t) + \sin(\omega_1 t + \phi - \omega_1 t)),$$
$$= A(t)(\sin(2\omega_1 t + \phi) + \sin(\phi)). \tag{23.33}$$

Then, it goes through the LPF which will filter the $\sin(2\omega_1 t + \phi)$ term. We thus get the Q-part as

$$Q = A(t)\sin(\phi). \tag{23.34}$$

For the I part, the readout signal is mixed with phase-shifted LO signal, $2\cos(\omega_1 t - \pi/2) = 2\sin\omega_1 t$, and we get:

$$V_I(t) = A(t)\sin(\omega_1 t + \phi)2\sin\omega_1 t,$$
$$= 2A(t)\frac{1}{2}(\cos(\omega_1 t + \phi - \omega_1 t) - \cos(\omega_1 t + \phi + \omega_1 t)),$$
$$= A(t)(\cos(\phi) - \cos(2\omega_1 t + \phi)). \tag{23.35}$$

Then, it goes through the LPF which will filter the $\cos(2\omega_1 t + \phi)$ term. We thus get the I-part as

$$I = A(t)\cos(\phi). \tag{23.36}$$

We can plot I and Q on the $I - Q$ plane to obtain the IQ blobs in Fig. 22.5. We can then distinguish the qubit states because ϕ depends on the qubit states.

23.5 Summary

In this chapter, we look into various microwave components in a superconducting qubit quantum computer. We see that typical and matured microwave skills and techniques are used. Important components include mixers, circulators, local oscillators, attenuators, and amplifiers. We also discuss the concept of noise and noise factors. Particularly, we emphasize the importance of lowering the overall noise factor by adding a quantum-limited amplifier as the first stage of an amplification chain. We also discuss the concept of parametric amplifier.

Problems

23.1 Noise Energy
If the system has a bandwidth of 10 GHz, what is the noise energy reaching the resonator? Hint: the noise energy is the bandwidth multiplied by the NSD.

23.2 Three-Port Network Properties
Prove that a three-port network cannot be lossless, reciprocal, and matched at the same time. Firstly, simplify the matrix in Eq. (23.15) by substituting the definitions of a matched network and a reciprocal network. Then try to show that it cannot be unitary. This only proves one case. Now try to prove other combinations. It will be easier if you can form equations and show that the number of equations is more than the number of variables. Note that the elements in the scattering matrix are generally complex. Therefore, it has two real numbers.

23.3 Circulator Constraints
Derive Eqs. (23.20) to (23.25) based on the fact that the matrix is unitary for a lossless network.

23.4 Circulator S-Matrix
Prove Eq. (23.28) by using the same approach we used to prove Eq. (23.26).

23.5 I-Q Demodulation and Low Pass Filter

An LPF can be modeled as an integration in the time domain. Try to apply integration to Eqs. (23.33) and (23.35) and show that you will get Eqs. (23.34) and (23.36), respectively, if it is integrated over multiples of $\frac{2\pi}{\omega_1}$.

References

1. Hiu Yung Wong, Prabjot Dhillon, Kristin M. Beck, and Yaniv J. Rosen. A simulation methodology for superconducting qubit readout fidelity. *Solid-State Electronics*, 201:108582, 2023.
2. Behzad Razavi. *RF Microelectronics*. Pearson, 2011.
3. B. Yurke, L. R. Corruccini, P. G. Kaminsky, L. W. Rupp, A. D. Smith, A. H. Silver, R. W. Simon, and E. A. Whittaker. Observation of parametric amplification and deamplification in a josephson parametric amplifier. *Phys. Rev. A*, 39:2519–2533, Mar 1989.
4. C. Macklin, K. O'Brien, D. Hover, M. E. Schwartz, V. Bolkhovsky, X. Zhang, W. D. Oliver, and I. Siddiqi. A near–quantum-limited josephson traveling-wave parametric amplifier. *Science*, 350(6258):307–310, 2015.
5. Eunjung Cha, Niklas Wadefalk, Per-Åke Nilsson, Joel Schleeh, Giuseppe Moschetti, Arsalan Pourkabirian, Silvia Tuzi, and Jan Grahn. 0.3–14 and 16–28 ghz wide-bandwidth cryogenic mmic low-noise amplifiers. *IEEE Transactions on Microwave Theory and Techniques*, 66(11):4860–4869, 2018.

Chapter 24
Design of an Integrated Superconducting Qubit Chip

24.1 Introduction

In this chapter, we will use the theories that we have learned from the previous chapters to design transmon qubits on an integrated superconducting chip. The purpose is to highlight some of the design procedures and numerical calculations. We will talk about feedline design, resonator design, and qubit design. Then we will talk about feedline-resonator coupling and qubit-resonator coupling. We will also briefly mention the design of flux tunable transmon qubit.

24.1.1 Learning Outcomes

Be aware of the design tools available for superconducting qubit design; be able to use the equations from the previous chapters to design qubits; appreciate the role of feedline and resonator in a superconducting qubit system.

24.1.2 Teaching Videos

- Search for Ch24 in this playlist
  - https://tinyurl.com/3yhze3jn
- Other videos
  - https://youtube.com/playlist?list=PLnK6MrIqGXsIS_97Nt-8R6uNrJXxMiOro

24.2 Layout

Figure 24.1 shows a typical layout of a superconducting qubit chip. This is generated by Qiskit Metal [1]. Although this is just a dummy design, it reveals a few important components and aspects of a superconducting qubit circuit.

Qiskit Metal can then export the GDS layout file which can be used for fabrication. GDS layout is a bird view of the chip which has conductive metal patterned on top of a resistive layer. It is common that the circuit is built on highly resistive silicon. For example, Fig. 24.2 shows the GDS in the transmon region. The green regions represent the existence of conductive metal (such as niobium which is superconducting at the operation temperature). The black regions represent the absence of conductive metal and thus are insulating. In Fig. 24.2, the cross section under cutline $A - A'$ is also shown. It should be noted that, besides highly resistive silicon, other types of substrate are also used. For example, oxide on silicon is used in [2].

24.2.1 Transmon Qubit

In the circuit in Fig. 24.1, there are two transmon qubits. Each has a shunt capacitor. Figure 24.2 shows the layout of the lower right one with the two unused coupling ports removed. The layout of the Josephson junction (JJ) is not included but its location is indicated. This is a regular transmon qubit with one Josephson junction. The reader can refer to the fabrication of JJ in Sect. 16.3. The green area is covered by niobium (Nb) which has a thickness of about 100 nm–300 nm. Nb is superconducting at the operation temperature. The Josephson junction will be

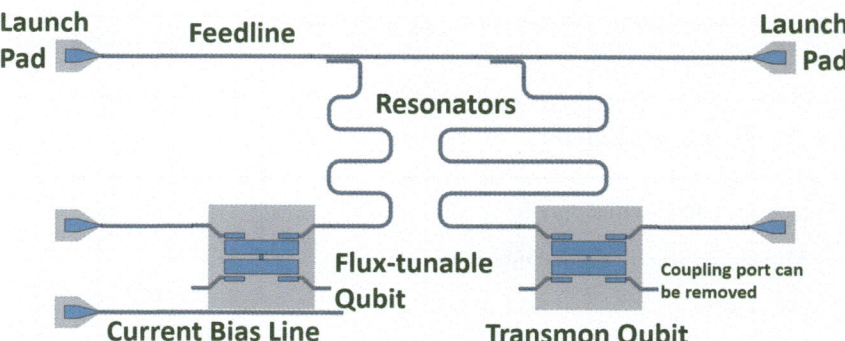

Fig. 24.1 Schematic showing what a typical transmon qubit circuit looks like. Note that this is not an optimized one. This is generated by Qiskit Metal

24.2 Layout

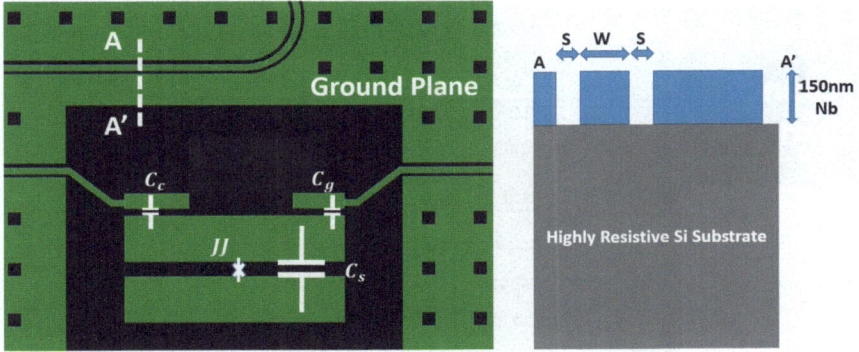

Fig. 24.2 Left: Transmon layout which is a bird view of the chip. Right: The $A - A'$ cross section (side view)

formed by Al/AlO$_x$/Al and connected to the two capacitive plates of the shunt capacitor, C_s.

The transmon qubit is coupled to the "outside world" through two other capacitors, C_g and C_c. C_g capacitively couples the transmon to a launch pad which will be wire-bonded to a **printed circuit board (PCB)** trace. This is the same gate capacitor as in Fig. 21.1, and it allows single-qubit gate manipulations of the qubit. C_c couples the qubit to the resonator which allows the qubit to pull the resonator's resonant frequency based on its state ($|g\rangle$ / $|e\rangle$). This is the same C_c as the one in Fig. 22.3.

It should be noted that it is not necessary to have two separate paths for readout and manipulation. One may also just use the readout path to manipulate the qubit. In that case, the manipulation pulse goes through the resonator to interact with the qubit.

24.2.2 Feedline and Resonator

Figure 24.3 shows the readout path layout. It is composed of a feedline and a resonator. Both the feedline and the resonator are **coplanar waveguide (CPW)**. It has a signal line (conductive niobium in green) with a width of w, which is separated from the ground plane by a spacing of s. The cross section of a CPW is also shown in Fig. 24.2. The spacing refers to the absence of conductive material on the silicon substrate (black).

The resonator is coupled to the feedline *inductively*. This can be seen by the fact that the conductive line of the resonator CPW is shorted to the ground plane at the coupling region. The voltage is close to zero in the coupling region but the current is maximum. Therefore, it is coupled to the feedline CPW inductively through the magnetic field. The other end of the resonator is capacitively coupled to the

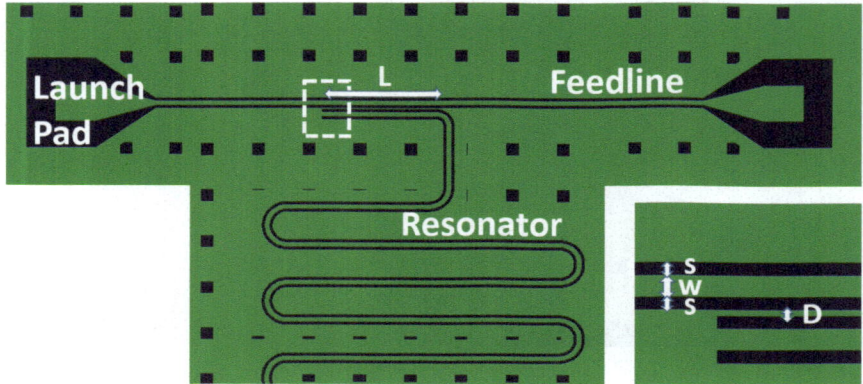

Fig. 24.3 Layout of feedline and resonator. The inset shows the coupling region between the feedline and the resonator, corresponding to the dashed box in the main figure

transmon as shown in Fig. 24.2 through C_c. Therefore, it is open at the other end with zero current and maximum voltage. As a result, this is a **quarter-wavelength** ($\lambda/4$) **resonator** [4].

In the readout process, a readout pulse is launched from the left launch pad which is wire-bonded to a PCB trace. The pulse will reach the right launch pad without being interrupted if its frequency is not the same as the resonator's resonant frequency. On the other hand, it will be interrupted at the resonant frequency of the resonator which is modulated by the qubit state. Note that this scheme is different from the case in Fig. 22.3. But both of them enable dispersive readout and the phase of the pulse is changed depending on the qubit state.

24.3 Design and Numerical Examples

24.3.1 Design Constraints

We now want to design a single qubit with the following constraints. Firstly, the qubit frequency is set to be $\omega_{01} = \omega_q = 2\pi \times 4.5$ GHz and the resonator frequency is set to be $\omega_r = 2\pi \times 7$ GHz. Secondly, there are three other constraints that need to be met in order to have a well-behaved transmon qubit:

$$\alpha > 0.1\omega_{10}, \tag{24.1}$$

$$\frac{E_J}{E_c} > 50, \tag{24.2}$$

$$R_{n,\min} < R_n < R_{n,\max}. \tag{24.3}$$

24.3 Design and Numerical Examples

It is desirable to have a large enough anharmonicity, α (Eq. (24.1)). However, sometimes it is difficult to reach 10% of the qubit frequency. We will try to have it close to 200 MHz. We also want to have a large enough $\frac{E_J}{E_c}$ ratio so that it is insensitive to charge noise (Eq. (24.2)). Finally, the Josephson junction needs to have its critical current or Josephson inductance within the fabrication range (Eq. (24.3)) supported by the manufacturer. This is often given by the foundry as the minimum ($R_{n,\min}$) and maximum ($R_{n,\max}$) normal state resistance which is related to I_c through Eq. (16.27).

24.3.2 Design of Feedline and Resonator

24.3.2.1 CPW Design

Firstly, we need to design the feedline and the resonator to have the same characteristic impedance, Z_0, as the system has. Here we assume $Z_0 = 50\,\Omega$. To do this, we can go through analytical calculations or simulations. One can find that $s = 5.8\,\mu\text{m}$ and $w = 10\,\mu\text{m}$ will give the required characteristic impedance by using the tools in [5] or [6]. Reference [6] is based on [7].

24.3.2.2 $\lambda/4$-Resonator Design

We need to design a $\lambda/4$-resonator with $\omega_r = 2\pi \times 7\,\text{GHz}$. Since an electromagnetic wave will have a shorter wavelength in matters than in a vacuum, we need to find the effective relative dielectric constant, ϵ_{eff}, so that we can find the wavelength and, thus, the length of the CPW. This can be performed by using simulations. For example, in [8], the effective relative dielectric constant for the EM fields for metals on the top of a silicon substrate is extracted to be 6.1. Therefore, the length of the resonator is found to be

$$\begin{aligned}
L &= \frac{\lambda_{\text{matter}}}{4}, \\
&= \frac{\lambda_{\text{vacuum}}}{4\sqrt{\epsilon_{eff}}}, \\
&= \frac{c}{4f\sqrt{\epsilon_{eff}}}, \\
&= \frac{3 \times 10^8}{4 \times 7 \times 10^9 \times \sqrt{6.1}}\,\text{m}, \\
&= 4.338\,\text{mm},
\end{aligned} \qquad (24.4)$$

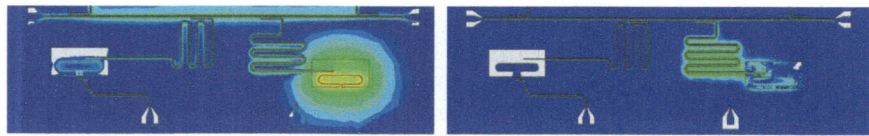

Fig. 24.4 Energy distributions at two eigenmodes of a quantum chip in which there are two qubits (left and right). Each of them is coupled to the top feedline through the respective resonator. Left: Eigenmode at the qubit frequency of 4.5 GHz. Right: Eigenmode at the resonator frequency of 7 GHz

where we have used $\lambda_{\text{vacuum}} = \lambda_{\text{matter}} \sqrt{\epsilon_{eff}}$. We can then layout the resonator (with the help of Qiskit Metal) with a length of 4.338 mm. A serpentine structure (or meander) can be used to make more efficient use of the area (Fig. 24.1). To verify the design, one can then perform **eigenmode** simulation which calculates the resonant frequency and **quality factor, Q,** of the resonator as demonstrated in [9]. Figure 24.4 plots the energy distributions at two of the eigenmodes of a part of a quantum chip. At the designed resonator frequency, $\omega_r = 7$ GHz, most of the energy concentrates at the resonator which shows that the design is successful.

24.3.2.3 Coupling and Q-Factor

Another important design aspect of the resonator is its **coupling strength** to the feedline. If the coupling is too small, it will take a long time to read the qubit and the qubit might have changed its state due to decoherence. The strength of coupling is usually gauged by the quality factor of the resonator. The quality factor of a resonator, Q_r, can be thought of as the number of oscillations, N, that occur before the energy is dissipated (in this case, through the coupling with the feedline to the outside world). Usually, Q_r is between 5,000 and 10,000. Therefore, the time, T_r, it takes to lose its energy is

$$\begin{aligned} T_r &= \frac{2\pi}{\omega_r} N, \\ &= \frac{2\pi Q_r}{\omega_r}, \\ &= \frac{2\pi \times 5000}{2\pi \times 7 \times 10^9} \text{ s}, \\ &= 714 \text{ ns}. \end{aligned} \quad (24.5)$$

This means that, approximately, a reading pulse in the order of hundreds of nanoseconds can be used to read out the qubit. This is short enough as we expect the decoherence time of a qubit to be more than 50 μs.

24.3 Design and Numerical Examples

The quality factor depends on the distance D between the resonator and the overlapping length, L, (Fig. 24.3). We can use the analytical method [6], which is demonstrated in [10], to find the appropriate distance and overlap. With $D = 3\,\mu m$ and $L = 230\,\mu m$, one will get a Q_r in the desired range. Of course, finite-element simulation as mentioned earlier needs to be conducted to confirm the results [9].

24.3.3 Qubit Design

24.3.3.1 Capacitor Design

Now we will design the qubit to satisfy the constraints given in Eqs. (24.1)–(24.3). Firstly, to satisfy the requirement of anharmonicity, based on Eqs. (17.8) and (19.19), we have:

$$|\alpha| = E_c,$$
$$= \frac{1}{2}\frac{e^2}{C}. \quad (24.6)$$

Therefore, to target an anharmonicity of at least 200 MHz, we need to have

$$\frac{1}{2}\frac{e^2}{C} > \hbar \times 2\pi \times 200\,\text{MHz},$$
$$C > \frac{1}{2}\frac{e^2}{\hbar \times 2\pi \times 200\,\text{MHz}},$$
$$C > 96.6\,\text{fF}. \quad (24.7)$$

C is the total capacitance felt by the transmon. It includes C_g, C_s, and C_J (Fig. 21.1). We can perform finite-element simulations [11] to obtain them. One can change the distance between the metal plates and their overlaps (Fig. 24.2; note that here we lumped C_c into C_g). For C_J, which is the intrinsic Josephson junction capacitor, it usually is given by the foundry. A reasonable value is 8 fF. As an example, $C_s = 88.2$ fF and $C_g = 8.4$ fF. Therefore, $C = 104.6$ fF. This gives $E_c = 1.22 \times 10^{-25}$ J and $\frac{\alpha}{\hbar} = 185$ MHz. Although not ideal, this is still close to 200 MHz. The reason why we do not try to reduce C further to obtain a higher α is because we also need to meet the other two constraints.

It is also instructive to mention that it is often to express energy in terms of frequency. For example, we can say that the qubit energy is 4.5×10^9 GHz. This means that its true energy is $\hbar \times 2\pi \times 4.5 \times 10^9$ GHz. This is why we also often just say $\alpha = 185$ MHz.

24.3.3.2 Josephson Junction Design

Now we need to find E_J and thus I_c based on Eq. (19.18) and the given qubit frequency:

$$E_{10} = \hbar \omega_q,$$
$$= \sqrt{8 E_c E_J} - E_c. \qquad (24.8)$$

Therefore,

$$\hbar \omega_q = \sqrt{8 E_c E_J} - E_c,$$
$$4.5 \,\text{GHz} = \sqrt{8 \times 185 \,\text{MHz} \times E_J} - 185 \,\text{MHz},$$
$$E_J = 14.8 \,\text{GHz}. \qquad (24.9)$$

In the second line, we use the convention to express energy in terms of frequency without $\hbar$ as mentioned at the end of the previous section. Then,

$$\frac{E_J}{E_c} = \frac{14.8 \,\text{GHz}}{185 \,\text{MHz}} = 80 > 50, \qquad (24.10)$$

which satisfies the constraint in Eq. (24.2). Finally, we need to make sure the E_J calculated can be realized in the fabrication process. We will first convert it to the critical current I_c using Eq. (16.21) and then the normal state resistance, R_n, using Eq. (16.27). Therefore,

$$R_n = \frac{\pi \Delta}{2 I_c e},$$
$$= \frac{\pi \Delta}{2 \frac{2\pi E_J}{\Phi_0} e},$$
$$= 9.6 \, k\Omega, \qquad (24.11)$$

where we have used the value of Δ from Eq. (16.29). The R_n obtained falls in the range available in most fabrication processes.

24.3.3.3 Quality Factor

One should also perform an eigenmode simulation to estimate the quality factor of the qubit. Figure 24.4 (left) shows the energy distribution at eigenmode with a frequency of about 4.5 GHz using the design from the previous section. It is clear that the energy concentrates at the qubit. Therefore, the design is considered to be correct. Note that in most electromagnetic wave simulators, quantum components

24.3 Design and Numerical Examples

such as the Josephson junction cannot be simulated. Therefore, Josephson inductance is modeled as a lumped inductor (usually between 1 nH and 20 nH). Moreover, the quality factor of the transmon, Q_q, can be simulated and it needs to be in the order of 10^6–10^7.

Example 24.1 Find the energy relaxation time due to the coupling of the qubit to the outside world for $Q_q = 10^6$ and $Q_q = 10^7$.

Based on Eq. (24.5)

$$\begin{aligned} T_q &= \frac{2\pi}{\omega_q} N, \\ &= \frac{2\pi Q_q}{\omega_q}, \\ &= \frac{2\pi \times 10^6}{2\pi \times 4.5 \times 10^9} \text{ s}, \\ &= 222\,\mu\text{s}. \end{aligned} \qquad (24.12)$$

This is still too small because it is similar to the qubit decoherence time (about $50\,\mu\text{s}$–$200\,\mu\text{s}$). If $Q_r = 10^7$, it will increase to 2.2 ms and will have minimal impact on the decoherence time. ∎

24.3.4 Qubit-Resonator Interaction

As mentioned in Sect. 22.4, qubit readout relies on qubit-state-dependent dispersive shifts of the resonator's resonant frequency. We need to create a large enough **cross-Kerr**, χ, based on Eq. (22.22). For the given ω_q and ω_r, we can change the coupling energy g to achieve the desired cross-Kerr. 2χ should be similar to the **full width at half maximum (FWHM)**, Δ_f, of the resonator. This can be estimated using

$$\begin{aligned} \Delta_f &= \frac{1}{T_r}, \\ &= \frac{\omega_r}{2\pi Q_r}, \\ &= \frac{2\pi \times 7 \times 10^9}{2\pi \times 5000} \text{ Hz}, \\ &= 1.4\,\text{MHz}. \end{aligned} \qquad (24.13)$$

Therefore, we want to have $\chi \approx 0.7\,\text{MHz}$. We can use Eq. (22.12) which is copied here for convenience:

$$g = e \frac{C_c}{C_1 C_r} \left(\frac{E_J}{2E_c}\right)^{\frac{1}{4}} \sqrt{\frac{C_r \omega_r \hbar}{2}}. \qquad (24.14)$$

We need to find C_r which is the effective lumped capacitance of the resonator. The effective capacitance of a *quarter wavelength resonator* (Chap. 6 in [4]) is

$$C_r = \frac{\pi}{4\omega_r Z_0}. \qquad (24.15)$$

The effective inductance (although not used here) is

$$L_r = \frac{1}{\omega_r^2 C}. \qquad (24.16)$$

The capacitance is found to be 357 fF. The cross-Kerr is then found to be 0.51 MHz where we used $C_1 = 96.2\,\text{fF}$ and $C_c = 84\,\text{fF}$. Although this is smaller than expected, we need to perform more rigorous calculations of cross-Kerr using methods such as **energy participation ratio (EPR)** [3] because the result from Eqs. (22.22) and (24.14) might not be accurate enough. Indeed, the cross-Kerr obtained through EPR simulation is about 1 MHz. An example of performing EPR calculation using finite-element simulation is shown in [12]. Eventually, we need an experiment to confirm our design.

24.3.5 Other Design Considerations

24.3.5.1 Flux-Tunable Qubit

In Fig. 24.1, there is a flux-tunable qubit at the lower left. The layout is the same as that in Fig. 24.2 except that it has two Josephson junctions in parallel. They enclose an area together with the two capacitive plates to allow external flux to control its effect on Josephson energy (Fig. 16.8). There are a few design considerations.

Firstly, we need to be able to tune the flux by passing a low enough current through the current bias line. The current needs to be low enough so that it does not exceed the critical current density of Nb which will cause Nb to leave the superconducting state and generate heat. Therefore, it needs to be close enough to the qubit to minimize the current required. The minimum flux it needs to provide is one flux quantum, Φ_0, as shown in Eq. (16.41) so that it can tune the effective E_{JT} from 0 to maximum.

Example 24.2 How many magnetic quanta can a wire carrying 1 mA generate through a 40 μm × 20 μm area which is given to be effectively 70 μm away?

Using **Biot-Savart law**

$$B = \frac{\mu_0 I}{2\pi R},$$
$$= \frac{4\pi \times 10^{-7} H/m \times 0.001 A}{2\pi \times 70 \times 10^{-6} m},$$
$$= 2.857 \times 10^{-6} T. \tag{24.17}$$

Therefore, the number of magnetic flux quanta is

$$n = \frac{BA}{\Phi_0},$$
$$= \frac{2.857 \times 10^{-6} T \times 40 \mu m \times 20 \mu m}{2.067 \times 10^{-15} \text{Wb}},$$
$$= 1.11. \tag{24.18}$$

This is enough to tune a flux-tunable Josephson junction. ∎

Secondly, the current bias line should not reduce the quality factor of the qubit, Q_q, too much. This requires simulations as discussed earlier. The layout in Fig. 24.1 is *not* optimal in this regard.

24.3.5.2 Flux Trapping

In the layout we have shown so far such as Fig. 24.3, there are holes (black insulating squares) in the ground plane. They are added so that parasitic magnetic flux such as those from the Earth's magnetic field can be trapped. The purpose is to avoid the parasitic magnetic flux damaging the superconductivity of the ground plane in case it exceeds the critical magnetic field. For silicon design engineers, this resembles the dummy filling we do after the layout of a circuit.

24.4 Summary

In this chapter, we apply the knowledge we have learned so far to design a superconducting transmon qubit circuit. We discuss the design of feedlines, resonators, and qubits. More importantly, we also discuss the importance of the feedline-resonator coupling and qubit-resonator coupling in the design process. We show that it is desirable to start with analytical calculations first. However, rigorous finite-element simulations are required to confirm calculation before tape-out. This is similar to a silicon tape-out process where we first perform hand calculation and

SPICE simulation. After that, rigorous post-layout simulation is required. Finally, experiments are required to verify the design.

Problems

24.1 Qiskit Metal

Install Qiskit Metal, Anaconda, and HFSS to perform simulations by following Lab 0) to Lab 5) in this playlist https://youtube.com/playlist?list=PLnK6MrIqGXsKpkN3nL1OlxW6Gr6srYt0C&feature=shared

24.2 Design of Resonator

We want to add one more qubit and resonator to share the same feedline together with the one described in this chapter. The constraint is $\omega_r = 2\pi \times 7.4\,\text{GHz}$. Find the length of the resonator.

24.3 Design of Qubit

Find the normal state resistance of the Josephson junction if $\omega_q = 2\pi \times 5.6\,\text{GHz}$. Assume it has the same capacitance as the qubit discussed in this chapter.

References

1. Qiskit metal | quantum device design & analysis (Q-EDA) 0.1.5. https://qiskit-community.github.io/qiskit-metal/. Accessed:2024-07-07.
2. M. Göppl, A. Fragner, M. Baur, R. Bianchetti, S. Filipp, J. M. Fink, P. J. Leek, G. Puebla, L. Steffen, and A. Wallraff. Coplanar waveguide resonators for circuit quantum electrodynamics. *Journal of Applied Physics*, 104(11):113904, 12 2008.
3. Z.K. Minev, Z. Leghtas, S.O. Mundhada, L. Christakis, I.M. Pop, and M.H. Devoret. Energy-participation quantization of josephson circuits. *npj Quantum Inf.*, 7(131), 2021.
4. David M. Pozar. *Microwave Engineering*. Wiley; 4th edition, 201.
5. Coplanar waveguide calculator. https://www.microwaves101.com/calculators/864-coplanar-waveguide-calculator, 2024. Accessed:2024-07-07.
6. Quality factor of a transmission line coupled coplanar waveguide resonator (misis.ru). https://smm.misis.ru/CPW-resonator-coupling/, 2024. Accessed:2024-07-07.
7. Ilya Besedin and Alexey P. Menushenkov. Quality factor of a transmission line coupled coplanar waveguide resonator. *EPJ Quantum Technology*, 5(1):2, 2018.
8. Extraction of effective dielectric constant and meshing - quantum chip design. https://youtu.be/ZkxK5f-JAqY, 2024. Accessed:2024-07-07.
9. Resonator eigen-mode and quality factor simulation using HFSS - quantum chip design. https://youtu.be/H71XHyAYDIY, 2024. Accessed:2024-07-07.
10. Analytical design of feedline and resonator coupling - quantum chip design. https://youtu.be/DsGiP1Hs-K0, 2024. Accessed:2024-07-08.
11. Using Q3D to simulate transmon capacitance from qiskit metal - quantum chip design. https://youtu.be/_52zA2XM0fo, 2024. Accessed:2024-07-08.
12. HFSS energy participation ratio for cross Kerr analysis - quantum chip design. https://youtu.be/PIdo-gQZk-0, 2024. Accessed:2024-07-08.

Chapter 25
Errors and Decoherence

25.1 Introduction

In this chapter, we will talk about various types of errors in quantum computers, including initialization errors, readout errors, gate errors, bit-flip errors, and phase errors. To understand errors, the quantum master equations need to be used. However, this is beyond the scope of this book and will not be covered. Then, we will discuss two of the most common methods to measure the decoherence times of a qubit, namely, the T_1 and T_2 times. We will use superconducting qubits as examples to demonstrate the steps and understand the results.

25.1.1 Learning Outcomes

Be able to describe different types of errors in a quantum computer; know the procedure to perform T_1 and T_2 measurements; understand and be able to explain the meaning of the graphs for extracting T_1 and T_2.

25.1.2 Teaching Videos

- Search for Ch25 in this playlist
  - https://tinyurl.com/3yhze3jn
- Other videos
  - https://youtu.be/QYqWQeRS6kY
  - https://youtu.be/u2ueQ0sVt3s

25.2 Errors

As we have seen, quantum computers are expected to be very susceptible to errors. This is because the state of a qubit is determined by the coefficients of the basis states (e.g., α and β in $|\psi\rangle = \alpha |0\rangle + \beta |1\rangle$). Any deviations from α and β result in errors. In this section, we will talk about different types of errors from the hardware perspective.

25.2.1 Readout Error

Qubit readout is an essential operation in any quantum algorithm. Due to noise, the readout cannot be 100% accurate (see Fig. 22.5), which results in **readout error**. To reduce readout errors, accurate calibration is very important. There are also various techniques to improve **readout fidelity** (e.g., [1]). Very often, multiple reading is used to reduce the effect of noise. To do so, the qubit needs to stay in its collapsed state after each reading. This requires the so-called **quantum nondemolition (QND)** measurement.

Figure 25.1 shows the readout process of a superconducting qubit. It plots the I and Q of each measurement (readout). Assume it is well-calibrated, every point with a positive Q is designated as $|e\rangle$ ($|1\rangle$) and every point with a negative Q is designated as $|g\rangle$ ($|0\rangle$). However, some states that were prepared as $|e\rangle$ have negative Q, and some states that were prepared as $|g\rangle$ have positive Q. Assuming 0.5% of the $|e\rangle$ and $|g\rangle$ states have this problem, the readout fidelity is 99.5%.

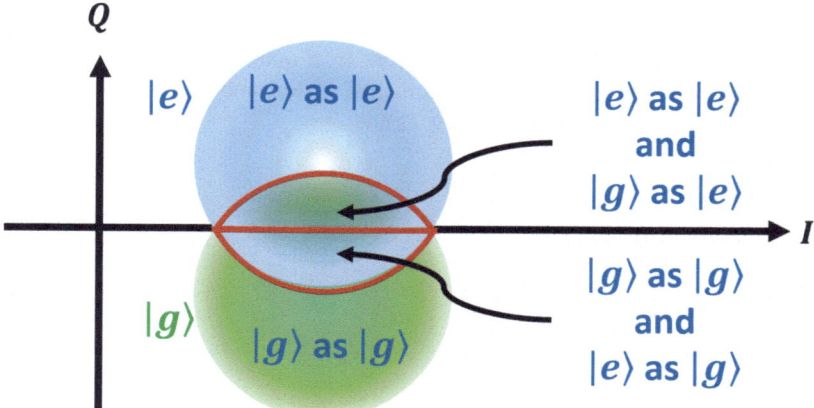

Fig. 25.1 $I - Q$ plot of a readout process. The true and measured states of each group of points are also shown

25.2.2 Initialization Error

Qubit initialization is a critical operation in a quantum computer and is one of the DiVincenzo criteria (Sect. 1.3, Chaps. 11 and 22). However, it is difficult to achieve 100% accuracy during qubit initialization. As discussed, one might use thermalization or active reset to initialize a qubit to its ground state, $|0\rangle$. If thermalization is used, we rely on the excited state decaying to the ground state by releasing energy. This is a random process and we cannot guarantee that the qubit will decay within a finite time. Moreover, a qubit at the ground state can also absorb energy from the environment and be excited to an excited state. Therefore, after the initialization process, there is a finite **initialization error** (e.g. 0.5%) and the initialization **fidelity** is not 100% (e.g., 99.5%).

If the initialization is performed using active reset, there are two sources of errors. Firstly, it needs to read the state of the qubit. The readout is not 100% accurate (Fig. 25.1), which causes the computer to apply a wrong operation. For example, the qubit may be at $|0\rangle$. However, it might read it as $|1\rangle$ ($|g\rangle$ as $|e\rangle$). It will then apply a π-pulse to try to "reset" it which will bring the ground state qubit to the excited state, $|1\rangle$ and cause an error. The following shows a sample code used to perform active reset [2]. A π-pulse is applied only if the I value is larger than a certain threshold.

```
measure(readout_pulse, resonator, ...)
save(I, I1+Q1)
with if_(I&gt;I_threshold):
    play(pi_pulse, qubit)
```

The second source of error is due to gate error. When a π-pulse is applied to bring $|e\rangle$ to $|g\rangle$, the pulse might not be perfect and can cause further errors.

Finally, it is worth mentioning that the initialization and measurement errors are difficult to decouple during characterization. Together, they are also called the **state preparation and measurement (SPAM)** error [3].

25.2.3 Gate Error

A quantum gate is just a microwave or laser pulse. As discussed in the previous chapters (e.g., Chaps. 12 and 21), a microwave pulse needs to have the right shape and frequency to achieve the desired gate action. The generation of the final pulse involves many steps and equipment (e.g., AWG and LO; see Sect. 23.3). Any error that occurs in the pulse generation process and its interaction with the qubit will be accumulated, resulting in **gate error**. However, with accurate calibration, this type of error may be corrected. Sometimes they are called the **coherence errors**.

However, when a gate is applied, it means that the qubit is also in contact with the external environment. **Incoherence errors** will occur after which the qubit might lose its purity and become a mixed state due to the entanglement with the environment (see Chap. 6 for the concept of mixed state). This type of gate error cannot be corrected through calibration.

25.2.4 Bit-Flip and Phase-Flip Errors

Bit-flip and **phase-flip** errors are errors categorized based on the type of change of a state. They can occur at any stage during quantum computing due to the interaction with the environment. A bit-flip error can be represented as

$$|0\rangle \longleftrightarrow |1\rangle. \tag{25.1}$$

A phase-flip error is also called a **sign-flip** error and can be represented as

$$\begin{aligned}|+\rangle &\longleftrightarrow |-\rangle, \\ \frac{|0\rangle + |1\rangle}{\sqrt{2}} &\longleftrightarrow \frac{|0\rangle - |1\rangle}{\sqrt{2}},\end{aligned} \tag{25.2}$$

where, on the computational basis ($|0\rangle / |1\rangle$), the sign is flipped. Figure 25.2 illustrates the bit-flip and phase-flip processes on the Bloch sphere. Bit-flip and phase-flip errors are the basic concepts used in quantum error correction [4].

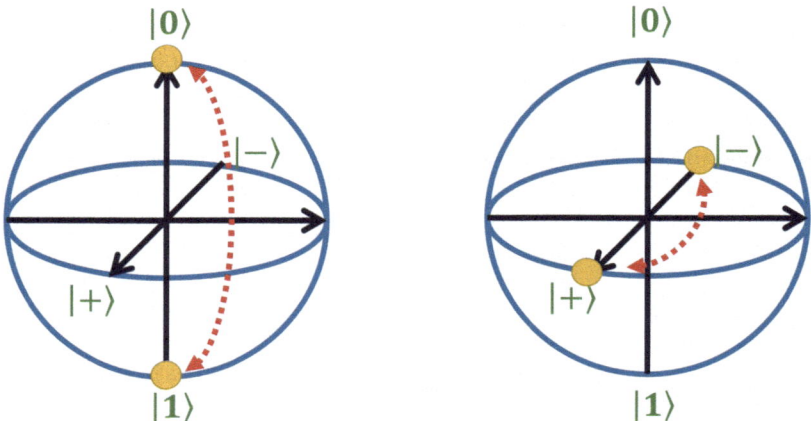

Fig. 25.2 Left: Illustration of the bit-flip error on Bloch sphere. Right: Illustration of the phase-flip error on the Bloch sphere

25.3 Decoherence Times

When a qubit interacts with the environment, errors will occur. This is a random process. To characterize the robustness of a qubit, two characteristic **decoherence times** are used, namely, the T_1 and T_2 times. The T_2 time is also called the T_ϕ time. T_ϕ is usually smaller than T_1 and limits the number of quantum gates that can operate before a qubit loses its information. However, it is difficult to measure T_ϕ directly because it is difficult to decouple the effect of T_1. Therefore, another characteristic time, T_2^*, is usually measured, and the following equation relates T_1, T_2^*, and T_ϕ [5, 6],

$$\frac{1}{T_2^*} = \frac{1}{2T_1} + \frac{1}{T_\phi}. \qquad (25.3)$$

25.3.1 T_1 Measurement

T_1 is also called the **energy relaxation time** or the **longitudinal relaxation time** [7, 8]. T_1 measurement is straightforward by exciting a qubit in the ground state to an excited state and observing its decay rate. To measure the T_1 of $|0\rangle\,/\,|1\rangle$ transition, the qubit is brought to the ground state $|0\rangle$ through active reset and then is excited to $|1\rangle$ by applying a π-pulse. The qubit is then measured at different delay times, t. For example, t changes from $1\,\mu s$ to $100\,\mu s$ at a step of $1\,\mu s$. For each delay time, many identical experiments will be performed (e.g., 1000) so that we can obtain the probability that it is still at the excited state, $|1\rangle$. For example, with 1000 identical experiments, for a delay time of $20\,\mu s$ after it is excited, 610 of them might give $|1\rangle$ in the measurement and 390 of them might give $|0\rangle$. Then, we say the probability of the excited qubit still being excited after $20\,\mu s$ is 0.61. The probability of measuring $|1\rangle$, $P_1(t)$, is then plotted against the delay time and fitted using

$$P_1(t) = P_1(0)e^{-\frac{t}{T_1}}. \qquad (25.4)$$

$P_1(0)$ is set to be 1. The fitting parameter T_1 is energy relaxation time. Figure 25.3 shows one of the possible experiments with fitting. T_1 is extracted to be about $40\,\mu s$.

25.3.2 T_2^* Measurement

T_ϕ is the **dephasing time** or **transverse relaxation time** [7, 8]. Since it cannot be measured directly, we will discuss the measurement of T_2^* which can be used to calculate T_ϕ using Eq. (25.3).

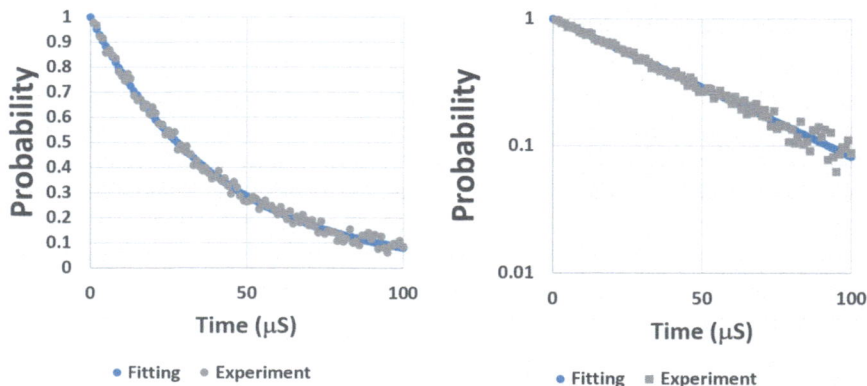

Fig. 25.3 T_1 relaxation time extraction experiment (simulation). T_1 is extracted to be about 40 μs using Eq. (25.4). Left: Linear scale. Right: Logarithmic scale

The method to measure T_2^* uses **Ramsey oscillation**. As shown in Fig. 25.4, firstly, the qubit is initialized through an active reset (Fig. 25.4a). Then a $\frac{\pi}{2}$-pulse for $|0\rangle / |1\rangle$ transition is applied with a detuning frequency, Δ. A $\frac{\pi}{2}$-pulse rotates the state about $-\hat{y}$ by $\pi/2$. This will bring the qubit to a superposition state of $\frac{|0\rangle+|1\rangle}{\sqrt{2}}$ (Fig. 25.4b). Due to detuning, the qubit will precess on the Bloch sphere (Fig. 25.4c). This is because of the $-\frac{\hbar\Delta}{2}\hat{\sigma}_z$ term in Eq. (21.19). After precessing for time t, another $\frac{\pi}{2}$-pulse is applied to rotate the state about $-\hat{y}$ by another $\pi/2$. Depending on the position of the state on the equator, the $\frac{\pi}{2}$-pulse will bring the state to $|0\rangle$ (e.g., Fig. 25.4d), $|1\rangle$ (if rotated by multiples of 2π), or a superposition state of $|0\rangle$ and $|1\rangle$. If a measurement follows, the probability of measuring $|0\rangle$, $P_1(t)$, oscillates as a function of t. When there is phase decoherence, the Bloch vector shrinks (Fig. 25.4e, f), and the amplitude of the oscillation decreases as a function of t.

To perform the experiment, t is varied from 1 μs to 80 μs in steps of 1 μs. For each delay time, many identical experiments will be performed (e.g., 1000) so that we can obtain the probability of it being $|1\rangle$, $P_1(t)$, which will indicate its position on the equator before the last $\frac{\pi}{2}$-pulse is applied and also the Bloch vector's length. As the vector precesses more on the equator, $P_1(t)$ changes from 0 to 1 periodically. If there is decoherence, the vector will shrink. $P_1(t)$ still oscillates but its amplitude decreases. For a dephasing process with a single time constant, the decay of the amplitude can be modeled as

$$P_1(t) = a + be^{-\frac{t}{T_2^*}} \cos(2\pi \Delta t + \phi). \tag{25.5}$$

where ϕ, a and b are fitting parameters and a and b are expected to be $\frac{1}{2}$ in theory. Figure 25.5 shows one of the example with $T_2^* \approx 20$ μs.

25.3 Decoherence Times

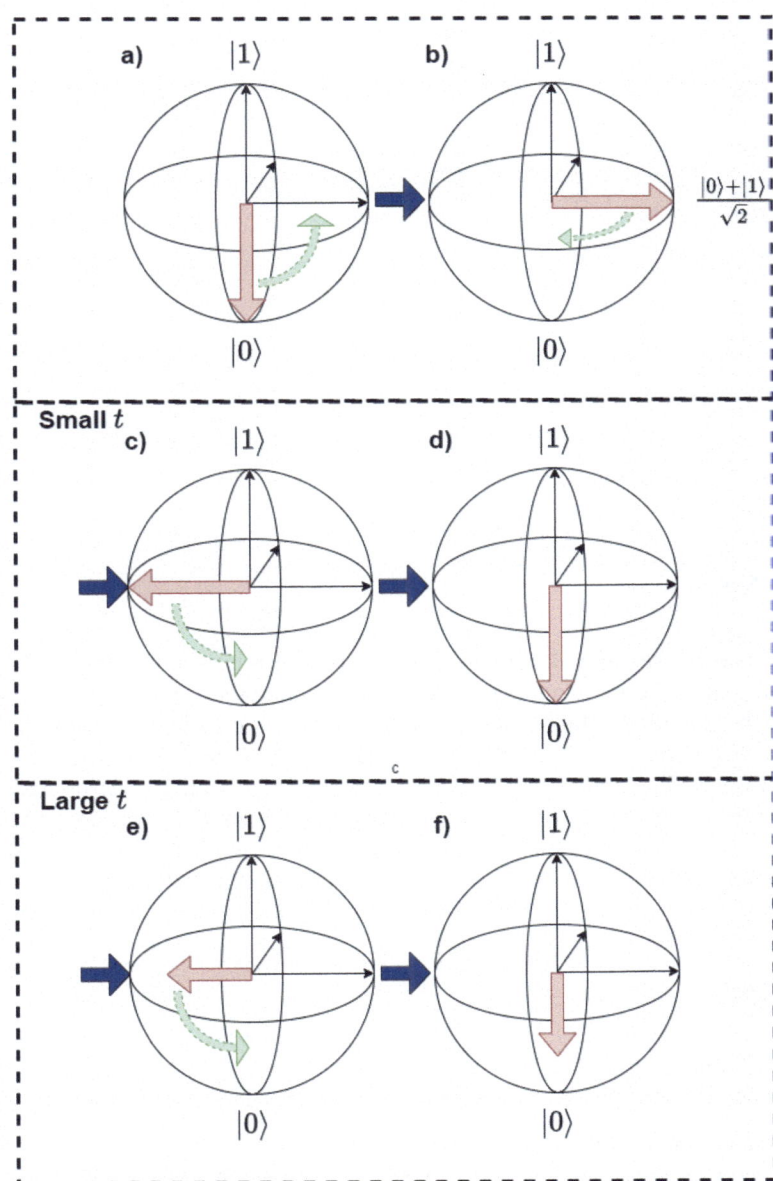

Fig. 25.4 Bloch spheres depicting the Ramsey measurement process. Pink arrows represent the Bloch vectors. Green arrows represent the subsequent operations on the Bloch vector. t is the delay time. From (**b**), due to detuning, the vector will precess to (**c**), or (**e**). When t is small (**c**), dephasing is not much. (**e**) is the case when t is large and the Bloch vector has shrunk noticeably

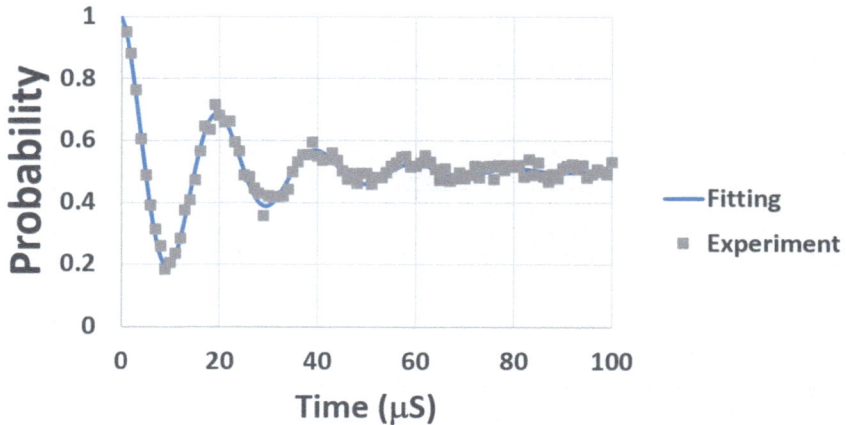

Fig. 25.5 T_2^* phase relaxation time extraction experiment (simulation). T_2^* is extracted to be about 20 μs using Eq. (25.5)

25.4 Summary

In this chapter, we study different types of errors in a quantum computer. There are initialization, readout, and gate errors. Depending on how the state is changed, they can be classified as bit-flip and phase-flip (sign-flip) errors. We characterize the qubit quality using two characteristic times, namely, the T_1 and T_2 (T_ϕ) decoherence time. However, T_2 cannot be measured directly. So, T_2^* is usually measured and T_2 is found using Eq. (25.3). We study how to extract T_1 and T_2^* experimentally. The theory of decoherence is a study of open quantum systems which is out of the scope of this book. Readers can read [9] if they are interested in the next step.

Problems

25.1 T_2 Time
Find T_2 based on the parameters extracted in Figs. 25.3 and 25.5.

25.2 Decoherence Time
Both T_1 and T_2 reduce the coherence of a qubit. Assume the effective decoherence time is $(\frac{1}{T_1} + \frac{1}{T_2})^{-1}$ Assume that T_1 is 300 μs, sweep T_2 from 50 μs to 500 μs, and plot the effective decoherence time as a function of T_2.

25.3 T_2^* Measurement
Try to draw the evolution of a Bloch vector that has precessed for 45° following the approach in Fig. 25.4.

References

1. Alistair W. R. Smith, Kiran E. Khosla, Chris N. Self, and M. S. Kim. Qubit readout error mitigation with bit-flip averaging. *Science Advances*, 7(47):eabi8009, 2021.
2. How can I do an active reset? https://www.quantum-machines.co/faq/how-can-i-do-an-active-reset, 2024. Accessed:2024-07-26.
3. Benjamin Nachman and Michael R. Geller. Categorizing readout error correlations on near term quantum computers, 2021.
4. Peter W. Shor. Scheme for reducing decoherence in quantum computer memory. *Phys. Rev. A*, 52:R2493–R2496, Oct 1995.
5. Matthew Reagor, Wolfgang Pfaff, Christopher Axline, Reinier W. Heeres, Nissim Ofek, Katrina Sliwa, Eric Holland, Chen Wang, Jacob Blumoff, Kevin Chou, Michael J. Hatridge, Luigi Frunzio, Michel H. Devoret, Liang Jiang, and Robert J. Schoelkopf. Quantum memory with millisecond coherence in circuit qed. *Phys. Rev. B*, 94:014506, Jul 2016.
6. H. Y. Wong, K. Beck, V. M. Iaia, and Y. J. Rosen. Study of phase method in tantalum superconducting qubit t2* measurements. In *IEEE International Conference on Quantum Computing and Engineering (QCE24)*, 2024.
7. M. A. Nielsen and I. L. Chuang. *Quantum Computation and Quantum Information: 10th Anniversary Edition*. Cambridge University Press, 2011.
8. Christopher J. Foot. *Atomic Physics*. Oxford University Press, 2005.
9. Francesco Campaioli, Jared H. Cole, and Harini Hapuarachchi. Quantum master equations: Tips and tricks for quantum optics, quantum computing, and beyond. *PRX Quantum*, 5:020202, Jun 2024.

Appendix A
Resources

The following provides a list of resources for this book.

Second Part of this Book: I might write a second volume for this book. Please check the information here:

- https://github.com/hywong2/Quantum_Computing_Architecture

Email me if you have any questions or suggestions.

- Email: intro.qc.wong@gmail.com

Introduction to Quantum Computing Hardware and Architecture: Teaching videos on YouTube.

- https://youtube.com/playlist?list=PLnK6MrIqGXsKpkN3nL1OlxW6Gr6s-Yt0C
- https://youtube.com/playlist?list=PLnK6MrIqGXsL1KShnocSdwNSiKnBodpie

Quantum Chip Design using HFSS: Teaching videos on superconducting chip design.

- https://youtube.com/playlist?list=PLnK6MrIqGXsIS_97Nt-8R6uNrJXxMiOro

Typos and Corrections: Any typos and corrections can be found here.

- https://github.com/hywong2/Quantum_Computing_Architecture

Quantum Algorithm Videos: Teaching videos on YouTube corresponding to the book, *Introduction to Quantum Computing—from a Layperson to a Programmer in 30 Steps*.

- https://youtube.com/playlist?list=PLnK6MrIqGXsJfcBdppW3CKJ858zR8P4eP

Index

A
Absolute anharmonicity, 260
Action, 171
Active reset, 4, 322
Additional phase shift, 165
Addition operation, 14
Adjoint, 26
Ambegaokar-Baratoff formula, 228
Amplifiers, 328
Angular frequency vector, 138, 284
Angular momentum, 94
Anharmonicity, 213, 260, 261, 314
Annihilation, 303, 310
Annihilation operator, 186
Anti-parallel spin state, 165
Application-specific integrated circuits (ASICs), 4
Arbitrary waveform generators (AWGs), 329
Area vector, 95
Artificial atom, 314
Attenuators, 328
Available power gain, 337
Azimuthal angle, 59

B
Bardeen-Cooper-Schrieffer pairs (BCS pairs), 218
Bare qubit frequency, 318
Basis vectors, 17
BCS theory, 218
Biot-Savart law, 157, 353
Bit-flip, 358
Bloch sphere, 59
Bloch sphere surface, 76
Bloch vector, 85
Bohr magneton, 97, 103
Bose-Einstein distribution, 219
Bosons, 218
Branch flux, 242

C
Cauchy sequence, 16
Cavity quantum electrodynamics (c-QED), 312
Change of basis, 18
Charateristic impedance, 204
Charge dispersion, 264
Charge noise, 247
Charge qubit, 274
Charge qubit regime, 264
Charging energy, 145, 164, 238
Charging energy of Cooper pair, 238
Chemical potential, 219
Circuit quantum electrodynamics (circuit-QED), 314
Circulator, 333
Coherence errors, 357
Coherent state, 197
Collapse of the state, 20
Commutation, 61
Completeness, 184
Complex vector space, 14
Conjugate transpose, 19, 26
Conjugate variables, 185, 205
Constitutive relations, 202
Continuous basis, 183
Controlled NOT (CNOT), 51

Controlled phase shift gate, 158
Control qubit, 51
Cooper pair box, 236
Cooper pair number operator, 291
Cooper pairs, 218, 238
Coplanar waveguide (CPW), 345
Coulomb blockade, 145, 164
Coupling energy, 301, 310
Coupling strength, 348
Creation operators, 186, 303, 310
Critical current, 222
Critical current density, 222
Critical magnetic field, 219
Critical temperature, 216
Cross Kerr, 319, 351
Current law, 202
Current-phase relation, 223

D

Decoherence times, 359
Decoupled, 42
Degeneracy, 98
Degenerated, 147
Degenerated states, 98
Degrees of freedom (DOFs), 58
Density matrix, 72, 76, 79
Dephasing time, 7, 359
Detuning, 137, 284, 297
Detuning frequency, 317
Diagonalizable, 27
Diagonalization, 28
Dilution refrigerator, 6
Dirac delta function, 184
Dispersive readout, 317
Dispersive regime, 317
Distance, 16
DiVincenzo's criteria, 143
Dolan bridge, 220
Dot/inner/scalar product, 99
Dot product, 15
Down-converted, 333
Dual correspondence, 19, 26

E

Effective dispersive shift, 319
Effective Josephson junction energy, 232
Eigenbasis, 27
Eigenenergies, 44
Eigenmode, 348
Eigenvalue, 18, 183

Eigenvectors, 27, 183, 185
Electric charge, 205
Electromagnetic (EM) wave, 210
Electrons, 96
Electron spin resonance, 120, 157
Embedded, 59
Energy eigenstates, 189
Energy participation ratio (EPR), 352
Energy relaxation time, 359
Ensemble of pure states, 77
Entangled, 35
Entanglement, 53
Entanglement gate, 157
Envelope function, 290
Equations of motion, 171, 175
Euclidean vector space, 14, 15
Euler rotations, 60
Excited state, 102
Expectation values, 35, 82

F

Fault-tolerant quantum computing, 144
Fermi-Dirac distribution, 150, 218
Fermi level, 150, 219
Fermions, 218
Fidelity, 357
Field programmable gate arrays (FPGAs), 4, 328–329
First Josephson equation, 222
Fock states, 237, 291, 310
Friis equation, 337
Full width at half maximum (FWHM), 351

G

Gate error, 357
Generalized coordinates, 170, 176, 206, 236
Generalized momentum, 174, 176, 206, 236
Generalized Rabi frequency, 139
Generalized velocities, 170, 206, 236
Generator, 60
G-factor, 97
Ground state, 102
Gyromagnetic ratio, 95

H

H_0, 116–119, 127, 131, 133
H_1, 165
H_2, 165
Hadamard gate (H), 51

Hamiltonian, 4, 40–45, 48, 49, 140, 175, 176, 182, 204
Hamiltonian engineering, 55, 161
Hamiltonian mechanics, 169
Hamilton's equations, 175
Hamilton's principle, 171
Hermitian, 28, 183, 185, 186
H gate, 159
H_I, 116–119, 131, 133
$H_{JC,RF}$, 316
High electron mobility transistor (HEMT), 336
High-speed digital and analog electronics, 328
Hilbert space, 16
Hole, 98
H_{RF}, 134, 135, 137, 138, 140, 141

I
Idempotent, 29
Identity gate (I), 49
Incoherence errors, 357
Initialization, 149
Initialization error, 357
Inner product, 14, 75
Inner product of matrices, 72
Inner product space, 14, 75
In-phase, 296
Interaction Hamiltonian, 99
Inverse, 30
IQ blobs, 322
Isolators, 328, 333
Isotope, 156
iSWAP gate, 45, 289, 304, 307

J
Jaynes-Cummings Hamiltonian, 312
Josephson energy, 227
Josephson inductance, 225
Josephson junctions, 213, 220
Josephson parametric amplifier (JPA), 338
Josephson phase, 222

K
Kinetic energy, 171
Kinetic inductance, 226
Kirchhoff's current law (KCL), 243
Kirchhoff's voltage law (KVL), 202, 242
Kronecker delta, 21, 63
Kronecker delta function, 184

L
Laboratory frame, 283
Lagrange's equations, 171
Lagrangian, 171
Lagrangian mechanics, 169
Larmor frequency, 105, 275, 292
Larmor precession, 105, 275
LC tank, 208
Least significant bit (LSB), 52, 159
Legendre transform, 175, 207
Levi-Civita, 61
Linear space, 14
Line-width broadening, 6
Local oscillators (LOs), 328
Logical qubit, 144
Longitudinal relaxation time, 359
Lossless, 333
Lowering, 303, 313
Lowering operators, 132
Low-pass filter (LPF), 331, 340

M
Macroscopic quantum object, 219
Magnetic flux, 205
Magnetic flux quantum, 223
Magnetic moment, 95
Manhattan-style, 220
Matched, 334
Matrix, 79
Matrix differential equation, 41
Matrix exponential, 41
Matrix mechanics, 40
Mechanical SHO, 208
Meissner effect, 216
Microprocessors, 329
Mixed-state, 77
Mixers, 328
Momentum operator, 185
Momentum vector, 184
Most significant bit (MSB), 52, 159
Multiplicative identity, 14

N
Native entanglement gate, 157, 297
Newtonian mechanics, 171
Noise factor (NF), 336
Noise power spectral density (NSD), 331
Noncommutative, 185
Nonlinear inductance, 226
Nonlinear inductor, 228
Non-zero off-diagonal elements, 42

Norm, 16
Normalized vector, 21
NOT, 34, 49, 133, 303
Number operator, 188, 237

O
Offset charge, 246, 250
1-Qubit gate, 5
Operators, 26, 183
Orthogonal, 76
Orthonormal, 15, 183
Orthonormal basis, 21
Out-of-phase, 296
Over complete, 197

P
P, 28
Pauli exclusion principle, 218
Pauli matrices, 60, 274
Pauli spin matrix, 104
Pauli vector, 73
Perfect conductor, 216
Perturbation, 117
Phase-flip, 358
Phase operator, 291
Phase shift gate (UPS,Φ), 50, 51, 108
Phonon, 218
Photon, 210, 310
Physical qubit, 144
Planck constant, 40
Plane, 322
Polar angle, 59
Position operator, 183
Position vector, 183
Positive semi-definite (PSD), 79
Potential energy, 171
Precess, 275
Principle of least action, 171
Printed circuit board (PCB), 345
Probability density, 193
Project, 193
Promote, 182
Pure state, 76

Q
Quadrature, 329
Quality factor (Q), 348
Quantization, 185
Quantum bit (qubit), 4
Quantum dot (QDOT), 144, 152
Quantum gate, 47, 109

Quantum-limited amplifier (QLA), 336
Quantum mechanical tunneling, 195, 210
Quantum noises, 6
Quantum nondemolition (QND), 356
Quantum non-demolition (QND) measurement, 320
Quantum operators, 182
Quantum parametric amplifier, 9
Quantum point contact transistor, 152
Quantum supremacy, 144
Quarter-wavelength, 346
Quasiparticles, 219
Qubit energy, 268
Qubit frequency, 292
Qubit initialization, 4
Qubit manipulation, 4
Qubit readout, 5, 317

R
Rabi frequency, 123, 126, 131, 282
Rabi oscillation, 115, 122, 279, 311
Raising, 303, 313
Raising operators, 132
Ramsey oscillation, 360
Readout, 152, 157, 310
Readout error, 356
Readout fidelity, 356
Real vector space, 14
Real vector space of 2 ×2 Hermitian matrices, 72
Reciprocal, 334
Reduced magnetic flux quantum, 223, 237
Reduced Planck constant, 40
Relative anharmonicity, 261
Relative phase, 51
Resonant frequency, 203
Resonator, 346
Rotating frame, 126, 133, 134, 139, 140, 303
Rotating wave approximation (RWA), 120, 126, 283, 295, 304, 316

S
Scalars, 14
Scattering matrix (S), 333
Schrödinger equation, 40, 306
Second Josephson equation, 223
Second quantization, 310
Self-adjoint, 28
A set of universal quantum gates, 5
S gate, 50, 111
Signal-to-noise ratio (SNR), 9, 336
Sign-flip, 358

Index

Silicon spin qubit, 146
Similarity transformation, 33
Simple harmonic oscillator (SHO), 173, 182, 208, 268
Single electron transistor (SET), 157
Single-side band (SSB), 331
Spectral selectivity, 162
Spin, 96
Spin angular momentum, 60, 96
Spin angular momentum operator, 114
Spin-half, 96
Spin magnetic moment, 96
Spin magnetic number, 96
Spin-orbit (SO) coupling, 146
Spin-to-charge conversion, 152
Split transmon qubit, 231
Spontaneous emission, 311
Stark effect, 162
State preparation and measurement (SPAM), 357
Stimulated emission, 311
Superconducting quantum interference device (SQUID), 231
Superconductor, 216
Superposition, 51
Sweet spot, 253, 261

T
Target qubit, 51
Tensor product, 22, 35
T gate, 50, 111
Thermalization, 4, 149, 322
Thermal noise, 6
Three-port network, 333
Throughput, 9, 144
Time independent, 138
Torque, 107
Trace, 63, 74
Traceless, 63
Transmon, 254
Transmon qubit, 261, 276
Transmon regime, 265
Transverse coupling, 305
Transverse relaxation time, 359
Traveling wave parametric amplifier (TWPA), 8, 338

Tunable, 231
Tunneling, 164
2-Qubit gate, 5

U
U, 33, 47, 48, 297
U^{-1}, 33
U_{Cphase}, 157, 160–162, 165, 297
$U_{CPS,\pi}$, 157–160
Uncertainty principle, 6, 185, 338
Unitary, 48, 333
Unitary matrix (U), 30
Universal sets of quantum gates, 65
U_{NOT}, 49, 50
Up-conversion, 330
$U_{PS,\Phi}$, 108, 160
U_{RF}, 134–137, 303
U_{XOR}, 51–53, 65, 158, 161

V
Vacuum energy, 189, 311
Vacuum Rabi oscillation, 311
Vector axioms, 14
Vectors, 14
Vector space, 14

W
Wavefunction, 194
Weak link, 222
Well-characterized two-dimensional Hilbert space, 149

X
XOR, 51

Z
Zeeman splitting, 147
Zero-point energy, 189
Zero-point fluctuation, 187, 209
Zero-point fluctuation of the charge number, 269
Z gate, 50, 111

The manufacturer's authorised representative in the EU is Springer Nature Customer Service Centre GmbH, Europaplatz 3, 69115 Heidelberg, Germany. If you have any concerns regarding our products, please contact ProductSafety@springernature.com

Printed and bound by CPI Group (UK) Ltd, Croydon, CR0 4YY
26/03/2026
02078967-0002